The ELFNET Book on Failure Mechanisms, Testing Methods, and Quality Issues of Lead-Free Solder Interconnects

Günter Grossmann · Christian Zardini
Editors

The ELFNET Book on Failure Mechanisms, Testing Methods, and Quality Issues of Lead-Free Solder Interconnects

Springer

Editors
Günter Grossmann
EMPA
Überlandstrasse 129
Dübendorf CH-8600
Switzerland
e-mail: guenter.grossmann@empa.ch

Prof. Christian Zardini
ENSEIRB
Talence Cedex
France

ISBN 978-0-85729-235-3 e-ISBN 978-0-85729-236-0

DOI 10.1007/978-0-85729-236-0

Springer London Dordrecht Heidelberg New York

British Library Cataloguing in Publication Data
A Catalogue record for this book is available from the British Library

Cover design: eStudio Calamar S.L.

Printed on acid-free paper

Springer is part of Springer Science+Business Media (www.springer.com)

Contents

List of Contributors

E. Auerswald, Micro Materials Center Berlin and Chemnitz, Fraunhofer IZM, Volmerstr. 9 B, 12489, Berlin, Germany

I. Baylakoglu, TÜBİTAK UZAY, Ankara, Turkey

M. H. Biglari, Mat-Tech BV, Ekkersrijt 4605, Eindhoven, The Netherlands

A. Brewin, National Physical Laboratory, Teddington, Middlesex, UK

J.-Y. Deletage, Laboratoire IMS, 351 cours de la libération, 33405, Talence, France

R. Dudek, Micro Materials Center Berlin and Chemnitz, Fraunhofer ENAS, Technologie-Campus 3, 09126, Chemnitz, Germany

G. Grossmann, Swiss Federal Laboratories, EMPA, Überlandstrasse 129, Dübendorf CH-8600, Switzerland

E. Hedin, ERICSSON AB, Stockholm, Sweden

M. J. M. Hermans, Materials Science and Engineering Joining and Mechanical Behaviour, Delft University of Technology, Mekelweg 2, Delft, The Netherlands

C. Hunt, National Physical Laboratory, Teddington, Middlesex, TW11 0LW, UK

J. K. Kivilathi, Department of Electrical and Communications Engineering, Helsinki University of Technology, P.O. Box 3000, Otakaari 7A, 02015 TKK, Finland

R. Labie, IMEC, Kapeldreef 75, B-3001, Leuven, Belgium

K. Lambrinou, IMEC, Kapeldreef 75, B-3001, Leuven, Belgium

T. Laurila, Department of Electrical and Communications Engineering, Helsinki University of Technology, P.O. Box 3000, Otakaari 7A, 02015 TKK, Finland

T. T. Mattila, Department of Electrical and Communications Engineering, Helsinki University of Technology, P.O. Box 3000, Otakaari 7A, 02015 TKK, Finland

P.-E. Tegehall, Swerea IVF AB, Box 104, SE-431 22, Mölndal, Sweden

B. Vandevelde, IMEC, Kapeldreef 75, B-3001, Leuven, Belgium

A. Vicenzo, Dipartimento di Chimica, Materiali, Ingegneria Chimica "Giulio Natta", Politecnico di Milano, Via L. Mancinelli 7, 20131, Milano, Italy

V. Vuorinen, Department of Electrical and Communications Engineering, Helsinki University of Technology, P.O. Box 3000, Otakaari 7A, 02015 TKK, Finland

C. Zardini, Laboratoire IMS, 351 cours de la libération, 33405, Talence, France

C. Zou, NPL, Hampton Road, Teddington, Middlesex, TW11 0LW, UK

Chapter 1
Deformation and Fatigue of Solders

Günter Grossmann

In this chapter, a brief introduction in the definition of cyclic load as well as in the deformation mechanisms and their corresponding degradation mechanisms is given. For more in-depth information, there is ample literature available.

1.1 Deformation

1.1.1 Time-Independent Deformation

Every material deforms if stress is applied. The deformation can be either elastic, which means that the deformation is reversible as soon as the load is removed or instantaneous plastic if the stress is above the so-called yield stress.

The elastic deformation has its origin in the stretching of interatomic bonds.

The bonds are a combination of attractive and repulsive forces with the net force being zero when the bond is at equilibrium. Clearly, when the atoms are far apart, the attractive part must dominate and when they are very close together, the repulsive part is greater; both contributions decay away as the separation increases. The combination of attractive and repulsive forces manifest in the general atomic force (F)–atomic spacing (r) graph (Fig. 1.1).

A tensile force will lengthen the interatomic bonds and will be opposed by the attractive part of the interatomic force. As soon as the minimum of the F–r graph is reached, the external force will dominate and the bond will break.

G. Grossmann (✉)
EMPA, Überlandstrasse 129, Dübendorf, CH-8600, Switzerland
e-mail: guenter.grossmann@empa.ch

G. Grossmann and C. Zardini (eds.), *The ELFNET Book on Failure Mechanisms, Testing Methods, and Quality Issues of Lead-Free Solder Interconnects*,
DOI: 10.1007/978-0-85729-236-0_1,

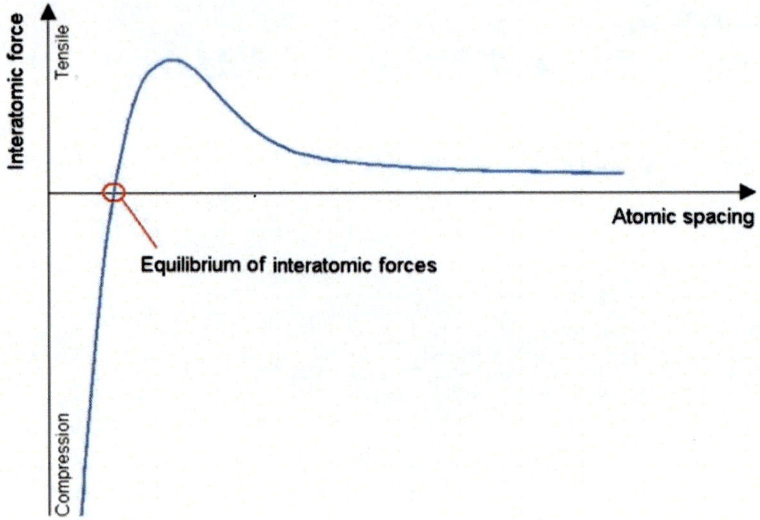

Fig. 1.1 Atomic force–atomic spacing graph

This behaviour can be illustrated with a spring. As a load is applied, the spring extends following a linear law

$$F = C \cdot \Delta l \tag{1.1}$$

F load, C spring constant, Δl deformation.

In construction materials, this relationship is expressed as

$$\sigma = E \cdot \varepsilon \tag{1.2}$$

σ stress, E modulus of elasticity or Young's modulus, ε strain or under shear load.

$$\tau = G \cdot \gamma \tag{1.3}$$

τ shear stress, G shear modulus, γ shear strain.

The elastic deformation is principally time independent. This means the specimen deforms as soon as a load is applied.

As soon as the spring is stretched above a certain amount, it will relax the maximum elastic amount but also the spring will show a remaining deformation if the load is removed. This behaviour is visible in most metals. If a metal is strained more than 0.5%, the stress–strain behaviour is clearly no more a straight line but a curve and a not-recoverable deformation called plastic deformation takes place. The plastic deformation is caused by the movement of crystal layers above each other. As soon as a certain stress is exceeded, dislocations in the crystal lattice will be activated to move through the crystal lattice and thus two plains of atoms are moving across each other. The point where the elastic deformation gradually merges to the plastic deformation is called the proportional limit. However, this

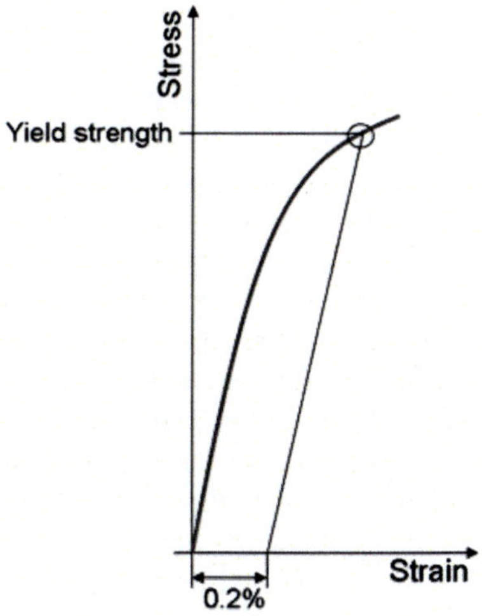

Fig. 1.2 Definition of yield strength

point cannot be defined precisely. To overcome this problem, a line with the same slope as the elastic line is drawn, originating from a given strain, usually 0.2%. The intersection of this line with the curved line of the stress–strain diagram is called the yield strength (Fig. 1.2).

1.1.2 Time-Dependent Deformation

There is also a time-dependent elastic deformation. This means that under a constant load, a material deforms and after the load is removed, the deformation goes back to 0 over time. This behaviour is called anelastic deformation.

If a material under load experiences an elevated temperature, time-dependent plastic deformation takes place, which means that the material shows an ongoing deformation under constant load. This deformation is called creep.

In contrast to the spring example, a material that creeps can be compared with a wet chewing gum. If one applies a constant load to such a specimen, it will constantly deform until it breaks. Varying the load results in differences in the deformation rate. The heavier the load, the faster the chewing gum deforms. If the specimen is cooled down, the deformation will come to an end as soon as the chewing gum is frozen.

How fast a material deforms if a certain load is applied depends on how close it is to its solidus temperature. How close the temperature of the application is to the solidus temperature is expressed with the homologues temperature

$$T_{\mathrm{H}} = \frac{T_{\mathrm{A}}}{T_{\mathrm{S}}} \tag{1.4}$$

T_{A} temperature of application (K), T_{S} temperature of solidus (K).

Note that absolute temperatures in Kelvin are to be used.

As soon as a material surpasses a T_H of 0.5, creep becomes an important part of the deformation history.

In creep, two deformation mechanisms are dominant: grain boundary sliding and dislocation creep. In grain boundary sliding (GBS), the crystals that make up the metal slide above each other along their grain boundaries. Since the grains fit only in one configuration, they have to move on complex paths combining rotation and lateral movement in all three dimensions to come around each other. As soon as the stress applied exceeds a certain value, dislocations will start to glide through the grains (DC), thus deforming the grains. Again in both cases the deformation causes microvoids to form that have to be filled up by diffusion.

Thus, GBS occurs if low stress results in a slow deformation, DC is the dominant deformation mechanism if large stress is applied to a specimen.

In reality, GBS and DC never occur on their own. Also, with a low overall stress that would activate GBS only, local stress build-up can cause local DC.

In contrast to the elastic deformation, creep is time dependent as already mentioned earlier. Thus, the load does not cause a certain deformation but a deformation rate.

$$F = f\left(\frac{\Delta l}{\Delta t}\right) \tag{1.5}$$

F force, Δl elongation, Δt time; or as stress–strain.

$$\sigma = f(\dot{\varepsilon}) \tag{1.6}$$

σ normal stress, $\dot{\varepsilon}$ strain rate; or under shear load

$$\tau = f(\dot{\gamma}) \tag{1.7}$$

τ shear stress, $\dot{\gamma}$ shear strain rate.

This means the more load one applies to a specimen, the faster it deforms, and the longer one waits, the more the specimen deforms; in other words, every deformation can be achieved with every load—it is just a question of time. Thus, it is senseless to speak of a yield strength when characterising soft solder: Since Creep is the main deformation mechanism in soft solders due to their high homologues temperature even at room temperature (Table 1.1), every deformation rate will result in another yield strength.

If a creep test is done, where a constant load is applied to the specimen and the deformation is monitored, one sees that the strain often varies over time (Fig. 1.3). After an initial elastic deformation, primary creep with a varying strain rate takes place followed by a regime with mostly constant creep rate called secondary creep. In the end, the creep rate starts to increase due to damage in the material until the specimen fails.

If the creep rate of metals is plotted against the stress applied in a double logarithmic plot, one sees that over a limited interval, the measurement points are located on a straight line.

Table 1.1 T_H at 20°C of some soft solder alloys

Alloy	T_L	T_H (20°C)
$SnPb_{36}Ag_2$	183°C, 456 K	0.643
$SnAg_{3.8}Cu_{0.7}$	217°C, 490 K	0.598
$SnAg_{3.5}$	221°C, 494 K	0.593
$SnCu_{0.7}$	227°C, 500 K	0.586

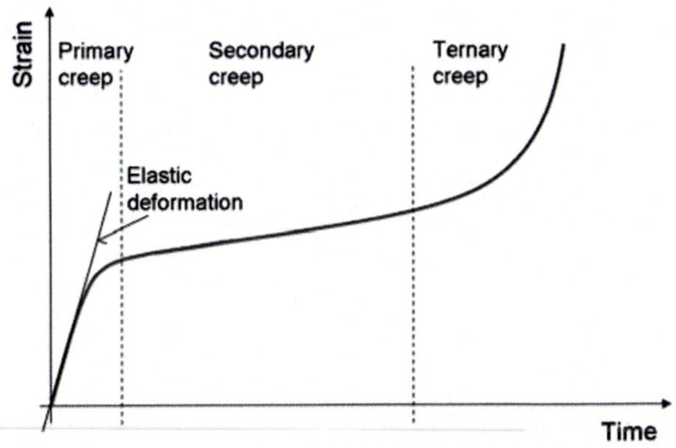

Fig. 1.3 Creep behaviour with constant load

Norton [1] surveyed creep in steel and described the nearly linear relationship of the secondary creep in the double logarithmic plot with

$$\dot{\gamma} = A\left(\frac{\tau}{G}\right)^n \exp\left[\frac{Q}{RT}\right] \tag{1.8}$$

$\dot{\gamma}$ creep rate, A constant, τ shear stress, G shear modulus, n exponent, Q activation energy, R gas constant, T temperature in K

where τ is normalised with the shear modulus G in order to eliminate the units of τ and an Arrhenius term is used to account for the influence of the temperature.

However, in solder, over an extended measurement range of stress vs. strain rate, one sees that the measurement points are no more located on a straight line (Fig. 1.4).

Several models have been used to describe this behaviour.

A very popular model is the approach of Garofalo [2] who approximated the curve where the points are located with a sinh-relationship:

$$\dot{\gamma} = C_1[\sinh(C_2\tau)]^n \tag{1.9}$$

$\dot{\gamma}$ creep rate, C_1, C_2 constant, τ shear stress, n exponent.

Another possibility has been proposed by Hart [3] who treated the system of grain boundary sliding (GBS) and matrix creep (MC which is the same as DC) with a rheological analogue (Fig. 1.5). Hart modelled GBS with a linear viscous law and MC by a power law.

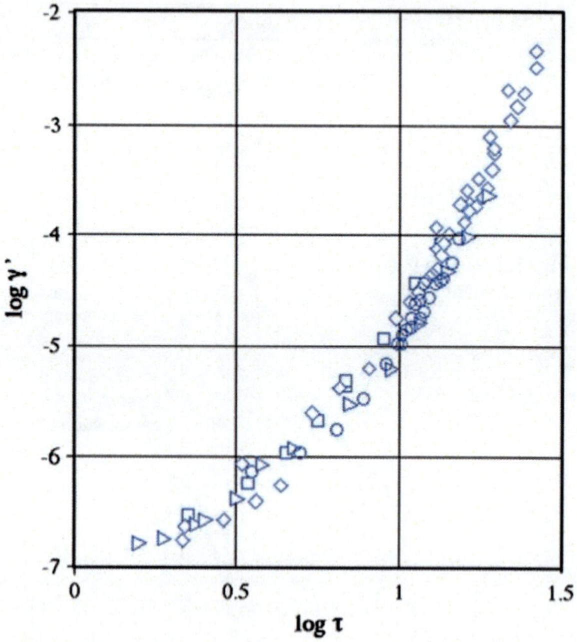

Fig. 1.4 Measurement of creep of $SnPb_{36}Ag_2$ at room temperature

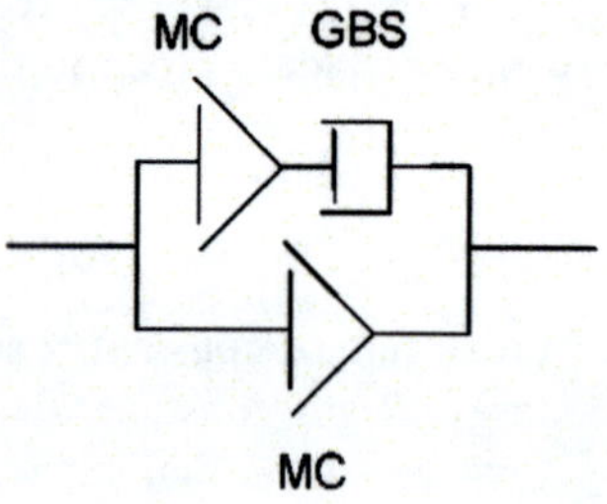

Fig. 1.5 Analogous model of Hart

The system of Fig. 1.5 can be modelled with

$$\sigma = \sigma_0[y + (1 - y)Z]\dot{\varepsilon}^{\mu} \tag{1.10}$$

σ stress, $\dot{\varepsilon}$ strain rate, σ_0 material constant, μ material constant, y parameter for the grain size; with

$$Z = \frac{\sigma_b}{\sigma_m} \tag{1.11}$$

σ_b stress in the grain boundaries, σ_m stress in the matrix.

Another constitutive equation proposed by Weber and Grossmann [4] models the deformation behaviour of soft solder as the sum of the two deformation mechanisms GBS and DC, both expressed with a Norton equation:

$$\dot{\gamma} = A\left(\frac{\tau}{G}\right)^n \exp\left[\frac{-Q_{GBS}}{RT}\right] + B\left(\frac{\tau}{G}\right)^m \exp\left[\frac{-Q_{DC}}{RT}\right] \quad (1.12)$$

$\dot{\gamma}$ strain rate, τ shear stress, G dynamic shear modulus, Q_{GBS} activation energy GBS, Q_{DC} activation energy DC, n exponent GBS, m exponent DC.

1.2 Degradation Due to Cyclic Load

1.2.1 Definition of a Cyclic Load

A regular cyclic load is defined by its stress amplitude σ_a, its mean stress σ_m as well as the upper stress level σ_u, the lower stress level σ_l and the period of a cycle t (Fig. 1.6) with

$$\sigma_a = \sigma_u - \sigma_l \quad (1.13)$$

σ_a stress amplitude, σ_u upper stress level, σ_l lower stress level or with the stress quotient R:

$$R = \frac{\sigma_l}{\sigma_u} \quad (1.14)$$

σ_l lower stress level, σ_u upper stress level; and

$$f = \frac{1}{t} \quad (1.15)$$

f frequency, t time for 1 period.

A cyclic load can be single-levelled with σ_m and σ_a constant or multi-levelled with varying σ_m and σ_a. The frequency f is rarely varied since if f is varied, the evaluation of the resistance of a material against fatigue becomes multidimensional and extremely complex.

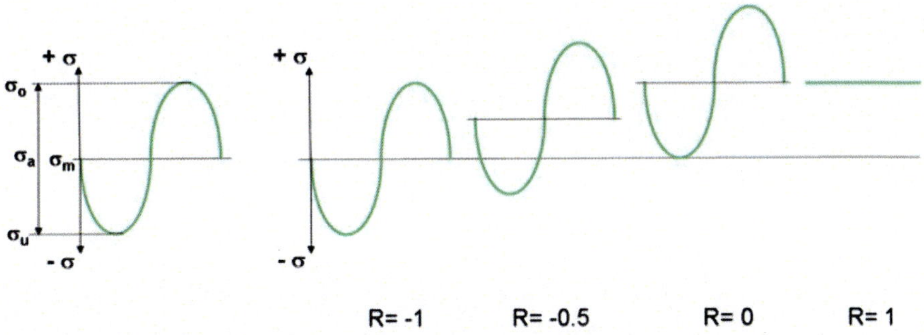

Fig. 1.6 Definition of a cyclic load

1.2.2 Fatigue Testing

The resistance of a material against cycle fatigue is evaluated by cyclic fatigue testing with two kind of testing set-up: In one set-up, σ_m is held constant and σ_a varies and in the second set-up, both, σ_m and σ_a, are varied.

1.2.2.1 σ_m Constant, σ_a Varying

Usually σ_m is 0, which means that the specimen encounters cyclic tensile and compressive stress. By varying σ_a, the average number of cycles to failure (N_f) is determined for each level of σ_r (Fig. 1.7). The resulting curve of $N_f = f{\cdot}(\sigma_a)$ is called Wöhler's curve.

Of course there is a considerable spread in the measurements of Wöhler's curve. Thus, the failure probability has to be included in the measurements to describe the resistance of a material against high-cycle fatigue (Fig. 1.8).

1.2.2.2 σ_m and σ_a Varying

If both parameters are varied, there are 2 ways to look at the results. The straightforward one is to draw the average Wöhler's curve for various σ_m in one diagram (Fig. 1.9).

The other possibility is to fix a desired N_f for a specimen and plot the σ_a and σ_m combinations leading to this particular N_f in the Haigh diagram (Fig. 1.10).

1.2.3 Brittleness

The brittleness of a material is measured in the Charpy test with an impact tester where a pendulum hits a specimen that bears a notch (Fig. 1.11). The geometry of the specimen and the notch are standardised in EN10045.

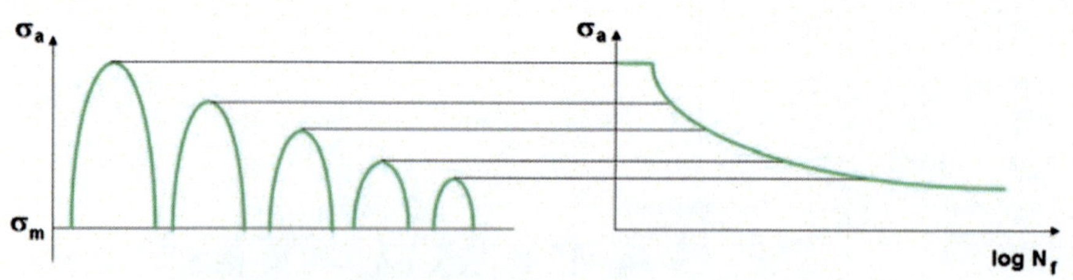

Fig. 1.7 Determination of Wöhler's curve [5]

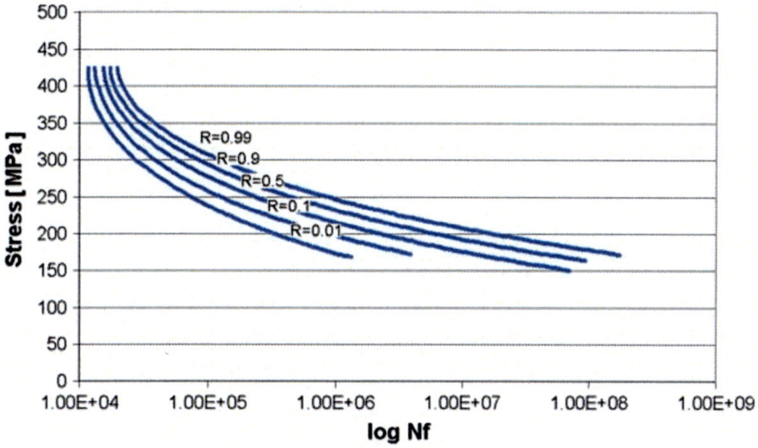

Fig. 1.8 Wöhler's curve of an Al-alloy with $\sigma_m = 0$ with the probability of occurrence as free parameter [6]

Fig. 1.9 Wöhler's curve for varying σ_m with $\sigma_{m1} > \sigma_{m2} > \sigma_{m3}$ [6]

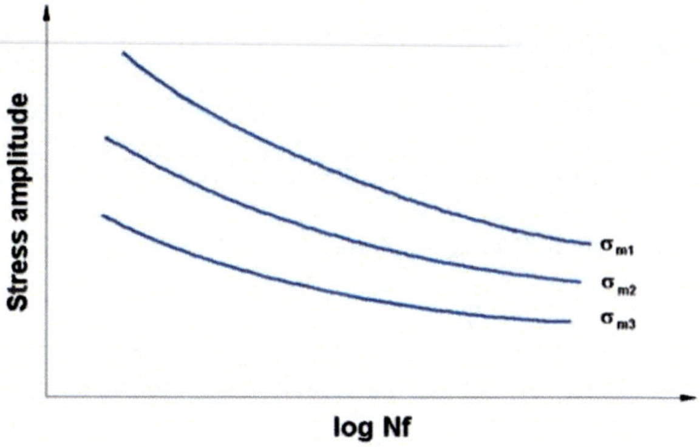

The energy loss of the pendulum given by the difference of height between the starting position and the end position of the pendulum (H–h) is a measure for the brittleness of the specimen. A ductile material will absorb more kinetic energy than a brittle material and thus *h* will be smaller if a ductile material is tested than for a brittle material. The brittleness of a material depends on the crystal structure, on grain size and on shape and size of particles to name a few influencing factors. Whether a material is brittle or not depends also on its temperature. Many materials show a transformation from ductile to brittle as a function of temperature called the transition temperature.

This transition from brittle to ductile is influenced by the crystal lattice. Cubic face-centred (fcc) materials such as Pb, Al or Cu retain their plasticity at low temperature where cubic body-centred materials such as Fe, Cr or W show a clear transition temperature (Fig. 1.12).

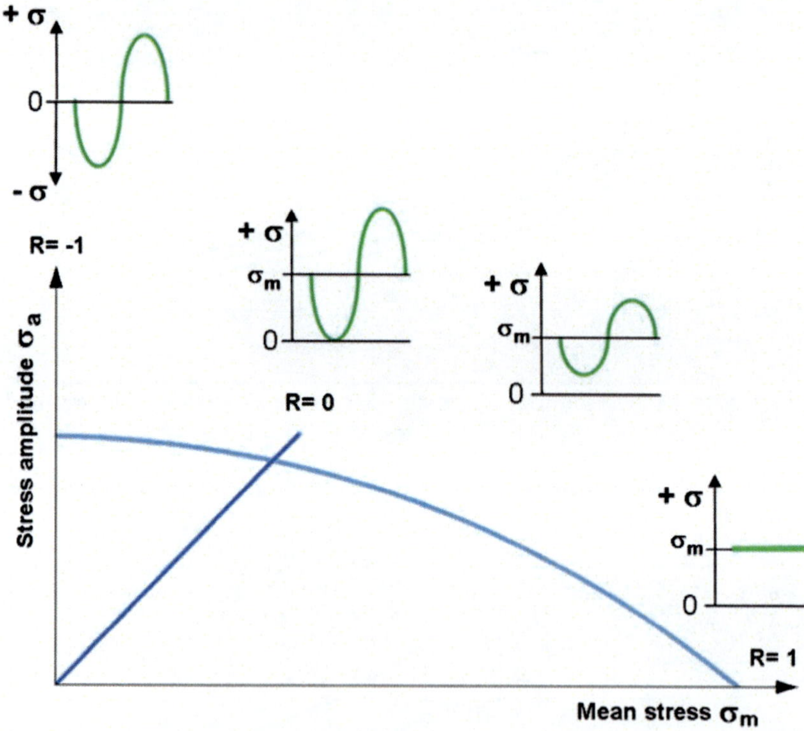

Fig. 1.10 Haigh diagram for a given N_f [6]

Various parameters influence the brittleness of solder, as discussed in Chap. 2.

There seems to be no widely accepted parameter or index (in ASTM, or ASM or Mil Handbook standards or elsewhere) for distinguishing brittleness from ductile failure. The best statement might come from the qualitative guidance offered by Charpy impact toughness measurements (either as a function of temperature or strain rate), where failures below the transition toughness level are considered as generally brittle and those above that as ductile. For tensile tests, it is common to report failure strains, and let readers make the judgement in the context it was presented. In industry, anything below a few % failure strain such as 5 or 10% (from a room temperature tension test, for example) would be considered to be brittle. Thus, there can be an "arbitrary" industry basis, with some understandable justification, to call a material brittle for failure strains less than a few percent.

Regarding shear failure strains, it is difficult to apply such a definition. Microscopically, some single crystals of ductile metals will fail by pure shear along slip planes when suitably oriented in tension tests, but in shear, they can exhibit very large strain to fracture, often like 20, 50%, due to accumulation of sliding on planes across the gage section.

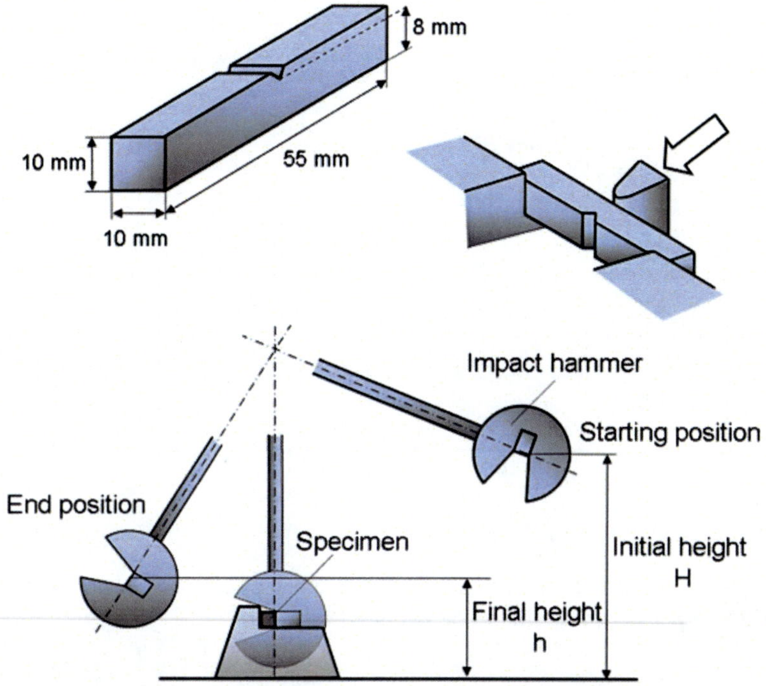

Fig. 1.11 Charpy test

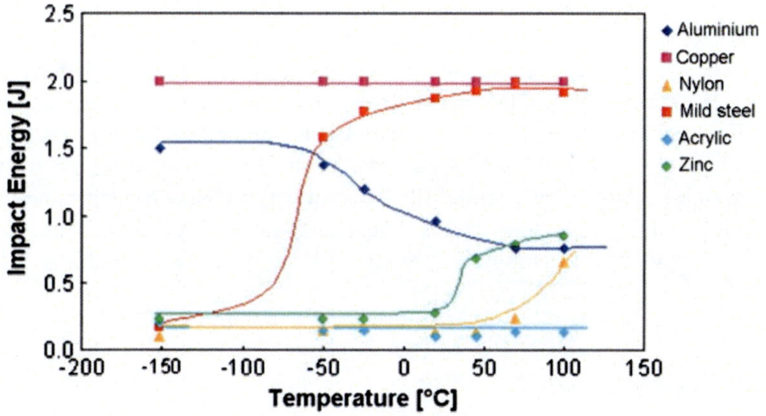

Fig. 1.12 Transition of some materials [7]

On the other hand, under biaxial pure shear load, one could have a material fail with an overall strain at rupture above 10%. However, taking the strain to failure in either stress direction, it could be that in one direction high strain occurs thus being

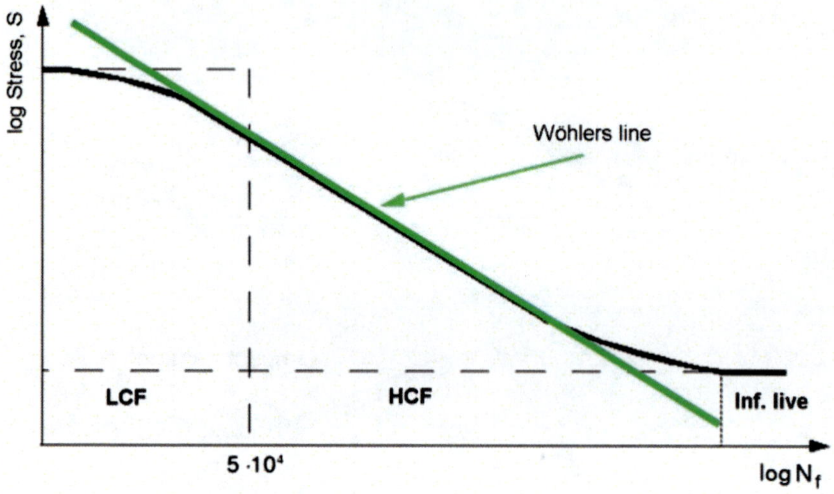

Fig. 1.13 Zones in Wohlers curve

characterised as ductile failure while in the other direction the material deforms only a few per cent and should be declared as brittle

1.2.4 Modes of Fatigue

In Wöhlers curve, three areas can be identified (Fig. 1.13). On the left side, one can see that the stress applied has nearly no influence on N_f. Degradation in this area appears early and is called low-cycle fatigue (LCF). In order to have a more precise definition, it is common sense that failures below 5×10^4 are assigned to LCF. If a specimen fails, later it is attributed to high-cycle fatigue (HCF). Some materials, such as steel, show a fatigue limit, a stress level below which no failures occur which is assigned as infinite life. This limit is not easy to determine because the limit occurs at low stress levels where tests usually run for a long time and where it is difficult to distinguish between a slow degradation and a real fatigue limit.

1.2.4.1 Low-Cycle Fatigue

As already mentioned, in LCF, the stress amplitude applied has no direct influence on N_f. This is due to the fact that in LCF, the degradation is strain driven. Be it time-independent or time-dependent strain. In slow cycles where time-dependent deformation takes place, not only the stress applied to a specimen has to be considered but also the time (the frequency) is important. This applies specially to ductile materials such as solder where creep is essential.

Each cycle will leave some microvoids because the grains only fit in one configuration and every plastic deformation due to GBS will cause voids due to the geometric mismatch. Along the grain boundaries, atoms will diffuse from areas with compressive stress to the voids and fill them up. The longer one waits, the more atoms will diffuse into the voids, and the higher the temperature, the faster the diffusion takes place. However, usually the cycle will be reversed before the voids are totally filled up, thus leaving a microvoid. With every cycle, new microvoids form and coalescent to large defects. The larger the deformation per cycle is, the larger are the microvoids that form. This means that in LCF, a degradation field develops where microvoids coalesce during the cyclic load to crack like separations. This degradation field corresponds to the strain filed in a material which means that degradation due to LCF can form locally in a volume. The occurrence of local degradation is well visible in lead-free solder after thermal cycling (Fig. 1.14).

The degradation due to LCF has been modelled by Coffin and Manson with

$$N_f \cdot \varepsilon_{pl}^{\alpha} = C \tag{1.16}$$

N_f number of cycles to failure, ε_{pl} plastic deformation per cycle, α degradation exponent, C constant as well as by Morrow with

$$N_f \cdot E^{\beta} = C \tag{1.17}$$

N_f number of cycles to failure, E_{pl} strain energy per cycle, β degradation exponent, C constant.

The strain energy can be derived from the area in the stress–strain hysteresis, which occurs in a material under cyclic load with plastic deformation. The main difference of the two models is the fact that in the Coffin–Manson approach, only the strain is the driving force for the degradation. The model of Morrow assumes that also the stress applied influences the degradation. With the approach of

Fig. 1.14 Degraded solder joint made of $SnAg_{3.5}Cu_{0.7}$ after 4,000 thermal cycles (−20°C/+120°C)

Morrow, this means in practical applications that short thermal cycles (e.g. in an accelerated test with steep temperature ramps and large temperature excursion) with large stress but small strain result in an equivalent degradation as in a slow cycle with reduced temperature excursion which, due to creep, will induce large strain with small stress because the area in the stress–strain hysteresis is the same in both cases. The Coffin Manson approach restricts the possibility of accelerated testing since in any test, the cycles must be long enough to allow the material to creep. Generally, it does not make a big difference which of the two models is used. Since it is easier do determine the strain per cycle than the strain energy, the Coffin–Manson relationship is more popular than Morrows equation. Assuming total relaxation during each thermal cycle in the solder joint of an electronic component, the strain can be estimated by:

$$\varepsilon_{\mathrm{pl}} = \frac{\Delta\alpha \cdot \Delta\vartheta \cdot l_0}{d} \tag{1.18}$$

$\Delta\alpha$ difference of the coefficients of thermal expansion between the component and the PCB, $\Delta\vartheta$ temperature swing, l_0 half-length of the component, d thickness of solder gap.

Of course this is only an estimate. The deformation field can only be determined with FEM simulation as it is outlined in Chap. 3. However, also the simulation has its limitations. It shows the deformation field for a given idealised geometry only. In reality, the geometry and also the structure of the solder, which influences the creep behaviour of the solder, change continuously as the degradation evolves. And, due to differences in the geometry and in the structure of the solder joints, the degradation evolves different in each joint. The consequence is evident in a component with two connectors such as a chip resistor. One of the solder joints will degrade faster than the other. This means that over time, the load-bearing area of one joint becomes smaller than the one of the other joint, which again means that in the end of the process, the weaker joint only will be deformed and degrade. As a consequence, in Eq. (1.18), the full length of the component has to be used for l_0 instead of its half-length.

However, for a first approximation of the live time of a solder joint, one can say that the estimate as shown above is often precise enough despite its shortcomings.

1.2.4.2 High-Cycle Fatigue

HCF is stress driven in contrast to LCF. If a load is applied to a homogeneous specimen, the stress is evenly distributed. However, if the bearing cross-section of the specimen contains a discontinuity, stress amplification at the tip will occur (Fig. 1.15). The maximum stress occurring at the tip can be calculated with

$$\sigma_{\max} = 2\sigma_0\sqrt{\frac{a}{\rho}} \tag{1.19}$$

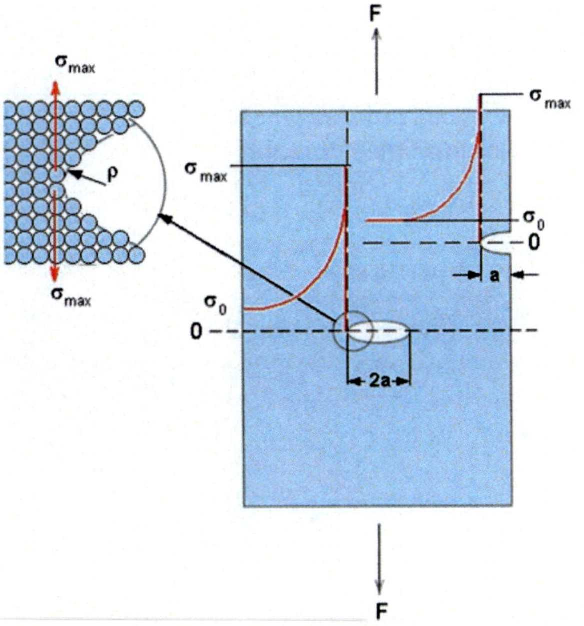

Fig. 1.15 Local overstress due to a discontinuity [6]

σ_{max} maximum stress, σ_0 average stress, a length of discontinuity, ρ radius of discontinuity.

Due to a certain plasticity in metals, σ_{max} is transformed to strain in a limited zone around the tip. The radius of this zone depends on the mechanical properties of the material. In a brittle material, the radius ρ is smaller than in a ductile material and because of the term $1/\rho$, σ_{max} becomes infinite if ρ is converging to 0, independently of the applied load. This maximum stress can be high enough to break up the bonds of the atoms that are right at the end of the tip. As a result, a crack will grow as long as a positive load is applied in a cycle even if the load is far below the yield strength of the material. In other words, the more brittle a material is, the more HCF becomes a thread. The progress of a crack due to high-cycle fatigue can be seen on the surface of a rupture. From an initial site, the stair-like striations indicate the path of the fatigue until the bearing area is so small that the yield strength of the material is exceeded and a forced rupture occurs (Fig. 1.16).

Because in the HCF zone, the Wöhler curve is very close to a straight line, the degradation has been modelled by Basquin with

$$\sigma \cdot N_f^{\alpha} = C \tag{1.20}$$

σ stress amplitude, N_f number of cycles to failure, α degradation exponent, C constant.

Fig. 1.16 High-cycle fatigue damage

Generally, high-cycle fatigue is not a concern for solder joints, since due to the ductility of the material, the critical radius will never converge to 0. With SnPb solder also the brittle interfaces with intermetallic compounds which are prone to high-cycle fatigue rarely see enough load to exhibit failure due to rapid cycling. Usually the leads of components fail before the solder joint shows significant degradation.

Lead-free solders on the other hand show some properties that make the occurrence of high-cycle fatigue in the solder joint more probable:

- The lower creep rate of lead-free solder causes larger stress in the intermetallic compound, thus provoking high-cycle fatigue in the interface of the solder with the substrate.
- Lead-free solder containing Ag shows a clear transition temperature as shown at IMEC with a mini-Charpy tester on specimen measuring 10 × 10 × 55 mm (Fig. 1.17). The outcome of the investigation was that SnPb shows a gradual transition below −20°C. The Sn–Cu alloys show a clear transition between −140 and −120°C. The most interesting result was that Sn–Ag shows a transition temperature depending on the amount of Ag in the alloy. The alloy with 5% Ag had its transition between −60 and −20°C, which is in the range technical applications with lead-free solder are subject to.

The same damaging characteristic as in high-cycle fatigue (incremental growth of a crack due to local overstress in brittle materials) has been observed in tests on lead-free solders but the number of cycles to failure is not necessarily $>10^5$ cycles as expected in high-cycle fatigue. Thus, in solder joints, it might be more appropriate to talk about stress-induced fatigue rather than high-cycle fatigue.

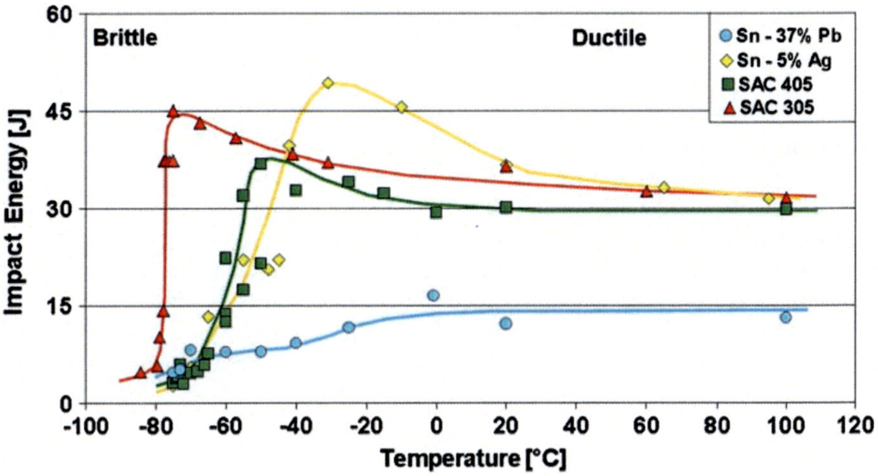

Fig. 1.17 Transition of solders [8]

This means that in certain mission profiles, high-cycle fatigue in lead-free solder might become important:

- Low temperature
- High Ag content
- Fast alternating load such as bending of a PCB due to vibration or drop. Tony Mattila at HUT Helsinki found that a PCB starts to vibrate after impact in a drop test with bending modes occurring at various natural frequencies of the specimen [9]. This is discussed in depth in Chap. 9.

1.2.4.3 More than One Degradation Mechanism

Of course usually more than one degradation mechanisms occur in a material in real-field applications. As it is implied in Eq. (1.12), GBS and DC are always effective in varying proportions. But also HCF and LCF might occur under combined vibration and thermal cycle load in a temperature range close to the transition temperature where creep is active. This has been modelled by Miner and Palmgren with

$$\sum_{i=1}^{n} \frac{n_i}{N_{fi}} = 1 \text{ at failure} \qquad (1.21)$$

n_i number of cycles with degradation mechanism i, N_{fi} number of cycles to failure with degradation mechanism i.

For the special case where different degradation mechanisms are active in one load cycle, Eq. (1.21) can be modified as proposed by Halford and Manson:

$$\sum_{i=1}^{n} \frac{f_i}{N_{fi}} = \frac{1}{N_f} \tag{1.22}$$

f_i fraction of a cycle with degradation mechanism i, N_f number of cycles all mechanisms combined.

In [10], it has been shown in a joined work between CRF and IMEC within the EC-IMECAT project that a combined load of temperature cycling and vibration, as encountered in automotive applications, has a more severe effect on the degradation of a solder joint than the same stress history applied sequentially, incorporating a considerable amount of stress-induced fatigue.

References

1. Norton FH (1929) The creep of steels at high temperatures. Mc. Graw-Hill, New York
2. Garofalo F (1963) Trans Metall Soc AIME 227
3. Hart EW (1967) Acta Metall 15:1545 ff
4. Grossmann G, Weber L (1999) The deformation behaviour of Sn62Pb36Ag2 and its implications on the design of thermal cycling tests. IEEE Trans Electron Packag Manufact 22:71
5. Bargel HJ, Schulze G (2004) Werkstoffkunde, 8th edn. Springer, Berlin
6. Callister WD Jr (2003) Materials science and engineering. Wiley, London
7. University of Cambridge, http://www.doitpoms.ac.uk/tlplib/BD6/results.php. Accessed June 2009
8. Lambrinou K et al (2009) A novel mechanism of embrittlement affecting the impact reliability of tin-based lead-free solder joints. J Electron Mater. doi:10.1007/s11664-009-0841-0
9. Marjamäki P, Mattila TT, Kivilahti JK (2006) A comparative study of the failure mechanisms encountered in drop and large amplitude vibration tests. In: Proceedings 56th electronic component and technology conference, San Diego, CA, May 30–June 2, pp 95–101
10. Vandevelde B et al FP5-CSG-IMECAT: highlights of a EC funded project on lead-free materials and assembly development technology, IPC Barcelona

Chapter 2
Factors Affecting the Bulk Embrittlement of Pb-Free Solder Joints

K. Lambrinou

Abbreviations

ASTM	American Society for Testing and Materials
bcc	Body-centred cubic
bct	Body-centred tetragonal
BSE	Back scattered electron
CAT	Crack-arrest temperature
CMOD	Crack-mouth opening displacement
CTOD	Crack-tip opening displacement
CVN	Charpy V-notch
DBTT	Ductile-to-brittle transition temperature
fcc	Face-centred cubic
FTE	Fracture transition elastic temperature
FTP	Fracture transition plastic temperature
hcp	Hexagonal close-packed
IMC(s)	Intermetallic(s) or intermetallic compound(s)
LLD	Load line displacement
NDT	Nil ductility temperature
Pb-free	Lead-free
SAC	Sn–Ag–Cu
SE	Secondary electron
SEM	Scanning electron microscopy
Sn-based	Tin-based

K. Lambrinou
IMEC, Kapeldreef 75, 3001 Leuven, Belgium

Present Address:
K. Lambrinou (✉)
SCK-CEN, Boeretang 200, 2400 Mol, Belgium
e-mail: klambrin@sckcen.be

G. Grossmann and C. Zardini (eds.), *The ELFNET Book on Failure Mechanisms, Testing Methods, and Quality Issues of Lead-Free Solder Interconnects*,
DOI: 10.1007/978-0-85729-236-0_2,

Special Symbols

[]* Figure adapted from the reference given in []

[]§ From "The Science and Engineering of Materials", 5th edition, by Askeland and Phulé. © 2006 by Nelson. Reprinted with permission of Nelson, a division of Thomson Learning: www.tomsonrights.com. Fax: 800 730 2215

[]# From "The Science and Engineering of Materials", 1st edition, by Askeland. © 1984 by Wadsworth, Inc. Reprinted with permission of Brooks/Cole, a division of Thomson Learning: www.tomsonrights.com. Fax: 800 730 2215

[]& From "Deformation and Fracture Mechanics of Engineering Materials", 4th edition, by Hertzberg. © 1996 by John Wiley & Sons, Inc. Reprinted with permission of John Wiley and Sons, Inc

[]† From "Fracture and Fatigue Control in Structures: Applications of Fracture Mechanics", 3rd edition, by Barsom and Rolfe. © 1999 by ASTM International. Reprinted with permission of ASTM International

[]∝ Reproduced with permission of Emerald Group Publishing Ltd. © 2000 by MCB University Press

2.1 Introduction

The fracture behaviour of all materials is affected by several factors, some of which are 'intrinsic', i.e. reflecting the material constitution, and some 'extrinsic', i.e. relating to the processing or service conditions. Examples of 'intrinsic' factors are the material composition, crystal structure, and microstructure, while examples of 'extrinsic' factors include the cooling rate during processing, as well as the temperature, strain rate, or constraint of the material during service. Changes in the processing or service conditions often alter the crystal structure, microstructure, or even composition of the material (e.g. by contamination), which might affect its fracture behaviour. Electronic materials, like the lead-free (Pb-free) solders studied in the framework of ELFNET, are no exception to the above general rules of thumb. It has already been observed, for example, that the fracture behaviour of tin-based (Sn-based) solders changes from ductile to brittle under solder-specific service conditions. This type of brittle fracture occurs in the solder joint unlike the typical brittle failures occurring in the intermetallic layers that are present between solder joint and bond pad; therefore, it is often referred to as 'bulk solder embrittlement'. This chapter attempts to identify the factors affecting the bulk embrittlement of Sn-based Pb-free solder joints, as this type of embrittlement may occur during the solder joint life cycle, depending on the requirements of the 'mission profile'. Identifying these factors and understanding how they facilitate embrittlement is the first step in finding ways to limit, delay, or even avoid bulk solder joint embrittlement during service. However, due to the limited available data, the factors affecting the bulk embrittlement of Pb-free solder joints will be addressed by referring to data acquired from studying other metallic materials with similar behaviour, e.g. ferritic steels.

2.1.1 Comparing Ductile to Brittle Fractures

It is common knowledge that *brittle fractures* occur with little or no plastic deformation prior to failure and are characterised by very limited energy absorption. This is the type of failure that occurs at extremely high speeds (up to 2,000 m/s or 7,000 ft/s in steels) [1]. An example of the stress–strain behaviour of a brittle material subjected to tension is given in Fig. 2.1a, where one may observe the essential absence of strain (i.e. deformation) before failure of the material at the ultimate tensile strength, σ_{tens}.

Contrary to brittle fractures, ductile fractures are characterised by appreciable plastic deformation prior to failure and high energy consumption [1]. When a ductile material is subjected to tension, as shown in Fig. 2.1b, an appreciable amount of strain usually results from loading the material above its yield strength, σ_{yield}. The fact that ductile materials deform a lot before failing, and that this failure is relatively slow, makes ductile fractures more appealing than brittle ones in the eyes of the design engineer.

Two notions very closely related to the type of material failure (i.e. ductile vs. brittle) are those of toughness and ductility. *Toughness* is the ability of a material to resist fracture and represents the energy absorbed during failure [2, 3]. A brittle material, for example, absorbs small amounts of energy during fracture, while a tough/ductile material absorbs large energy amounts. For an unnotched tensile bar, the energy dissipated during failure can be estimated from the area under the stress–strain curve (Fig. 2.1), as follows [2]:

$$\text{Energy/volume} = \int_{0}^{\varepsilon_f} \sigma \mathrm{d}\varepsilon \tag{2.1}$$

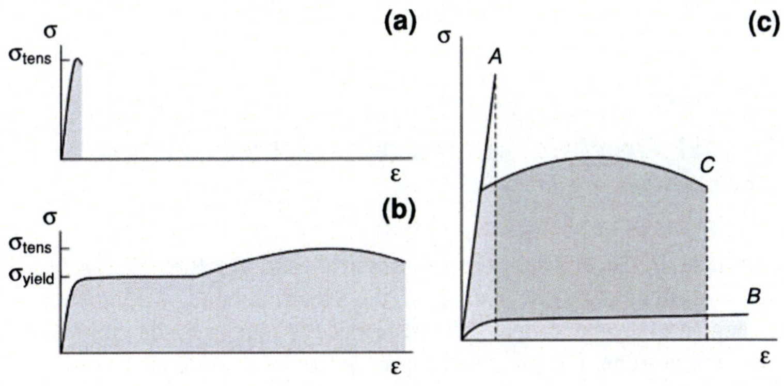

Fig. 2.1 Typical stress–strain curves of (**a**) brittle materials and (**b**) ductile materials [1]*. **c** Superposition of the stress–strain curves of a strong material with little capability for plastic flow (*A*), a material with low strength but high ductility (*B*), and a material with optimum combination of strength and ductility for maximum toughness (*C*) [2]*. * Figure adapted from the Reference in []. The meaning of symbol * is the same throughout this text

where σ is the stress, ε is the strain, and ε_f is the strain at failure.

On the other hand, *ductility* is the ability of a material to deform plastically without fracturing. Ductility is commonly expressed either as % elongation in the gage length, or % reduction in the area of a tensile specimen tested to failure [4]. It must be emphasised that maximum toughness requires a combination of strength and ductility [2], since high strength or good ductility alone cannot guarantee an enhanced toughness (Fig. 2.1c).

2.1.2 Factors Affecting the Type of Fracture

It is often erroneously believed that the type of fracture of a certain material is either ductile or brittle and that defining the fracture type is quite straightforward. This could, for example, be based on the fact that the deformation and fracture characteristics of materials depend on very material-specific properties, such as the nature of the electron bond, the crystal structure, and the degree of atomic order [2]. Another way to predict the type of failure of a material would be to consider its fundamental engineering properties, such as the yield strength, tensile strength, and tensile ductility [2], since high-toughness materials combine good strength and ductility. However, the job of a design engineer is not that simple, because some materials (like many metals) that fail in a ductile manner under mild service conditions might embrittle if subjected to low temperatures, high strain rates, or multiaxial stress states related to the presence of notches or defects [1–3]. The Sn-based Pb-free solders extensively addressed in this chapter fall into that category, for example. Moreover, some materials (like many ceramics) that fail in a brittle manner at mild ambient temperatures will certainly increase their ductility when exposed to sufficiently high temperatures; for ceramics, this change in fracture behaviour occurs usually above 1,000°C [5].

2.2 Crystal Structure, Microstructure, and Fracture Resistance

As mentioned in Sect. 2.1.2, the deformation and fracture characteristics of a material are affected by the nature of the electron bond, the crystal structure, and the degree of atomic order in the material. For example, the more rigidly fixed the valence electrons, the greater the propensity of a material to fail in a brittle manner [2]. In ionic solids, the valence electrons are tightly bound to their respective atoms inside the crystal [6], which explains their high propensity to brittle failure. In both covalent and metallic crystals, the valence electrons are shared between atoms [6]; however, their fracture behaviour is different due to the following reasons: in covalent solids, the valence electrons are shared between an atom and its nearest neighbours, while in metallic solids, the valence electrons are shared equally by all atoms in the material. The greater restriction in the movement of valence electrons experienced by covalent materials (e.g. diamond, silicon, carbides, nitrides, and silicates) accounts for their tendency to fail in a brittle manner, while materials held together by metallic bonds are the easiest to deform, failing usually in a ductile manner [2].

Materials with a low crystal symmetry, where slip is difficult, show a greater tendency to fail in a brittle manner. On the contrary, close-packed crystals characterised by high crystal symmetry are capable of considerable plastic deformation. For example, *face-centred cubic (fcc)* and *hexagonal close-packed (hcp)* crystals are less prone to brittle failure than *body-centred cubic (bcc)* crystals. Moreover, bcc metals show a much higher propensity to low-temperature embrittlement than fcc and hcp metals, as will be discussed more extensively later on. Figure 2.2 shows schematic representations of the fcc, bcc, and hcp crystal structures.

Another structural factor that affects the fracture resistance of a material is the degree of order in the atomic arrangement. It has been observed that the tendency for brittle failure increases as the degree of order in the atomic arrangement increases. In fact, the addition of a solute to a crystal lattice progressively suppresses plastic flow, as the solid solution that results from this addition changes gradually from disordered to short-range ordered first, and subsequently to long-range ordered [2].

The above-discussed theoretical aspects of the material structure have a great impact on the material's fracture resistance, but might be difficult to grasp and put

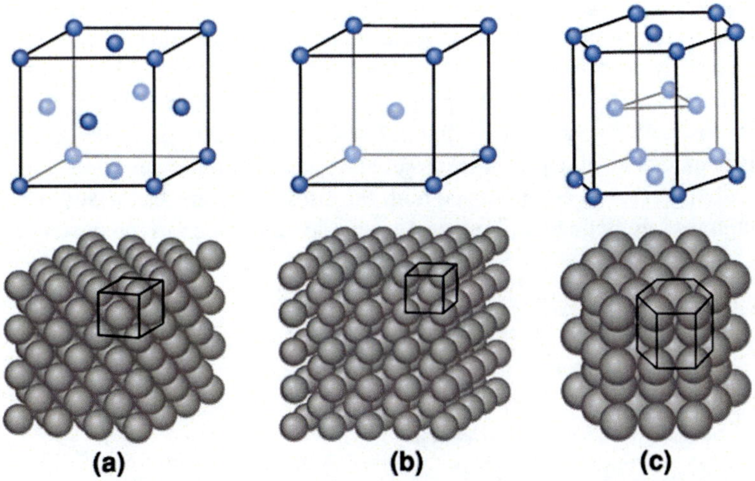

Fig. 2.2 **a** The face-centred cubic (fcc) crystal structure is shown as a reduced-sphere unit cell and an atomic aggregate [5]*. **b** The body-centred cubic (bcc) crystal structure is shown as a reduced-sphere unit cell and an atomic aggregate [5]*. **c** The hexagonal close-packed (hcp) crystal structure is shown as a reduced-sphere unit cell and an atomic aggregate [5]*

in practice by the design engineer. A simple question remains: What are the practical ways one could use to limit the tendency for brittle failure of a structural material? And if embrittlement cannot be avoided, as often happens at extreme service conditions, what are the available testing methods that can provide accurate information on the limitations of a specific material?

2.2.1 Elements of Fracture Mechanics

Before attempting to answer the above critical questions, it would be useful to introduce and put into perspective the following two concepts: (a) the stress-intensity factor and (b) the fracture toughness. It is common knowledge that the materials used to build any structure are not perfect, since they contain cracks, sharp notches, and discontinuities of various kinds. These defects create stress concentrations within the material, increasing its tendency to failure. Therefore, the design engineer needs a tool that allows the quantitative determination of the allowable stress levels in such structures, so as to avoid fracture during service. This tool is the scientific discipline of Fracture Mechanics, which envisages the intensity of stress concentration as the *driving force* to failure, and the material's fracture toughness as the *resistance force* against failure [1].

Depending on the mode of loading, the fracture of flaw-containing components may be described by a stress analysis based on principles of elasticity theory. Figure 2.3 illustrates the three major modes of loading, which are [1, 2]:

1. *Mode I*: the opening or tensile mode, where the crack surfaces move directly away from each other,
2. *Mode II*: the sliding or in-plane shear mode, where the crack surfaces slide over one another in a direction perpendicular to the crack leading edge, and
3. *Mode III*: the tearing or anti-plane shear mode, where the crack surfaces also move relative to one another, but in this case parallel to the crack leading edge. Since mode I loading is the most frequently encountered in structural applications involving cracked components, both the stress-intensity factor and the fracture toughness are defined with respect to that loading mode in the present text.

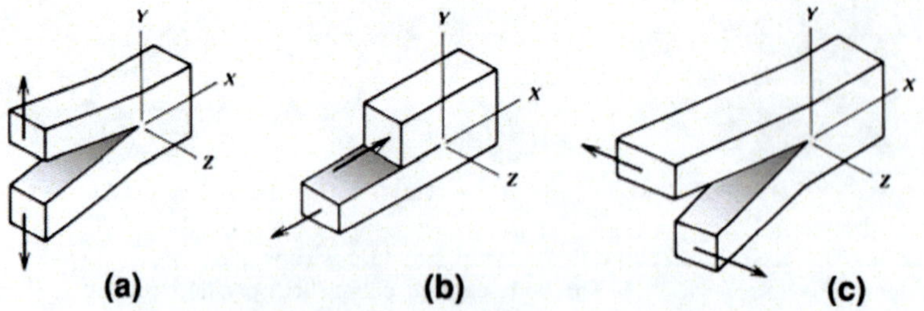

Fig. 2.3 The three basic loading modes: **a** Mode I, **b** Mode II, and **c** Mode III [1, 2]*

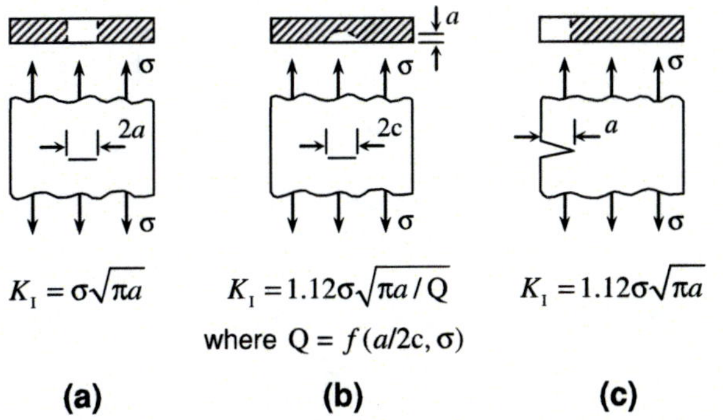

Fig. 2.4 Equations relating the stress-intensity factor, K_I, with the stress, σ, and the crack size, a, for the following crack configurations: **a** through-thickness crack, **b** surface crack, and **c** edge crack. These equations describe the elastic-stress field in the vicinity of a crack tip inside a body subjected to tension [1]*

In linear-elastic fracture mechanics, the stress field ahead of a sharp crack subjected to mode I loading is characterised by the *stress-intensity factor*, K_I, which has units of $\text{MPa}\sqrt{\text{m}}$ or $\text{ksi}\sqrt{\text{in.}}$ [1]. This parameter is related to both the stress level, σ, and the flaw size, a. Figure 2.4 provides equations describing the elastic-stress field near the tip of cracks with different configurations, inside a body stressed in tension. In summary, K_I is affected by the specimen geometry, the applied load, as well as the size and shape of the flaw [1]. When the combination of σ and a results in a critical K_I value, unstable crack growth (related to brittle failure) occurs. This critical value of the stress-intensity factor at failure, K_{Ic}, is the so-called *fracture toughness*, and it is a material property [1]. The fracture toughness represents the resistance of a certain test specimen against crack propagation. It must be emphasised, however, that determining the fracture toughness of a structural material is not always an easy task, as the fracture toughness is influenced by the service temperature, loading rate, and degree of material constraint. Therefore, the K_{Ic} of a certain material can be determined for a specimen with a given thickness at a specified temperature and loading rate. In general, the fracture toughness decreases by decreasing the temperature, increasing the loading rate, and/or increasing the material thickness [1, 2]. Special reference will be made to each one of these factors later in the text, as they are also expected to affect the fracture behaviour of certain Pb-free solders.

2.2.2 Metallurgical Aspects of Fracture Resistance

It has already been mentioned that high toughness demands the combination of strength and ductility. However, metallurgists take this concept one step further and strive for the production of structural materials that exhibit both high strength

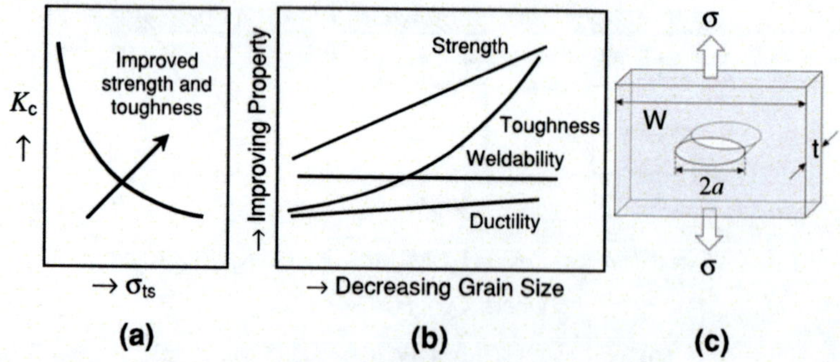

Fig. 2.5 **a** Inverse relation between the tensile strength, σ_{ts}, and the fracture toughness, K_c, of an alloy. Optimising the alloy properties would ideally result in a shift of the tensile strength–fracture toughness curve in the direction of the *arrow* [2]*. **b** Grain size refinement results in a simultaneous improvement in the strength and the toughness of the alloy. Reduction in the grain size does not affect adversely the alloy's weldability or ductility [2]*. **c** Through-thickness crack in a large plate subjected to a tensile stress σ. W is the plate width, t the plate thickness, and $2a$ the crack length [2]*

and fracture toughness. This is asking for a lot, since the strengthening of a material happens almost invariably at the detriment of its fracture toughness, as shown in Fig. 2.5a. Still, there are ways to improve both the strength and the fracture toughness of a material [2]; these ways should also be considered for the improvement of the in-service performance of commercially available Pb-free solders. The simultaneous improvement in strength and toughness may be achieved in one of the following ways:

1. By modifying the alloy chemistry and melting/processing process to improve the toughness; this could, for example, happen by removing or neutralising certain undesirable elements that degrade the toughness,
2. By optimising the microstructure and phase distribution to maximise the toughness, and/or
3. By refining the microstructure, or in other words, by opting for finer grain sizes.

Point (3) is quite obvious for all materials, and its effect on strength, toughness, and weldability is graphically presented in Fig. 2.5b. In order to provide a better insight into the benefits of microstructural refinement, the following could be said: strengthening occurs because smaller grains are likely to contain smaller flaws; smaller inherent flaws imply the need of a greater stress to achieve failure.

In terms of equations, Griffith [1, 2] suggested that the stress required for failure is given by:

$$\sigma = \sqrt{\frac{2E\gamma_s}{\pi a}} \quad \text{for plane-stress (biaxial stress) conditions} \tag{2.2}$$

and

$$\sigma = \sqrt{\frac{2E\gamma_s}{\pi a(1 - v^2)}} \quad \text{for plane-strain (triaxial stress) conditions} \tag{2.3}$$

where E is the material's modulus of elasticity, γ_S is the specific energy of the surface created during fracture, $2a$ is the crack length, and v is the Poisson's ratio of the material. Obviously, the larger the size of the flaw/crack, the lower the stress required to fracture the material. A model crack may be visualised in Fig. 2.5c. With respect to the toughening of a single-phase material by grain refinement, one could think that the finer the grain size, the more often will a crack cross grain boundaries during its propagation. Since grain boundaries tend to decelerate or even stop cracks, the travelling crack will spend more energy to propagate in a finer-grained material than in a coarser-grained one. The higher the energy absorbed during fracture, the tougher the material.

Examples of improving the properties of Pb-free solders based on the suggestions made in points (1) and (2) include the addition of special retardants to avoid tin pest on tin-based (Sn-based) solder alloys (in relation to point 1), and/or the control of the size, shape, and acuity (i.e. sharpness) of the intermetallic (IMC) particles in the solder bulk (in relation to point 2). Reference to the last two examples will be made later in the text. Before moving to the next section, however, it would be useful to briefly underline the role of IMC particles in the in-service embrittlement of Pb-free solder alloys.

Intermetallics or *intermetallic compounds* (IMCs) are discrete chemical compounds made by the mixing, in a certain proportion, of two or more metals with each other or with non-metals [5, 7]. An example of a widely known intermetallic compound formed in steels is cementite (Fe_3C). It is quite important to realise that the structure and properties of intermetallics are different from those of the constituent metals; for example, IMCs are characterised by enhanced atomic order and mixed (metallic and covalent/ionic) bonding. Moreover, their properties are a compromise between the typical properties of ceramics and those of metals: when compared to most metals, they exhibit higher hardness and refractoriness, together with poorer toughness and workability (and, hence, ductility). It has also been shown that the presence of brittle second-phase particles in an otherwise ductile matrix is a major cause of embrittlement [2]. Figure 2.6 illustrates this principle, showing that the probability of crack initiation at particles such as the IMCs in Pb-free solders is very high. The above indicates that the presence of IMCs in the bulk of Pb-free solder alloys is a major concern when it comes to the fracture behaviour of these alloys, as will be illustrated later in the text.

2.2.3 Elements of Dislocation Theory

Dislocations are very important, especially for metals and metallic alloys, since they provide a mechanism for plastic deformation, which may be regarded as the

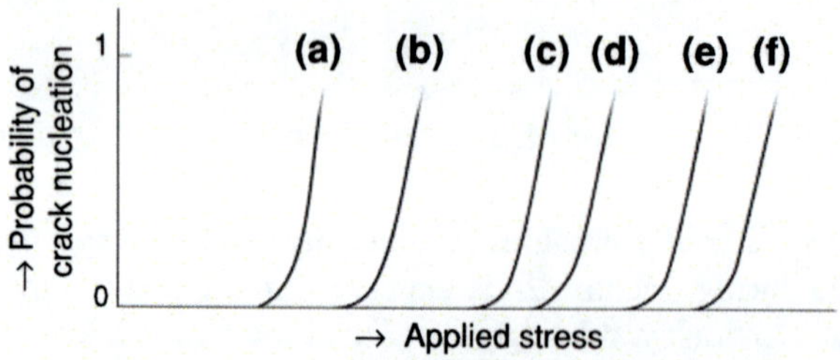

Fig. 2.6 Probability as a function of the applied stress that one of the following microcrack formation mechanisms is active: [*a*] embrittled grain boundaries, [*b*] brittle second-phase particles, [*c*] twin–twin intersections, [*d*] twin–grain boundary intersections, [*e*] slip-induced grain boundary opening (e.g. at triple points), and [*f*] slip band-slip band intersections. From a study conducted on high-purity iron-containing carbides as brittle second-phase particles [2]*

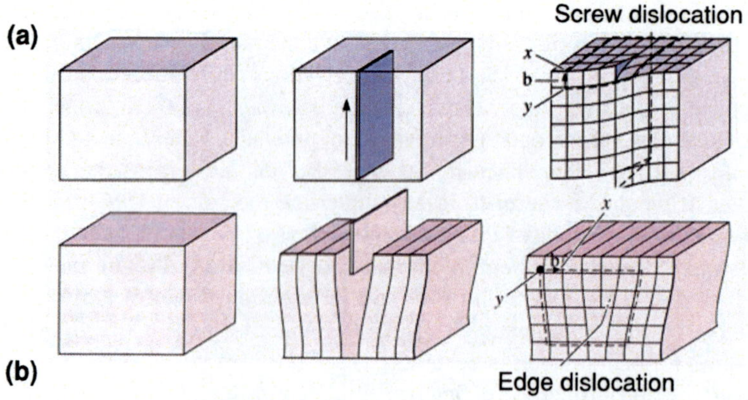

Fig. 2.7 **a** For the creation of a screw dislocation, the perfect crystal is cut and sheared one atom spacing; the screw dislocation is the line across which shearing occurs [7]§. **b** For the creation of an edge dislocation, the perfect crystal is cut and an extra plane of atoms is inserted; the edge dislocation is the bottom edge of the extra plane [7]§. In both types of dislocations, a Burgers vector (**b**) is required to close a loop of equal atom spacings around the dislocation. § From "The Science and Engineering of Materials", 5th edition, by Askeland and Phulé. © 2006 by Nelson. Reprinted with permission of Nelson, a division of Thomson Learning: www.tomsonrights.com. Fax: 800 730 2215. The meaning of symbol § is the same throughout this text

cumulative effect of the slip of numerous dislocations [7]. *Dislocations* are line imperfections in an otherwise perfect crystal, where they are usually introduced either during solidification or permanent (i.e. plastic) deformation [7]. There are three known types of dislocations: the screw dislocation, the edge dislocation, and the mixed dislocation. Figure 2.7 gives a visual impression of two of them: the screw dislocation and the edge dislocation.

Slip is the process by which a dislocation moves in a material, resulting in the deformation of this material [7]. Figure 2.8 illustrates the slip process for both an edge dislocation and a screw dislocation. Slip explains why the strength of a metal

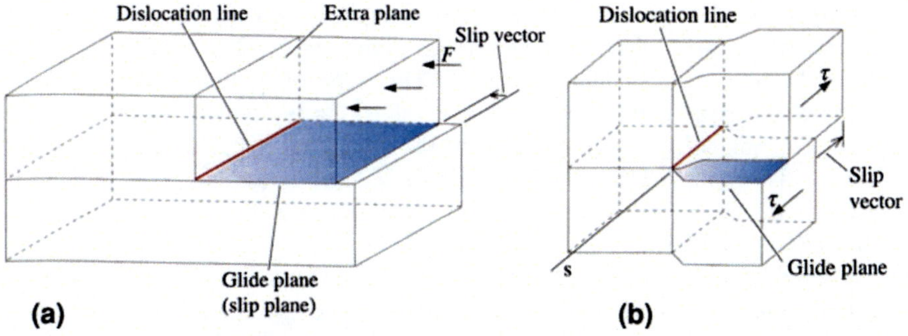

Fig. 2.8 Schematic representation of the slip line, slip plane, and slip (Burgers) vector for (**a**) an edge dislocation and (**b**) a screw dislocation [7]§

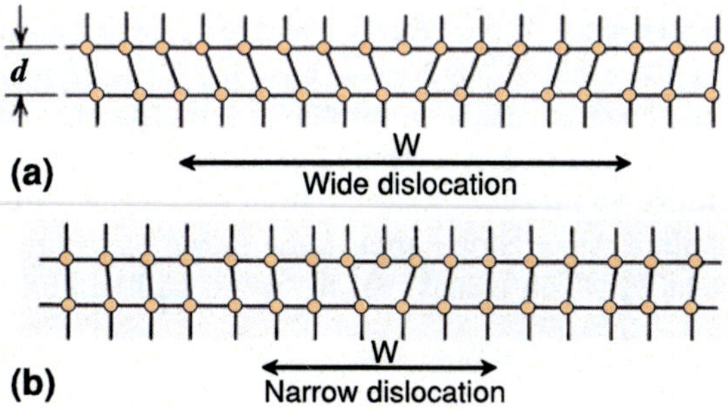

Fig. 2.9 Schematic representation of two edge dislocations with different width: (**a**) a wide edge dislocation, and (**b**) a narrow edge dislocation. *W* is the dislocation width [2]*

is much lower than the theoretical strength predicted from the metallic bond. In this sense, slip is responsible for the ductility and, therefore, the workability of metals. On the other hand, the presence of dislocations offers also the possibility of strengthening a certain metallic material by introducing obstacles (e.g. second-phase particles) that interfere with the dislocation motion [2, 8].

It has by now become clear that the movement of dislocations is a prerequisite for plastic deformation before fracture, and it should ideally occur even at extreme service conditions (e.g. low temperatures, high strain rates), if brittle failures were to be avoided. Unfortunately, the lattice resistance to dislocation movement changes with temperature, and this is the basis of the low-temperature embrittlement encountered in Sn-based Pb-free solders, which will be addressed later in the text. The resistance of a lattice to dislocation movement is expressed by the magnitude of the force required to move a dislocation through this lattice. This is known as the *Peierls–Nabarro force*, and the corresponding stress is the *Peierls–Nabarro stress* [2].

The magnitude of the Peierls–Nabarro stress decreases as the width of the dislocation, W, or the distance between similar planes, *d*, increases (Fig. 2.9). Since the interplanar spacing, *d*, increases with the planar atomic density, it is easier to have slip on close-packed planes. Moreover, the dislocation width, which represents the distance over which the lattice is distorted due to the presence of the dislocation, is large when the bonding forces are spherical in distribution and act along the line between atomic centres. This type of bonding characterises the close-packed structures of fcc and hcp crystals. On the other hand, highly directional bonding forces, as in the case of bcc, ionic, or covalent crystals, result in narrow dislocations. Therefore, the Peierls–Nabarro stress is low in fcc and hcp crystals, moderate in bcc crystals, and large in ionic or covalent crystals (e.g. ceramic materials) [2].

The Peierls–Nabarro stress is sensitive to the lattice thermal energy and, hence, to the ambient temperature [2]. Decreasing the temperature limits the thermal energy available for dislocation motion, which increases the Peierls–Nabarro stress, especially for crystals with a high starting Peierls–Nabarro stress (e.g. bcc). However, crystals with a negligible starting Peierls–Nabarro stress (e.g. fcc and hcp) experience little increase in their Peierls–Nabarro stress when the ambient temperature decreases. It is also quite important to realise that the temperature dependence of the Peierls–Nabarro stress is related to the temperature dependence of the yield strength of a material [2]. Before going deeper into the last phenomenon, it might be useful to introduce the yield strength and the (ultimate) tensile strength of a metallic material with the help of Fig. 2.10, which shows the stress–strain curve resulting from the tensile testing of an aluminium (Al) alloy to failure. Upon stress application in the elastic regime, the material deforms elastically, which means that the strain will be completely recovered as soon as the

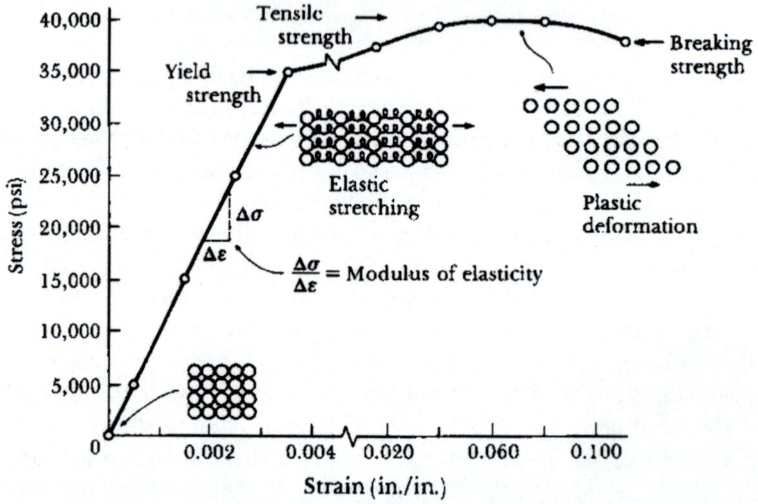

Fig. 2.10 The stress–strain curve resulting from the tensile testing of an Al alloy [9]#. # From "The Science and Engineering of Materials", 1st edition, by Askeland. © 1984 by Wadsworth, Inc. Reprinted with permission of Brooks/Cole, a division of Thomson Learning: www.tomsonrights.com. Fax: 800 730 2215. The meaning of symbol # is the same throughout this text

stress is withdrawn. The slope of the stress–strain curve in the elastic region is the material's *modulus of elasticity* or *Young's modulus, E.* As the applied stress increases, the material starts to deform both elastically and plastically. The critical stress value required to initiate plastic deformation is defined as the *elastic limit* of the material [3, 7]. In metallic materials, this is usually the stress required for dislocation motion or slip to be initiated. On the other hand, the stress level above which the relationship between stress and strain is not linear is the so-called *proportional limit* [3, 7]. In most materials, the elastic and proportional limits are quite close, but neither of them can be determined accurately since the measured values depend on the sensitivity of the equipment. Therefore, an *offset strain value* (typically, 0.002 or 0.2%) is defined, from which a line is drawn parallel to the linear portion of the stress–strain curve. The stress value corresponding to the intersection of this line and the stress–strain curve is the *0.2% offset yield strength* or *yield strength* [3, 7]. At stresses higher than the yield strength, plastic deformation occurs. The stress corresponding to the highest applied force is the *tensile strength* of the tested material. In ductile materials, the tensile strength is the stress at which *necking* begins. Necking refers to the formation of an area in the tensile specimen that is locally much more deformed (i.e. thinner) than the rest of the specimen.

The fact that the temperature dependence of the Peierls–Nabarro stress is related to that of the yield strength is the reason why the yield strength of fcc metals like Al, copper (Cu), and austenitic stainless steel alloys has a limited temperature dependence [2, 4, 9]. Figure 2.11a shows the temperature dependence of the tensile properties of an Al alloy [9]. This plot shows that as the temperature decreases, both the yield and tensile strength of the Al alloy increase. However, the

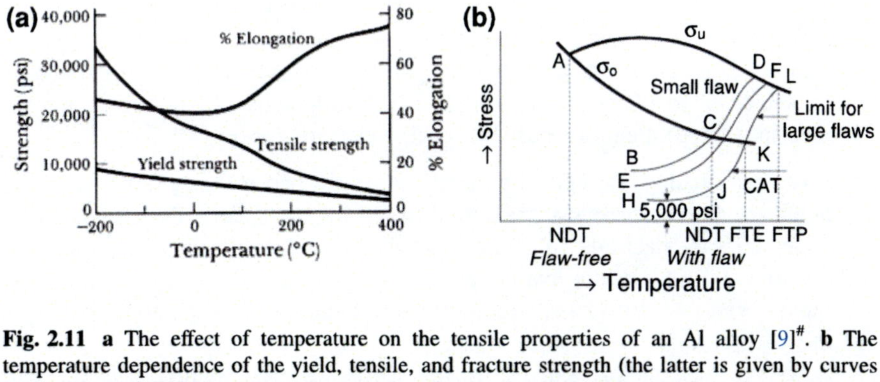

Fig. 2.11 **a** The effect of temperature on the tensile properties of an Al alloy [9][#]. **b** The temperature dependence of the yield, tensile, and fracture strength (the latter is given by curves BCD, EF, and HJKL) of a steel containing flaws of various sizes. CAT is the crack-arrest temperature curve: it gives the stress required for the unstable propagation of a large flaw as a function of temperature; fracture does not occur at any point on the right of that curve. FTE is the fracture transition elastic temperature, i.e. the temperature above which elastic stresses cannot propagate a crack. Above FTP, which is the fracture transition plastic temperature, the material behaves as if it were flaw-free; this means that a crack (irrespective of its size) cannot propagate in an unstable manner [8]

yield strength increases less rapidly than the tensile strength, which allows the Al alloy to retain some of its ductility at low service temperatures. The reason for the not-so-steep increase in the yield strength with lowering the temperature is related to the limited temperature dependence of the Peierls–Nabarro stress of this alloy. Examples of other fcc metals that for the same reason remain ductile at low temperatures include nickel (Ni), lead (Pb), and silver (Ag) [4].

On the contrary, the Peierls–Nabarro stress in bcc metals like iron (Fe), chromium (Cr), molybdenum (Mo), tantalum (Ta), tungsten (W), and ferritic stainless steel alloys rises rapidly with decreasing temperature [2, 4, 9]. This happens due to the strong temperature dependence of the Peierls–Nabarro stress, which becomes a large part of the yield strength in the low-temperature regime [2]. Figure 2.11b shows the temperature dependence of the tensile properties of a ferritic stainless steel alloy [8]. This plot shows that as the temperature decreases, the yield strength, σ_o, of the alloy increases faster than its tensile strength, σ_u. For an unnotched, flaw-free specimen, ductility is retained until a very low temperature, where $\sigma_o = \sigma_u$ (point A). This temperature is known as the *nil ductility temperature* (*NDT*); at temperatures lower than the NDT, fracture becomes 100% brittle [1, 2, 8]. One may also see that the presence of a small flaw ($a < 0.1$ in.) raises substantially the NDT of the steel from point A (flaw-free) to point C; for that particular case, the NDT was observed to increase by ~200°F, i.e. 110°C [8]! With materials like bcc metals, the presence of flaws/cracks/notches does not only influence the material's strength, but its toughness as well. The notch sensitivity of a material and the description of available testing methods to estimate it will be discussed in Sect. 2.3.

2.3 Impact Test: Work Principle and Interpretation of Results

The ability of a material to absorb energy in the presence of a sharp notch, often when subjected to impact, is known as *notch toughness* [1]. Materials may be divided into three categories based on their notch toughness [8]:

1. Low- and medium-strength fcc metals and most hcp metals exhibit very high notch toughness; these materials tend to fail invariably in a ductile manner.
2. High-strength materials ($\sigma_o > E/150$) with very low notch toughness; these materials fail in a brittle manner even when the applied stress is in the elastic range. High-strength steels, aluminium, and titanium (Ti; hcp) alloys belong to this category.
3. Low- and medium-strength bcc metals, ceramics, beryllium (Be; hexagonal), and zinc (Zn; hcp) fail in a brittle manner at low temperatures and in a ductile manner at high temperatures. In other words, their notch toughness is temperature dependent. The transition from ductile to brittle fracture for the metals in this category occurs at 0.1–0.2 of their absolute melting temperature, Tm; for ceramics, this transition occurs between 0.5 and 0.7 Tm.

The *notch sensitivity* of a certain material may be evaluated by comparing the absorbed energies of notched and unnotched specimens during impact [2, 9]. Since notches are often present in structural materials (due to poor machining, fabrication, or design), knowing the notch toughness of the materials used in any structure is very useful for the design engineer. Moreover, since the same engineer is interested in the worst possible combination of service conditions that might lead to failure, experimental testing methods were devised that intentionally suppress the ability of the material to deform plastically.

An example of such a test is the *Charpy V-notch impact test*, which (a) imposes high strain rates, and (b) creates a triaxial stress state by introducing a sharp notch to the test specimen. Furthermore, the Charpy V-notch (CVN) impact test can also be performed at low temperatures, limiting even further the ability of the test specimen to deform plastically. Figure 2.12 shows a Charpy impact test set-up, together with various possible test specimen configurations [9]. During testing, a heavy pendulum, starting at an elevation h_0, swings through its arc, strikes, and breaks the specimen, reaching a lower final elevation, h_f. Knowing the initial and final pendulum elevations allows the calculation of the potential energy difference. This difference is mainly the *impact energy* absorbed by the test specimen during fracture. For the Charpy test, this energy is expressed in joules (J) or foot-pounds (ft-lb). The ability of a material to withstand impact is often referred to as *impact toughness*; the impact toughness of a test specimen represents the energy absorbed during its fracture [7]. In this section, the factors affecting the impact toughness of a test specimen will be extensively addressed, so as to ensure (a) the correct evaluation of experimental results and (b) the understanding of the limitations of this testing method.

One could roughly divide the factors affecting impact toughness into (a) factors that are related with the service conditions and (b) factors that are related with the material microstructure. It is important to realise that the factors in these two categories are interlinked in more than one way. Examples of factors in the first category include temperature, loading/strain rate, and degree of material constraint; the response of a certain metal/alloy to these factors is largely determined

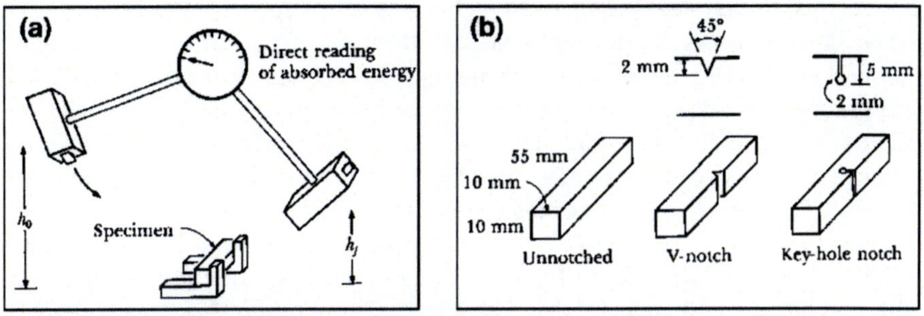

Fig. 2.12 **a** The Charpy impact test set-up [9]#. **b** Typical test specimen configurations designed for Charpy impact testing [9]#

by its crystal structure. For dense (i.e. non-porous) materials, factors belonging to the second category include the alloy composition as well as the volume fraction, size distribution and spacing of brittle second-phase particles, like the IMCs in Pb-free solders. At this point, one could also indicate that the volume fraction, size distribution, and spacing of second-phase particles are dictated by the alloy composition and the processing/service conditions experienced by the alloy. For example, the volume fraction of second-phase particles in an alloy is in principle defined by equilibrium thermodynamics. However, the size distribution and spacing of such precipitates is usually evolving in time, depending on the initial material microstructure (i.e. the microstructure at the end of processing) and the service conditions. The initial microstructure is greatly influenced by the chosen processing conditions, like cooling rate, temperature, and pressure; on the other hand, the service conditions are responsible for the growth and ripening of second-phase particles or the phase transformations that could take place in the material. For porous/cracked materials, the material response to impact is also dictated by the amount, spatial distribution, and orientation of material defects. Figure 2.13a illustrates the influence of the flaw size and flaw orientation on the probability of failure occurrence.

2.3.1 Temperature on Impact Toughness

As mentioned in Sect. 2.2.3, the temperature dependence of the Peierls–Nabarro stress, i.e. the stress required to move a dislocation through a crystal lattice, is more pronounced in bcc metals than in fcc or hcp metals. Since the Peierls–Nabarro stress and the yield strength are interrelated, the yield strength of bcc metals depends on temperature more strongly than the yield strength of fcc or hcp metals. The relative faster increase in the yield strength with respect to the tensile strength upon cooling is the main reason for the low-temperature embrittlement of bcc metals. Figure 2.13b shows the Charpy impact energy of several structural materials as a function of temperature. Some of the materials in this figure exhibit low impact toughness at all temperatures (e.g. the 75-ksi yield strength Al), while others show a high level of impact toughness at all temperatures (e.g. the 180-ksi yield strength steel). As may be seen from the same figure, the impact energy of some materials (Al alloys and high-strength steels) does not change with temperature; these are fcc alloys. On the other hand, the impact energy of some materials (e.g. the 40-ksi yield strength steel) changes dramatically with temperature: these are bcc alloys. Embrittlement occurs around a temperature known as the *ductile-to-brittle transition temperature (DBTT)*.

The exact location of DBTT on the temperature axis and the magnitude of embrittlement are affected by several factors that include the alloy composition, the amount, shape/sharpness, size and spatial distribution of brittle second-phase particles, the loading/strain rate, and the degree of material constraint. Figure 2.14a illustrates the change in the Charpy impact energy of steel as a

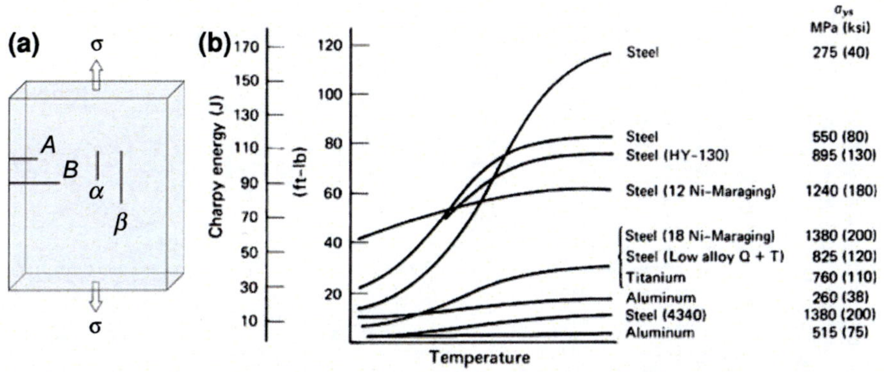

Fig. 2.13 **a** Arbitrary cracks in a solid body subjected to tension; the body contains cracks A, B, α, and β, where $\alpha = A$ and $\beta = B$. Since cracks α and β are parallel to the direction of stress, failure will occur in association with cracks A and B. The fact that $A < B$ increases the probability of failure originating from crack B [2]*. **b** Change of the Charpy impact energy of several engineering alloys with temperature [2] [&]. [&] From "Deformation and Fracture Mechanics of Engineering Materials", 4th edition, by Hertzberg. © 1996 by John Wiley and Sons, Inc. Reprinted with permission of John Wiley and Sons, Inc. The meaning of symbol [&] is the same throughout this text

function of its sulphur (S) content. The observed impact toughness change is attributed to the change in the volume fraction of sulphide inclusions in this steel. Figure 2.14b shows the effect of the shape and sharpness of graphite precipitates on the impact toughness of ferrous alloys: spherical precipitates are quite innocuous, while flake-like precipitates with sharp edges decrease the material's impact toughness. In fact, sharp and brittle second-phase particles, like several IMCs in Pb-free solders, act as *'internal notches'*, reducing the impact toughness of the otherwise ductile metallic matrix in which they have precipitated. These 'internal notches' play the role of stress concentrators inside the matrix material, facilitating the onset of brittle failure. Their impact on material properties is similar to the effect of external notches/cracks on the properties of notch-sensitive materials, like the ductile iron of Fig. 2.14b.

It has been mentioned more than once that the Peierls–Nabarro stress and the yield strength of a certain material are interrelated, and so are their temperature dependences. Since the Peierls–Nabarro stress expresses the resistance of a crystal lattice to dislocation motion, it would be interesting to see how the movement of dislocations is affected by the size and spatial distribution of second-phase particles. In general, second-phase particles in a metallic matrix are obstacles to dislocation motion [2]. After precipitation, the hardness and yield strength of the host alloy increases initially with time, but eventually decreases with prolonged annealing. This trend is demonstrated in Fig. 2.15a, which shows ageing curves of the 6061-T4 aluminium alloy [2]. These curves could be interpreted as follows: after solution treatment and quenching, the alloy is characterised by a maximum potential for solid solution strengthening, since most of the solute is present in the

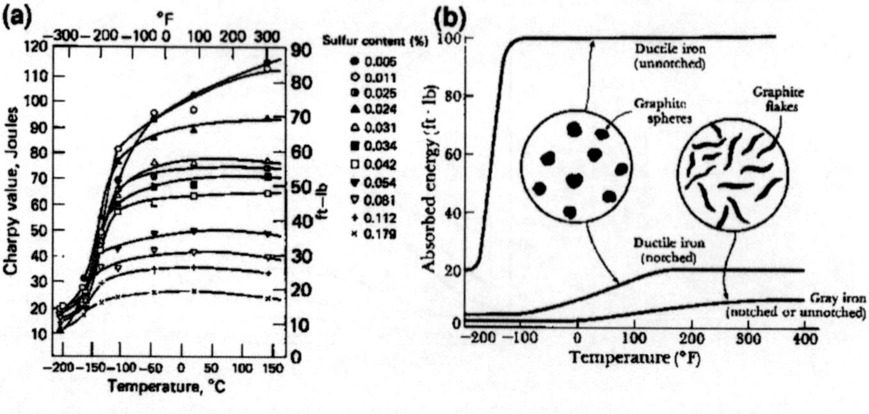

Fig. 2.14 **a** Effect of the *S* content on the Charpy impact energy of a steel plate [2]&. **b** The effect of notches (both external and internal) on the impact properties of two different ferrous alloys. The sharp graphite flakes in the grey iron act as internal notches, resulting in low absorbed energies during impact (i.e. brittle behaviour). On the other hand, the ductile iron contains spherical graphite nodules that do not act as internal notches, thus failing in a ductile manner. However, the presence of external notches reduces substantially the energy absorbed during the fracture of ductile iron [9]#

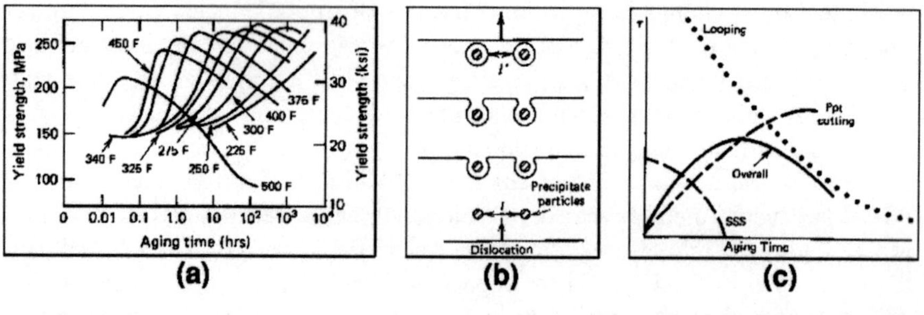

Fig. 2.15 **a** Ageing curves of the 6061-T4 aluminium alloy [2]&. **b** Dislocation looping around second-phase particles [2]&. **c** Schematic representation of the role of the major hardening mechanisms in the overall hardening response of an alloy [2]&. SSS stands for solid solution strengthening, and Ppt stands for precipitate

host lattice. Ageing will lead to the precipitation of second-phase particles, which on the one hand limits the effect of solid solution strengthening but on the other hand results in precipitation hardening (i.e. strengthening) the alloy. Precipitation hardening occurs through several mechanisms, like the interaction of dislocations with misfit strain fields around second-phase particles, particle cutting, and elastic modulus interaction effects. The extent of such hardening increases initially with time and, hence, with the precipitate particle size. Prolonged ageing, however, results in a wider inter-particle spacing, due to the Ostwald ripening (i.e. coarsening) of the precipitates. When this happens, dislocation looping occurs around the particles (Fig. 2.15b), a fact reflected in the weakening of the alloy, which simultaneously becomes more ductile; this weakening becomes worse with further ageing. The overall response of the alloy to ageing is characterised by maximum strength at intermediate ageing times and second-phase particle sizes (Fig. 2.15c).

The above discussion shows why the presence of IMCs in Pb-free solders alloys has a large impact on their fracture behaviour. Based on what has been said so far, it is expected that mainly IMC particles of a certain (intermediate) size will effectively 'pin' dislocations, limiting thus the capability of the host lattice for plastic deformation. Plastic deformation will be further suppressed with decreasing the temperature, especially for bcc metals, due to the increase in the Peierls–Nabarro stress. An enhanced Peierls–Nabarro stress implies that dislocation motion will occur at a higher stress level, so that the first dislocations to move will do so faster, since their velocity is proportional to the applied stress [8]. Therefore, the presence of IMC particles with the 'right' size and spacing in Pb-free solders stressed at low temperatures will cause dislocation coalescence and crack nucleation in the immediate vicinity of these IMCs, as suggested by Fig. 2.6.

Factors that should be of concern when one strives to control the effect of IMCs on the embrittlement of Pb-free solder alloys include (a) the volume fraction (as defined by the alloy composition), and (b) the shape, size distribution, and spacing of these IMCs (as defined by the microstructure after soldering and the 'mission profile' of the solder joint). In this chapter, these factors will be further discussed only with respect to the IMCs in the bulk solder. The IMC layers forming between solder joint and bond pads will be addressed separately in another chapter of this Handbook.

2.3.2 Strain Rate on Impact Toughness

The loading rates in most structures vary from slow to dynamic; with *slow loading rates*, the maximum load is reached in $\geq$10 s (usually, 10–60 s), while with *dynamic loading*, the maximum load is reached in $\leq$0.001 s [1]. *Intermediate loading rates* vary between the two extremes mentioned above, but typical loading rates in this category require $\sim$1 s to maximum load, as in the case of bridges subjected to truck loadings [1]. The common definition of loading rate is [1]:

$$\dot{K} = \frac{K_{\text{critical}}}{t} \tag{2.4}$$

where $\dot{K}$ = rate of increase in the stress-intensity factor, in MPa$\sqrt{\text{m}}$/s or ksi$\sqrt{\text{in.}}$/s, K_{critical} = critical stress-intensity factor (K_c, K_{Ic}, $K_{Ic}(t)$, K_{Id}), in MPa$\sqrt{\text{m}}$ or ksi$\sqrt{\text{in.}}$, and t = time required to reach K_{critical}, in *s*.

The three loading rate categories are defined as follows [1]:

Slow: $\dot{\varepsilon} \approx 10^{-5}\mathrm{s}^{-1}$
Intermediate: $\dot{\varepsilon} \approx 10^{-3}\mathrm{s}^{-1}$
Dynamic: $\dot{\varepsilon} \approx 10\,\mathrm{s}^{-1}$

where $\dot{\varepsilon}$ is the *strain rate* (i.e. strain per time unit) just ahead of the crack tip in a notched test specimen. Little change occurs in the fracture toughness of a material, unless the loading rate varies by at least one order of magnitude [1]. Figure 2.16a compares the three different loading rates with respect to the available time until failure.

In general, the fracture toughness of structural materials, especially bcc metals, increases with increasing temperature and decreasing loading rate. The first principle is illustrated as a general trend in Fig. 2.16b, while Fig. 2.17a shows fracture toughness data acquired from a 50-ksi yield strength steel [1]. Understanding Figs. 2.16b and 2.17a is easier using the following explanations: as mentioned in Sect. 2.2.1, the *fracture toughness* of a material is given by the *critical stress-intensity factor*, K_c. Specific service conditions, however, produce specific K_c values; an example is the K_{Ic}, which is the *mode I fracture toughness for slow loading under plane-strain conditions* [1]. The *mode I fracture toughness for intermediate loading rates under plane-strain conditions* is referred to as $K_{Ic}(t)$, where the time t to maximum load is given in the parenthesis [1]. Finally, the *mode I fracture toughness for dynamic loading under plane-strain conditions* is referred to as K_{Id} [1]. For a specific material, K_{Ic}, $K_{Ic}(t)$, and K_{Id} are known as the 'static', 'intermediate', and 'dynamic' fracture toughness, respectively, and can be determined as a function of temperature (Figs. 2.16b and 2.17a). Figure 2.16b shows that K_{Ic}, $K_{Ic}(t)$, and K_{Id} increase with the service temperature. However, at any given temperature, the dynamic fracture toughness is lower than both the intermediate and the static fracture toughness [1]; in general, the following relation holds: $K_{Id} < K_{Ic}(t) < K_{Ic}$.

At constant temperature, tests that aim at the determination of fracture toughness and are conducted at higher loading rates typically result in lower fracture toughness values, since $K_{Id} < K_{Ic}(t) < K_{Ic}$ (Fig. 2.17b). The effect of both

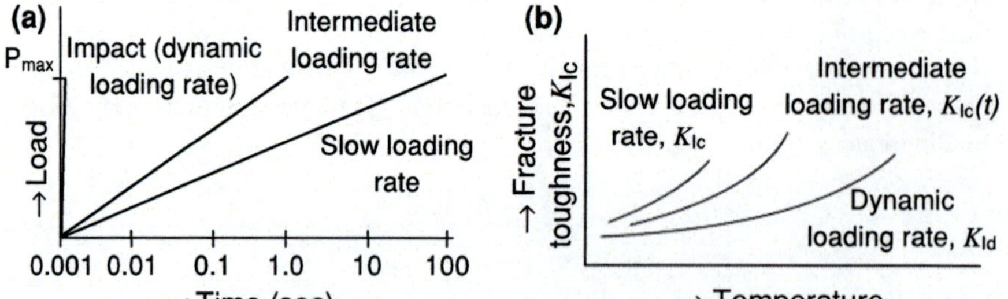

Fig. 2.16 **a** Dependence of the time to failure on the loading rate; P_{max} is the maximum load [1]*. **b** Effect of temperature and loading rate on the fracture toughness [1]*

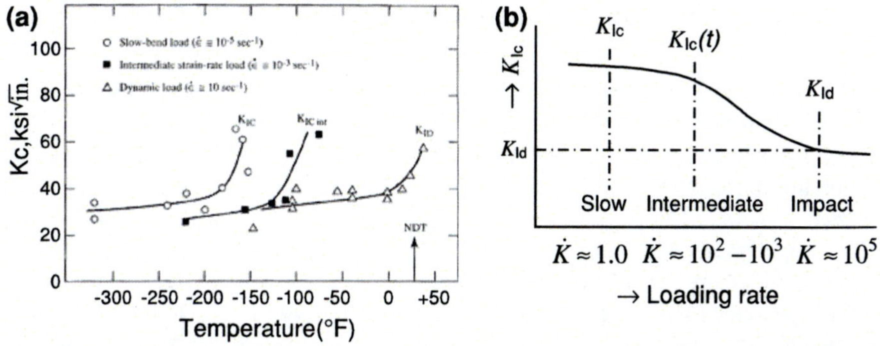

Fig. 2.17 **a** Effect of temperature and strain rate on the fracture toughness of a 50-ksi yield strength steel [1]†. **b** Effect of loading rate, $\dot{K}$, on the fracture toughness at constant temperature [1]*. † From "Fracture and Fatigue Control in Structures: Applications of Fracture Mechanics", 3rd edition, by Barsom and Rolfe. © 1999 by ASTM International. Reprinted with permission of ASTM International. The meaning of symbol † is the same throughout this text

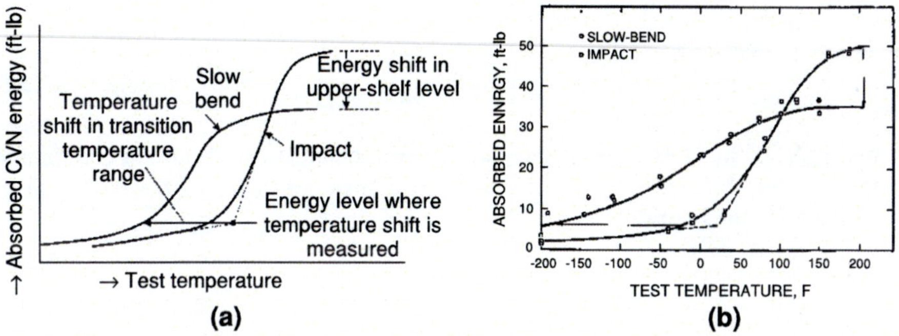

Fig. 2.18 **a** Schematic representation of the shift in the CVN transition temperature and the upper shelf level due to strain rate [1]*. **b** Slow bending and CVN impact test results acquired from an A36 steel [1]†

temperature and loading rate on the fracture behaviour of certain materials is observed in the Charpy V-notch impact test results, as shown schematically in Fig. 2.18a. Experimental results from an A36 structural steel are presented in Fig. 2.18b, which shows that the ductile-to-brittle transition of specimens tested at slow loading rates occurs at lower temperatures in comparison with the transition of specimens tested in impact. It should be pointed out that certain materials, like aluminium alloys, titanium alloys, and high-strength steels (yield strengths $\geq$ 140 ksi), do not exhibit loading-rate effects [1]. The response of these materials to impact may again be related to their crystal structure. For these materials, there would generally be no difference between the K_{Ic} and K_{Id} values that have been measured at the same test temperature [1].

2.3.3 Constraint on Impact Toughness

The concept of *constraint* refers primarily to the transition from a plane-strain to a plane-stress condition, due to changes in the specimen thickness. In linear-elastic fracture mechanics, *plane-strain* is the stress condition, where the strain is zero in a direction normal to both the axis of the applied tensile stress and the direction of crack growth. Plane-strain is related to *maximum constraint* and occurs in very thick specimens with deep cracks. On the other hand, *plane-stress* is the stress condition, where the stress is zero in the thickness direction. Plane-stress is related to *minimum constraint* and occurs in thin test specimens [1, 2]. Figure 2.19 shows schematically the plane-stress and plane-strain conditions. But why would the stress state of the test specimen influence its fracture resistance?

In order to answer this question, one must first understand that a region of plasticity is established near the crack tip of a notched specimen, whenever the stresses ahead of this tip exceed the yield strength of the material [2]. Figure 2.20a illustrates the onset of plastic deformation directly ahead of a crack tip, i.e. at an angle $\theta = 0$. At a certain distance r from the crack tip, the elastic stress will exceed the yield strength, σ_{ys}, of the material, and, at that particular distance, the following relation will hold [1, 2]:

$$\sigma_{ys} = \frac{K_I}{\sqrt{2\pi r}} \tag{2.5}$$

where K_I is the stress-intensity factor. For the plane-stress condition (i.e. biaxial stress state), the radius, r_y, of the plastic zone ahead of the crack tip may be given by the following expression [1, 2]:

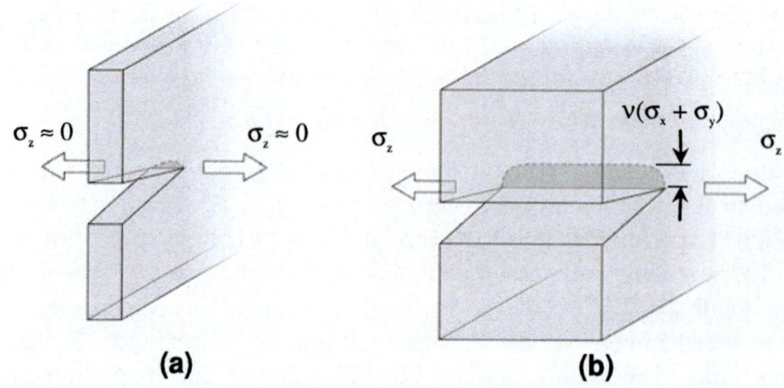

Fig. 2.19 Schematic representation of (**a**) the plane-stress condition for a thin sheet, and (**b**) the plane-strain condition for a thick plate [2]*. In the plane-stress condition, the through-thickness stress, σ_z, is close to zero, i.e. $\sigma_z \approx 0$. In the plane-strain condition, a through-thickness stress develops, which may be given by the expression $\sigma_z \approx \nu(\sigma_x + \sigma_y)$

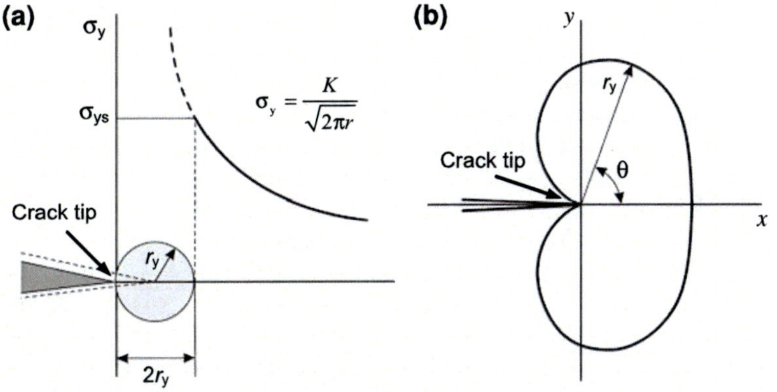

Fig. 2.20 **a** Onset of plastic deformation at the crack tip, where a plastic zone of radius r_y is established [2]*. **b** Boundary of the plastic zone ahead of a crack tip as a function of the angle θ with the horizontal [2]*

$$r_y \approx \frac{1}{2\pi} \frac{K_{IC}^2}{\sigma_{ys}^2} \tag{2.6}$$

For the plane-strain condition, the triaxial stress state ahead of the crack tip reduces the size of the plastic zone. In that case, the radius of the plastic zone is provided by the following expression [1, 2]:

$$r_y \approx \frac{1}{6\pi} \frac{K_{IC}^2}{\sigma_{ys}^2} \tag{2.7}$$

Equations 2.6 and 2.7 show that the size of the plastic zone ahead of a crack tip is smaller in the plane-strain condition than in the plane-stress condition. It must also be noted that the size of the plastic zone ahead of a crack tip varies also with θ, as shown in Fig. 2.20b.

Based on the above, it becomes clear that when the test specimen is thick in the direction parallel to the crack front, a large through-thickness stress, σ_z, will restrict the sample's plastic deformation in that direction. Since the fracture toughness of a material depends on the volume of material capable of plastic deformation prior to fracture and this volume depends on the specimen thickness, the fracture toughness K_c varies with the specimen thickness, as shown in Fig. 2.21a. For a thin sample, the plastic constraint at the crack tip is minimal, and the fracture toughness is, thus, maximum. As the specimen thickness increases, plane-strain conditions are established at the crack tip, increasing the plastic constraint and, thereby, decreasing the fracture toughness. When the specimen thickness exceeds a certain value, the fracture toughness stops decreasing; the lowest fracture toughness value is known as the *plane-strain fracture toughness*, K_{Ic}, and it is the conservative lower limit of material toughness that the design engineer can use for any given application [2]. One could summarise the above by

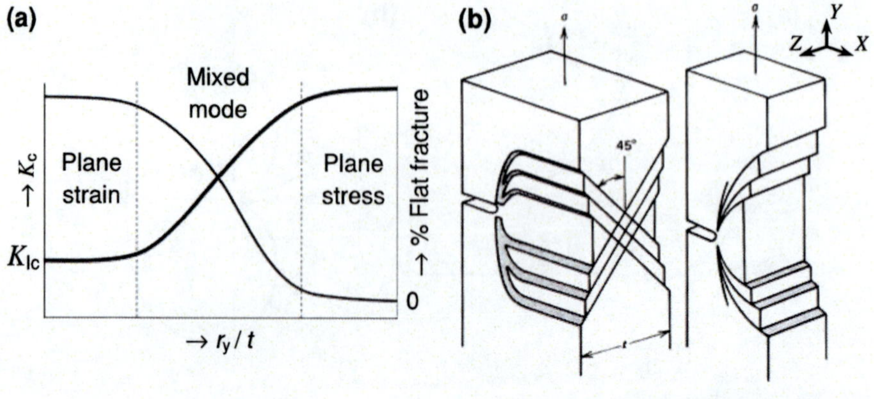

Fig. 2.21 **a** Effect of the plastic zone size-to-plate thickness ratio, r_y/t, on the fracture toughness and macroscopic appearance of the fracture surface. The fracture toughness curve is thicker than that related to the percentage of flat fracture. Plane-stress conditions are associated with maximum fracture toughness and a slant fracture. Plane-strain conditions are associated with minimum fracture toughness and a flat fracture [2]*. **b** Crack-tip deformation patterns in the plane-stress (*left*) and plane-strain (*right*) state [2][&]

saying that the plane-stress fracture toughness, K_{c}, is affected by both metallurgical factors and specimen geometry, while the plane-strain fracture toughness, K_{Ic}, depends only on metallurgical factors [2].

Since the effect of the stress state on the fracture toughness depends on the relation between the size of the plastic zone ahead of a crack tip, r_y, and the specimen thickness, t, it would be interesting to consider the change in the stress state in terms of the ratio r_y/t. Typically, when $r_y/t \geq 1$, plane-stress conditions prevail and the fracture toughness is high. On the other hand, plane-strain conditions exist when $r_y/t < 1/10$. In both cases, the yield strength of the material determines the thickness required for the establishment of plane-stress or plane-strain conditions, since it determines r_y according to Eqs. 2.6 and 2.7. This means that very thin plates of a material with high yield strength can still experience plane-strain conditions, whereas very thick specimens of a low yield strength material may never experience true plane-strain conditions [2].

The dependence of the fracture toughness on the stress state may also result in a transition of the fracture mode from ductile to brittle. As shown in Fig. 2.21a, the relative degree of flat-to-slant fracture depends on the stress state ahead of the crack tip. For plane-stress conditions and $r_y \geq t$, the fracture plane often assumes a ±45° orientation with respect to the load axis and sheet thickness (Fig. 2.21b). This happens because failure occurs on the planes containing the maximum resolved shear stress; these planes lie along ±45° lines in the *YZ* plane [2]. For plane-strain conditions and $r_y \ll t$, the plane of maximum shear is in the *XY* plane, and the fracture plane lies midway between the two maximum shear planes [2]. The fracture mode transition is observed with materials like aluminium, titanium, and certain steel alloys, as well as with a number of polymers [2]. In general, flat fracture surfaces imply plane-strain conditions and brittle failures, while slant or shear fracture surfaces are related to plane-stress conditions and more ductile failures [2].

2.3.4 Standard ASTM Fracture Test Methods

Several standard test methods of the American Society for Testing and Materials (ASTM) have been developed to measure the various critical stress-intensity factors for materials with different types of fracture behaviour. Figure 2.22 presents the general regions of the fracture behaviour of structural steels and the ASTM test methods used in each one of them.

The ASTM test methods mentioned in Fig. 2.22 are [1]:

1. K_{Ic}: Plane-strain fracture toughness value obtained at slow loading rates. The material constraint is maximum, leading to unstable brittle fracture with little or no deformation. ASTM test method E-399-90: Standard Test Method for Plane-Strain Fracture Toughness of Metallic Materials.
2. $K_{Ic}(t)$: Plane-strain fracture toughness value obtained at intermediate loading rates, where t is the time to maximum load. The material constraint is maximum, leading to unstable brittle fracture with little or no deformation. ASTM test method E-399—Annex A7: Special Requirements for Rapid-Load Plane-Strain Fracture Toughness $K_{Ic}(t)$ Testing.
3. K_{Ia} (K_{Id} or K_{ID}): Linear-elastic behaviour during dynamic/impact loading that results in fast, unstable brittle fracture. K_{Ia} is the plane-strain crack-arrest toughness, while K_{Id} or K_{ID} is the plane-strain fracture toughness under dynamic loading. K_{Id} and K_{Ia} are assumed to be equivalent, but there is no standard K_{Id} test method. ASTM test method E-1221-96: Standard Test Method

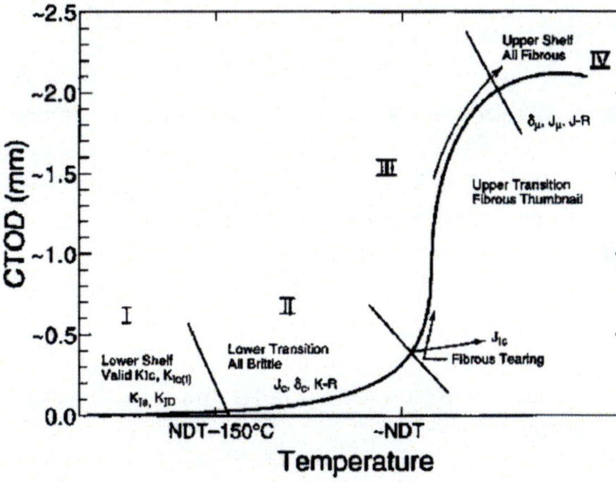

Fig. 2.22 General regions of the fracture behaviour of structural steels [1]†

for Determining Plane-Strain Crack-Arrest Fracture Toughness, K_{Ia}, of Ferritic Steels.

4. δ_c, J_c, K-R: Elastic–plastic plane-stress behaviour during slow loading, accompanied by plastic zone development, but without stable crack growth. Fast, unstable brittle fracture occurs.
   δ_c: ASTM Test Method E-1290-93: Standard Test Method for Crack-Tip Opening Displacement (CTOD) Fracture Toughness Measurement.
   K-R: ASTM Test Method E-561-94: Standard Practice for R-Curve Determination.
   J_c: ASTM Test Method E-1737-96: Standard Test Method for J-Integral Characterisation of Fracture Toughness.
5. J_{Ic}: Critical value of the J-integral that describes the stress–strain field ahead of a crack. J_{Ic} is a measure of the fracture toughness at the onset of slow, stable crack growth. The behaviour is non-linear elastic–plastic. ASTM Test Method E-813-89: Standard Test Method for J_{IC}, A Measure of Fracture Toughness.
6. δ_u, J_u, J-R: Elastic–plastic behaviour during slow loading, accompanied by slow, stable crack growth. Stable crack growth is either followed by brittle fracture (unstable crack growth) or continues until the separation of the test specimen.
   δ_u: ASTM Test Method E-1290-93: Standard Test Method for Crack-Tip Opening Displacement (CTOD) Fracture Toughness Measurement.
   J_u: This is a point on the J-R curve described next, as there is no single standard for J_u.
   J-R: ASTM Test Method E-1152-95: Standard Test Method for Determining J-R Curves.
7. J_c, J_{Ic}, J-R: New method to cover all J-integral test results in one standard. The behaviour is elastic–plastic, with or without stable crack growth. ASTM Test Method E-1737-96: Standard Test Method for J-Integral Characterisation of Fracture Toughness.
8. K, J, CTOD (δ): This standard replaces all previous test methods. A new fracture test method, called the Standard Test Method, has been developed for materials with an unknown type of fracture behaviour. For these materials, the needed type of test method is also unknown prior to testing. A bend or compact test specimen is tested, and the P-Δ_{CMOD} and P-Δ_{LLD} records are analysed either as K, J, or δ values, depending on the test records. CMOD is the Crack-Mouth Opening Displacement, while LLD is the Load Line Displacement. ASTM Test Method E-1820-96: Standard Test Method for Measurement of Fracture Toughness.
9. K_{Jc}: This test method determines the reference temperature, T_o, which characterises the fracture toughness of ferritic steels. These steels experience the onset of cleavage cracking either at elastic, or elastic–plastic K_{Jc} instabilities, or both. ASTM Test Method E-1921-97: Standard Test Method for the Determination of Reference Temperature, T_o, for Ferritic Steels in the Transition Range.

It falls beyond the scope of this text to explain in detail any of the above test methods. Before closing this section, however, the significance of the four regions in Fig. 2.22 must be briefly explained; these regions are known as the lower shelf (I), the lower transition (II), the upper transition (III), and the upper shelf (IV) region [1]. The behaviour in the *lower shelf region* is brittle, irrespective of the loading rate, and the fracture toughness changes little or not at all with the service temperature. Materials with such behaviour should best be avoided, since their toughness levels do not provide adequate safety margins, except under carefully monitored service conditions. The behaviour is nearly linear-elastic, and K_{Ic} could be used to describe the behaviour of the material in the largest part of this region. The behaviour in the *lower transition region* is elastic–plastic with increasing plastic zone sizes but without stable crack growth (ductile 'thumbnail') prior to the final brittle or mixed-mode fast fracture. Depending on the service conditions, such behaviour is satisfactory for many structural applications. The fracture toughness increases with temperature from approximately the linear-elastic K_{Ic} values to the fracture toughness values at the initiation of a fibrous 'thumbnail' visible to the naked eye. The behaviour in the *upper transition region* is elastic–plastic with large plastic zone sizes and increasing amounts of ductile tearing followed by unstable, mixed-mode fast fracture, but only after considerable deformation. Stable crack growth, i.e. 'thumbnail' behaviour, may occur; in that case, failure starts by stable ductile tearing (recognisable by the presence of a coarse fibrous 'thumbnail' on the fracture surface) and is followed by brittle crack propagation. The behaviour in the *upper shelf region* is generally yielding, and fracture is accompanied by fibrous ductile tearing that occupies the whole fracture surface. The exact point where the upper shelf region begins is ambiguous, due to the large data scatter in the upper transition region. This ambiguity is avoided, if the onset of the upper shelf region is defined at the temperature where fracture becomes 100% fibrous tearing. Figure 2.23 also presents the ductile-to-brittle transition in the fracture behaviour of a material with respect to the four regions discussed above.

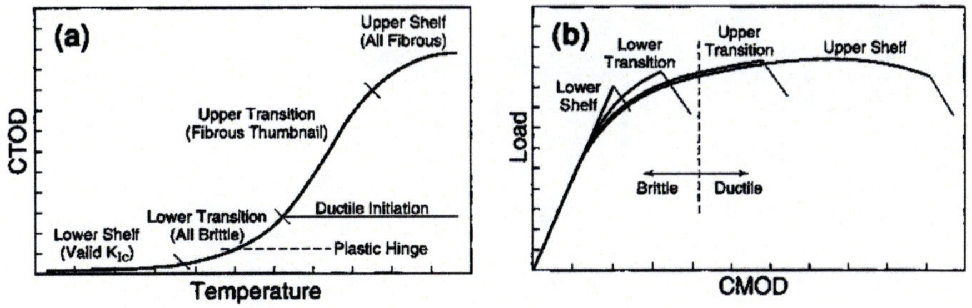

Fig. 2.23 **a** Curve showing the transition in Crack-Tip Opening Displacement (CTOD, i.e. the opening of the notch faces) with temperature [1]†. **b** Load versus Crack-Mouth Opening Displacement (CMOD) curve [1]†

2.3.5 Correlation of K_{Ic} and K_{Id} with CVN Energy

It has often been remarked that one of the main limitations of the CVN impact test is that it only gives a qualitative estimation of the toughness of a material subjected to impact loading, but it is unable to really determine the fracture toughness of the material [4]. In order to address this limitation, a number of methods have been devised [1]; this chapter will present a method to correlate K_{Id} and K_{Ic} with the CVN energy as an example. For the persons interested, additional information on alternative methods may be found in [1].

First, it should be noted that there is an ASTM standard test method dedicated to the impact testing of notched metallic bars. The latest revised version of this standard is the *ASTM Test Method E 23-06: Standard Test Methods for Notched Bar Impact Testing of Metallic Materials* [10]. Second, there seems to be a correlation between the CVN energy measured from the impact testing of a notched metallic material and the impact plane-strain fracture toughness, K_{Id}, of that material [1]:

$$\frac{(K_{\mathrm{Id}})^2}{E} = 5(\mathrm{CVN}) \tag{2.8}$$

where E is the material's modulus of elasticity, and K_{Id} is the impact plane-strain fracture toughness, which may be determined using the ASTM test method E-1221-96 (see Sect. 2.3.4). The validity of Eq. 2.8 is illustrated in Fig. 2.24a for various steel grades with yield strength varying in the 36–140 ksi range. The loading rate in all tests was the same, but the notch acuity varied; the latter was taken into account empirically by the factor of 5 [1]. In low-strength steels like

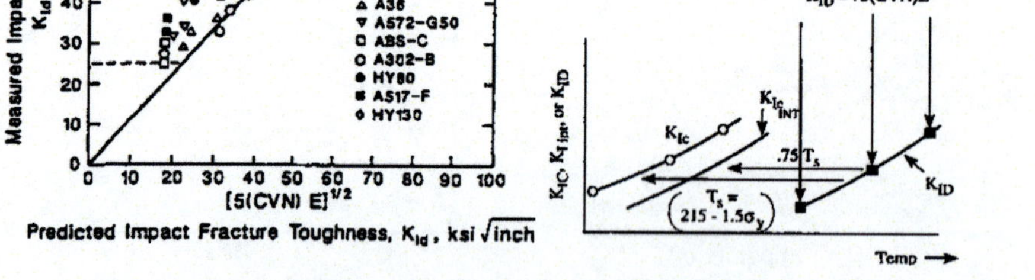

Fig. 2.24 **a** Correlation of the plane-strain impact toughness, K_{Id}, with the CVN energy for various grades of steel [1]†. **b** Relationship between CVN energy, K_{ID}, $K_{\mathrm{Ic}}(t)$, and K_{Ic} test results. In these plots, T_s is measured in °F, CVN in ft-lb, K_{ID} in psi$\sqrt{\text{in.}}$, σ_{ys} in ksi, and E in psi [1]†

the ones that produced the data in Fig. 2.24a, a shift in the CVN transition temperature is observed when the loading rate changes from slow bending to impact (see Fig. 2.18a). This shift may be calculated using the following expression [1]:

$$T_{\text{shift}} = 215 - 1.5\sigma_{\text{ys}} \tag{2.9}$$

where σ_{ys} is the 0.2% offset yield strength of the material.

Based on Eqs. 2.8 and 2.9, one may predict the K_{Ic} from CVN impact test results, following the steps of the procedure described below and shown in Fig. 2.24b [1]:

1. Start by testing CVN impact specimens in the lower transition region for the material of interest.
2. At each temperature where CVN values are available, calculate the corresponding K_{Id} values using Eq. 2.8.
3. Shift the K_{Id} values to K_{Ic} values using Eq. 2.9.
4. If fracture toughness values at intermediate loading rates are required, e.g. $K_{\text{Ic}}(t) = K_{\text{Ic}}(1)$ at $t = 1$ s, use ¾ of the K_{Id}-to-K_{Ic} shift, as shown in Fig. 2.24b.

2.3.6 Other Impact Testing Considerations

A problem that often characterises Charpy V-notch impact test results is that the ductile-to-brittle transition of several materials occurs over a whole range of temperatures and not at a specific temperature. In such a case, the definition of the DBTT becomes a challenge. In order to overcome this problem, people have defined the following criteria for the determination of the DBTT [1, 2]:

1. The ductile-to-brittle transition occurs at a certain level of CVN energy, like the 13.5, 20, or 27 J energy level (equivalent to 10, 15, or 20 ft-lb, respectively).
2. The ductile-to-brittle transition occurs at a certain level of lateral expansion on the compression side of the test specimen (Fig. 2.25a).
3. The ductile-to-brittle transition occurs at a certain amount of fibrous (i.e. ductile) or cleavage (i.e. brittle) fracture as measured on the fracture surface of the test specimen (Fig. 2.25b).

The problem with having more than one criterion does not only lie in which criterion to choose. Unfortunately, transition temperatures defined using the energy absorption, ductility, or fracture surface appearance criteria do not agree for the same material! An example is given in Table 2.1, which shows that the transition temperatures of different steels defined by the 20-J energy criterion and the 0.38-mm (15-mil) lateral expansion criterion are in reasonable agreement, but they are both lower than the transition temperature defined by the 50% fibrous fracture criterion.

In contrast to the results presented in Table 2.1, Fig. 2.26 presents CVN test results from a low-strength steel, showing that the transition temperatures defined

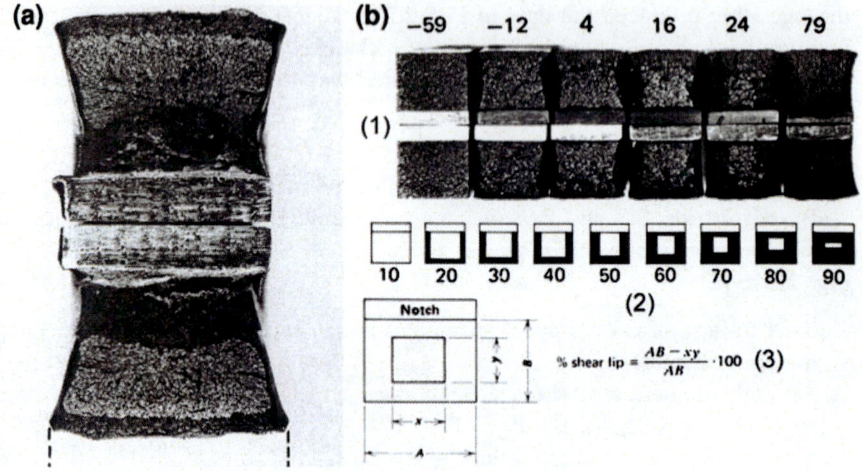

Fig. 2.25 **a** Measurement of lateral expansion at the compression side of a Charpy bar [2]&. **b** Transition in the appearance of the fracture surface of A36 steel Charpy bars as a function of temperature. This figure shows (1) an actual fracture series of A36 steel test specimens, (2) a standard comparison chart showing the percentage shear lip, and (3) the formula used for the calculation of the percentage shear lip [2]&

Table 2.1 Transition temperature data from selected steels [2]

Material	$\frac{\sigma_{ys}}{\sigma_{ts}}\left(\frac{MPa}{MPa}\right)$	DBTT (°C)		
		20 J	0.38 mm	50% fibrous
Hot-rolled C–Mn steel	$\frac{210}{442}$	27	17	46
Hot-rolled, low-alloy steel	$\frac{385}{570}$	−24	−22	12
Quenched and tempered steel	$\frac{618}{688}$	−71	−67	−54

σ_{ys} is the yield strength of the material, σ_{ts} is its tensile strength

by the 15 ft-lb energy, 20-mil lateral expansion, and 50% fracture appearance criteria coincide at ~30°F. This indicates that the right choice of criterion plays an important role in getting reasonable results, but this choice is rarely known a priori. It should also be emphasised that the transition temperature value will also be influenced by other factors, like the specimen size and shape, and the sharpness of the notch [2], so one needs to be very cautious when interpreting impact test results.

One of the factors that merits more consideration is the specimen size, and, in particular, the specimen thickness. As mentioned in Sect. 2.3.3, changes in the specimen thickness affect the degree of material constraint and can be responsible for a transition in the stress state from plane-stress (thin specimens) to plane-strain (thick specimens) conditions. It is already clear that the determined fracture toughness in each one of these conditions is different, with the plane-strain fracture

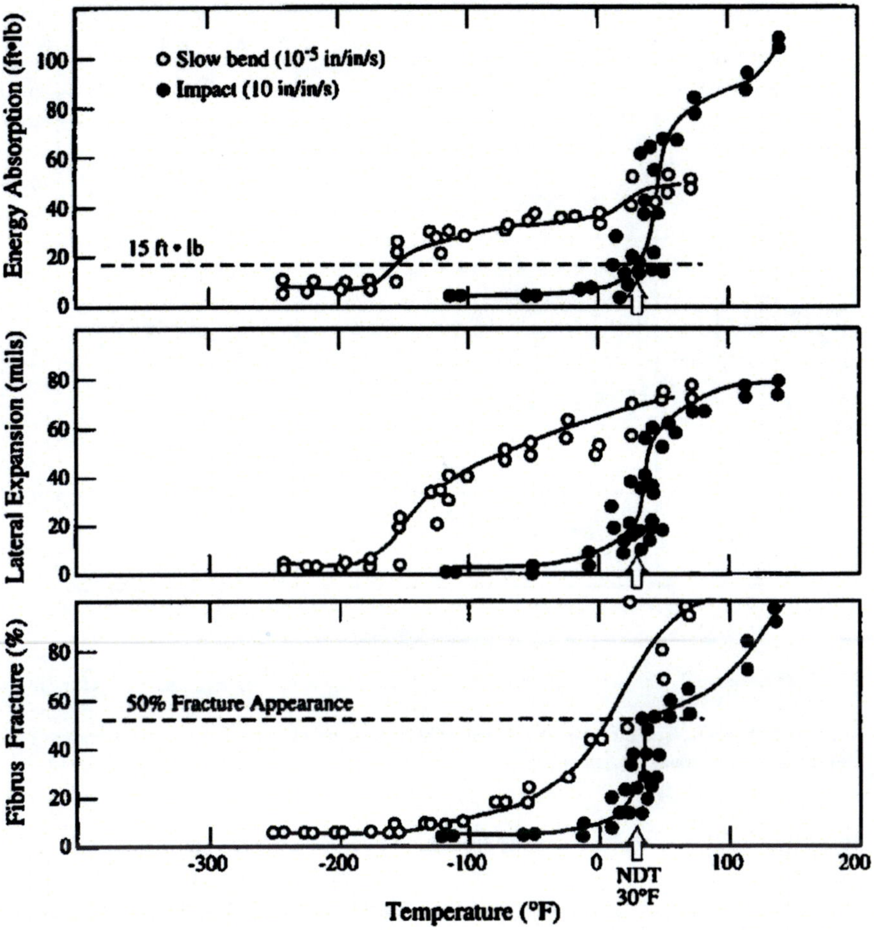

Fig. 2.26 CVN energy absorption, lateral expansion, and fibrous fracture from the impact and slow-bend testing of standard CVN specimens of a low-strength structural steel [1]†

toughness being lower than the plane-stress one (Fig. 2.21a). What may not be clear, however, is that the experimentally determined DBTT often depends on the thickness of the test specimen [2]. This phenomenon is also related to the transition from plane-stress to plane-strain conditions, which is likely to occur as the text specimen dimensions change. For example, Fig. 2.27 shows that the ductile-to-brittle transition temperature of A283 steels increases with the thickness of the Charpy test specimens [2]. This figure shows that the absorbed energy per 2.5 mm of sample thickness, or the percentage of shear fracture, changes at a certain temperature, indicating the transition in the fracture behaviour of these steels from ductile to brittle. As the specimen thickness increases, the DBTT is observed to increase until a limiting value, which most likely corresponds to true plane-strain conditions in the material. Studies like the above point out that one needs to keep in mind the

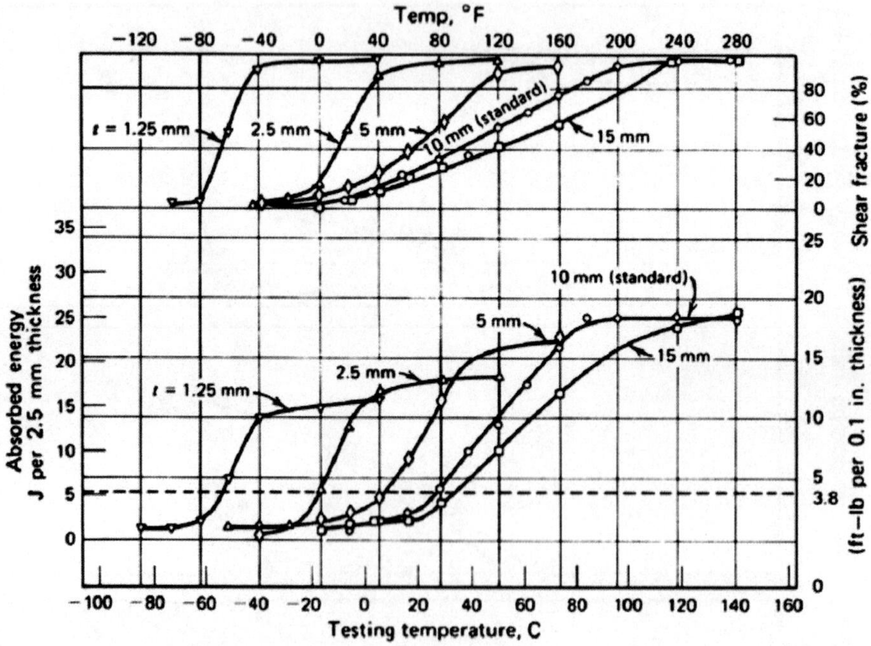

Fig. 2.27 Adjusted energy vs. temperature and shear fracture vs. temperature curves for 38-mm-thick plates of A283 hot-rolled carbon manganese steel tested in impact using CVN specimens of various thicknesses. The DBTT is defined at an absorbed energy level of 5.2 J per 2.5 mm (3.8 ft-lb/0.1 in.) of specimen thickness [2][&]

potential DBTT dependence on the specimen thickness when interpreting the impact test results of certain materials. For all the reasons explained in this chapter, the Sn-based Pb-free solder alloys, for example, are expected to exhibit a similar dependence of their DBTT on the test specimen dimensions.

2.4 Fracture Behaviour of Sn-Based Pb-Free Solders

This section reports and discusses experimental results related to the fracture behaviour of commercial Sn-based Pb-free solder alloys, focusing on the conditions under which bulk embrittlement is promoted. The section starts by presenting results acquired from the impact testing of specimens with different geometries and with one of the following compositions (all in wt%): SAC 305 (Sn-3%Ag-0.5%Cu), SAC 405 (Sn-4%Ag-0.5%Cu), Sn-5%Ag, Sn-0.7%Cu-0.1%Ni, Sn-0.7%Cu, 99.99% Sn, and, for the sake of comparison, Sn-37%Pb (near-eutectic Sn–Pb alloy). Apart from the energy absorbed during impact testing, a fractography study of selected samples tested in impact is presented; this study was conducted using scanning electron microscopy (SEM). In order to investigate the effect of IMC particle size changes on the fracture behaviour of Sn-based Pb-free solders, SAC 405 specimens were subjected to post-casting annealing in the 150–175°C range prior to impact testing; the results of this study are also presented here. Furthermore, SEM was utilised to study the interaction of cracks and IMC particles in solders SAC 405 and Sn-0.7%Cu-0.1%Ni; the cracks formed and propagated during routine thermal cycling of these solders. Finally, this section presents the results of a study of the embrittlement of Sn-based solders due to the formation of "tin pest"; for this purpose, Sn-0.5%Cu ingots were aged at −18°C for a period of up to 2 years.

2.4.1 Ductile-to-Brittle Transition

2.4.1.1 CVN Impact Testing of Bulk Solder Samples

As already mentioned in the introduction of this section, bulk specimens of various commercial Sn-based Pb-free solders and of the Sn-37%Pb solder were tested according to the CVN impact test methodology described in ASTM standard E 23-06 [10]. Standard specimens of $10 \times 10 \times 55\ mm^3$ were made from alloys SAC 305, SAC 405, Sn-5%Ag, and Sn-37%Pb, while smaller specimens of $5 \times 5 \times 55\ mm^3$ were made from alloys Sn-0.7%Cu-0.1%Ni, Sn-0.7%Cu, and 99.99% Sn, due to the limited availability of the last three alloys [11, 12]. The experimental procedure is described in [11, 12], and the impact test results are discussed later in this section. Due to the two different test specimen dimensions, the impact test results are presented in two plots, following the guidelines of the ASTM standard E 23-06 [10], which recommends the separate presentation of impact results obtained from specimens of different size or shape.

Before commenting on the results obtained from the CVN impact testing of the aforementioned Sn-based Pb-free solders, it is worthwhile mentioning that Sn has the *body-centred tetragonal (bct)* crystal structure [13]. This Sn form, also known as *β-Sn*, transforms at 13.2°C to *α-Sn*, which has the *diamond cubic* crystal structure. The transformation of β-Sn-to-α-Sn (i.e. the notorious *"tin pest"*) is very sluggish, requiring a large incubation period (about 18 months) below 13.2°C. Therefore, even at low service temperatures, β-Sn is the crystalline form of reference for all Sn-based Pb-free solders. It should be pointed out that the bct crystal structure resembles greatly the bcc structure in terms of crystal symmetry and directionality of bonding forces, as illustrated in Fig. 2.28. This means that some of the properties of pure Sn and Sn-based alloys are expected to be comparable to the respective properties of bcc metals: one of them is the existence of a rather sharp transition from ductile to brittle failure at a given DBTT.

In agreement with this expectation, Figs. 2.29 and 2.30 show that the tested Sn-based Pb-free solders demonstrate a sharp ductile-to-brittle transition in the −20 to −130°C range. In contrast to the Sn-based solders, the near-eutectic Sn–Pb solder shows a very gradual and rather weak transition between 0 and −80°C. This

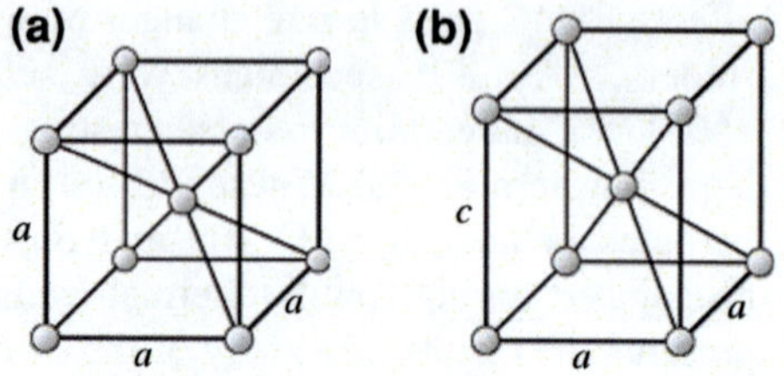

Fig. 2.28 Schematic representation of (**a**) the body-centred cubic (bcc) crystal structure, and (**b**) the body-centred tetragonal (bct) crystal structure

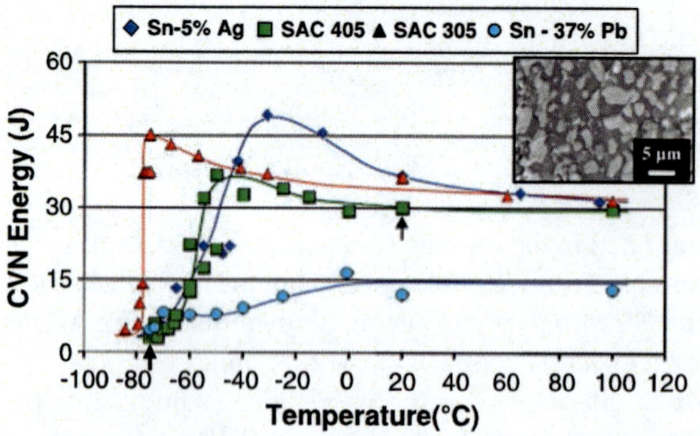

Fig. 2.29 Change of the CVN energy of Sn-37%Pb and various Sn-based solders as a function of temperature [11, 12]*. The specimen dimensions were 10 × 10 × 55 mm^3. The notch depth was 2.5 mm. The inset shows the microstructure of the Sn-37%Pb solder, where the Pb-rich phase appears bright and the Sn-rich phase grey. Data from imec

gradual and not-so-sharp overall transition in fracture behaviour is easy to understand, taking into account that the microstructure of the near-eutectic 63%Sn-37%Pb alloy consists of dendritic grains of a Pb-rich solid solution in a matrix of Sn [14]. The fracture behaviour of a 'composite' material like the Sn-37%Pb solder is expected to be a compromise between the fracture behaviours of its constituents, i.e. the Sn-rich solid solution (bct) and the Pb-rich solid solution (fcc). The fact that bcc and bct alloys tend to exhibit a ductile-to-brittle transition, while fcc alloys do not, explains the fracture behaviour of the Sn-37%Pb solder demonstrated in Fig. 2.29. The microstructure of the Sn-37%Pb solder is also shown in Fig. 2.29. Other comments that could be made on the test results of Figs. 2.29 and 2.30 are:

1. The DBTT of the Ag-containing alloys increases with the Ag content (Fig. 2.29). This is presumably related to the increase in the volume fraction of Ag_3Sn IMCs in the solder. These IMCs are believed to act as internal 'notches' with high stress intensity in their immediate vicinity, facilitating locally the solder embrittlement and shifting the DBTT towards higher temperatures.

2. The nearly pure Sn solders (i.e. Sn-0.7%Cu-0.1%Ni, Sn-0.7%Cu, and 99.99% Sn) exhibit a substantially lower DBTT in comparison with the two SAC solder alloys and the Sn-5%Ag one. To be precise, the DBTT of the very Sn-rich solders is around −130°C, while the lowest DBTT of the Sn-Ag-(Cu) solders is around −80°C (for the SAC 305). This could be attributed to a number of reasons: first, the very Sn-rich solders have a very limited amount of IMCs. Second, the very Sn-rich test specimens were smaller (5 × 5 × 55 mm^3) than the Sn-Ag-(Cu) test specimens (10 × 10 × 55 mm^3). This means that the stress state in the smaller specimens might be closer to plane-stress than to plane-strain. As has already been said, the plane-strain toughness is lower than the plane-stress one, which implies that the initiation of brittle failure is easier for plane-strain than for plane-stress conditions. This is reflected in the increase in the DBTT with the specimen thickness, as shown in Fig. 2.27.
3. The notch size in the nearly pure Sn impact test specimens was not the same: for the Sn-0.7%Cu alloy, the notch was 1.3 mm, while for the other two alloys, the notch was 2.5 mm. Interestingly, one may notice that the energy absorbed during the impact failure of Sn-0.7%Cu specimens was higher than the energy absorbed by the other two very Sn-rich alloys (Fig. 2.30). This confirms the notch sensitivity of Sn-based alloys, pointing out that the effective control of the bulk embrittlement of Sn-based solders requires control over the presence of notches in these materials. This automatically implies control over the shape, sharpness, size distribution, and spacing of the IMC particles, which act as 'internal notches'.

Figures 2.31 and 2.32 show SEM micrographs of the fracture surface of selected specimens that have been tested in impact. The chosen specimens, made of the SAC 405 and 99.99% Sn solders, are indicated in Figs. 2.29 and 2.30 with arrows. As may be seen from Figs. 2.31 and 2.32, the fracture behaviour of both the 99.99% Sn and the SAC 405 solders changes drastically as the test temperature decreases. For example, the fracture of the 99.99% Sn solder at room temperature appears to be completely ductile, and the study of its fracture surface reveals boundaries of Sn grains that have been clearly 'stretched' during

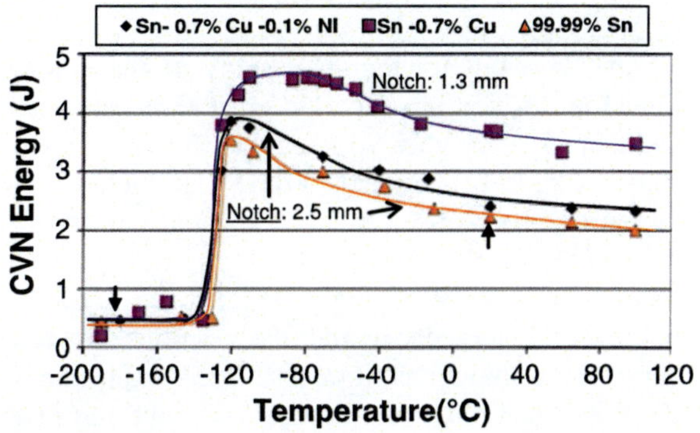

Fig. 2.30 Change of the CVN energy of various Sn-based solder alloys as a function of temperature [11, 12]*. The specimen dimensions were 5 × 5 × 55 mm^3. The notch depth was either 1.3 mm or 2.5 mm. Data from imec

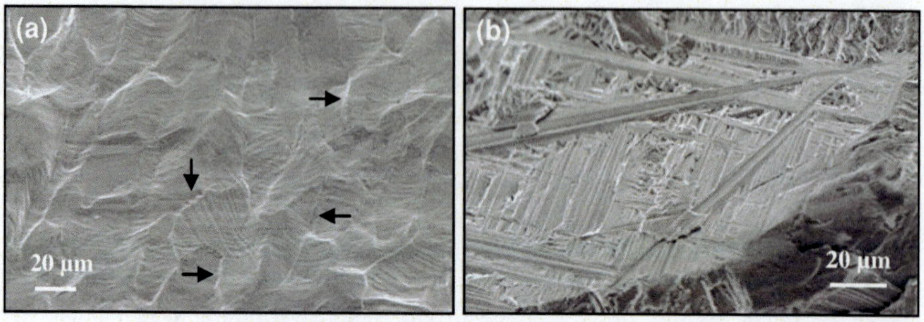

Fig. 2.31 Secondary electron (SE) detector images of the fracture surfaces of impact test specimens made of the 99.99% Sn solder. **a** Specimen tested at room temperature. The *arrows* indicate Sn grain boundaries. **b** Specimen tested at −190°C. Data from imec

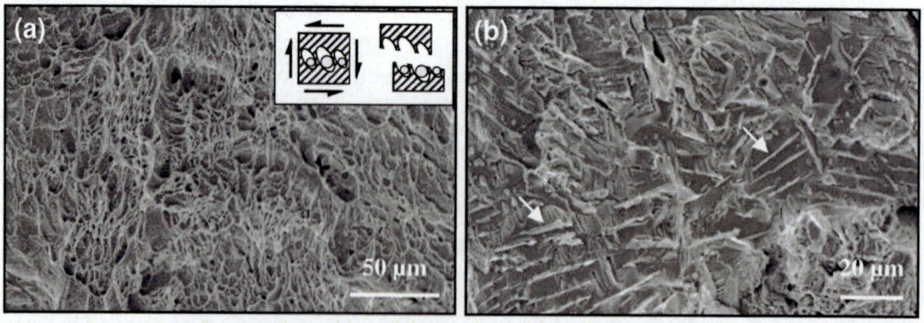

Fig. 2.32 SE detector images of the fracture surfaces of impact test specimens made of the SAC 405 solder. **a** Specimen tested at room temperature. The fracture surface is characterised by the presence of many 'dimples'. The idealised drawing in the inset shows two stages in the dimple-formation process; the dimples form around the hard, brittle IMCs in the soft, ductile Sn matrix and are slightly elongated since the material is sheared during the CVN impact test. **b** Specimen tested at −75°C. The flat grey areas are Sn, while the bright particles indicated by *arrows* are IMCs. Data from imec

testing (Fig. 2.31a). The same solder fails in a completely brittle manner at −190°C, and the study of its fracture surface reveals the cleavage-like fracture of Sn (Fig. 2.31b).

The thorough fractography study of the commercial Sn-based solders mentioned in this text has revealed great similarities between their fracture behaviour and that of nearly pure Sn (i.e. 99.99% Sn). This was not unexpected, as the above solders consist of a dispersion of IMC particles in a matrix of Sn. For example, the fracture surface of a SAC 405 specimen tested in impact at room temperature is typical for the failure of a soft, ductile material containing hard, brittle second-phase particles (Fig. 2.32a). In such a case, failure is believed to occur by microvoid coalescence, and the resulting fracture surface contains numerous depressions, known as 'dimples'. The dimples are created during the stressing of the solder material, which leads to decohesion of the IMC particles from the Sn matrix and the formation of microvoids around them (inset of Fig. 2.32a); the coalescence of these microvoids leads to the formation of cracks and, eventually, to the failure of the material [2]. The fracture behaviour of SAC 405 is clearly brittle when the material is tested at −75°C, as shown in Fig. 2.32b; the solder embrittlement is mainly the outcome of the low-temperature embrittlement (under conditions of dynamic loading and in the presence of IMC 'internal notches') of the bct Sn matrix.

2.4.1.2 Impact Testing of Solder Joints

This section discusses the results of the fractography study performed on Sn-based solder joints tested in impact. The test specimens consisted of bumped SuperBGA® packages mounted on FR-4 circuit boards; for this purpose, solder spheres of 750 μm in diameter made of various Sn-based solder alloys were soldered onto Cu bond pads. The specimens were tested in impact using a miniaturised Charpy test set-up built at imec; this customised set-up and the impact test results are presented elsewhere [15, 16]. Figure 2.33 shows SEM images of a SAC 405 solder joint that was tested in impact at room temperature. The fracture of this joint was clearly ductile as it was preceded by appreciable plastic deformation and necking (Fig. 2.33a), while the fracture surface revealed the presence of dimples caused by the IMC particles in the Sn matrix (Fig. 2.33b).

The study of the solder joint surface (indicated by arrows in Fig. 2.33a), which was subjected to tension during impact testing, proved also to be quite interesting, as it revealed the multiple brittle fractures of close-to-the-surface Cu_6Sn_5 IMC particles (Fig. 2.34). These IMCs, being brittle in nature and, hence, prone to failure when subjected to tension, have been forced to break at several sites along their length, so as to follow the deformation of the ductile Sn matrix in which they were embedded. The sharp external notches created in the solder joint by such IMC failures might facilitate the in-service brittle failure of the joint, especially

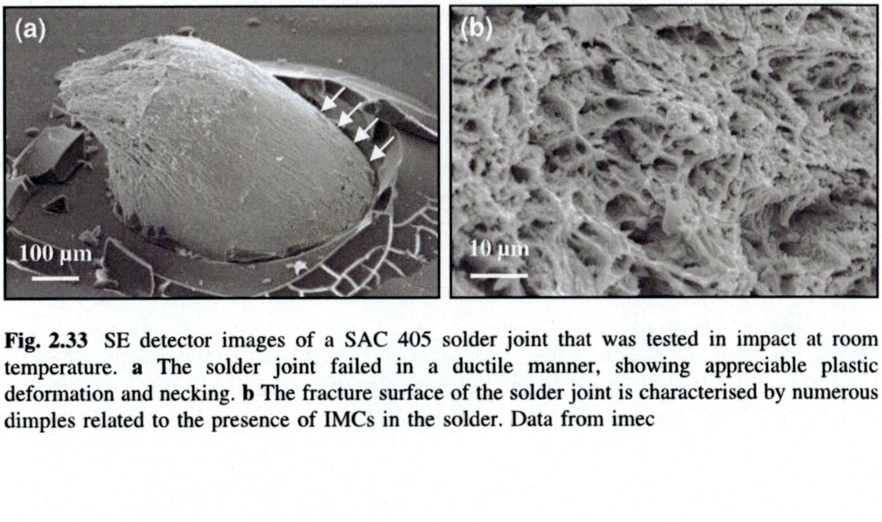

Fig. 2.33 SE detector images of a SAC 405 solder joint that was tested in impact at room temperature. **a** The solder joint failed in a ductile manner, showing appreciable plastic deformation and necking. **b** The fracture surface of the solder joint is characterised by numerous dimples related to the presence of IMCs in the solder. Data from imec

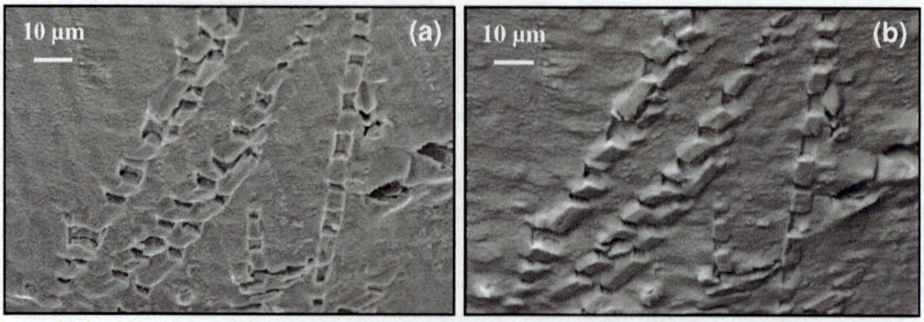

Fig. 2.34 Additional SEM micrographs of the SAC 405 solder joint shown in Fig. 2.33. **a** SE detector image of the area on the solder joint surface, which is indicated by arrows in Fig. 2.33a. This area, which has been subjected to tension during the impact testing of the joint, contains several fragmented Cu_6Sn_5 IMC particles. **b** Backscattered electron (BSE) detector image of the area shown in Fig. 2.34b. Data from imec

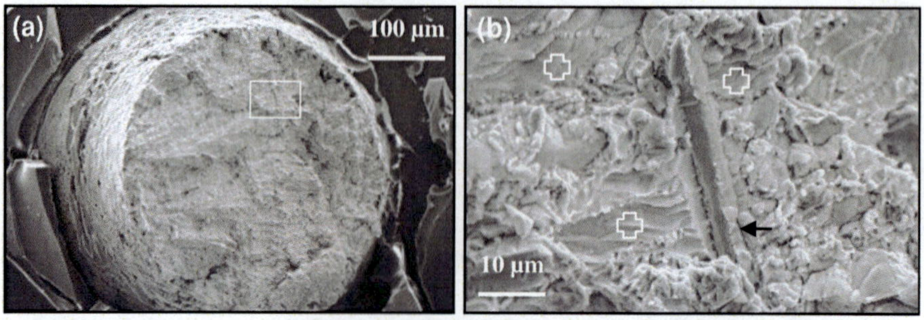

Fig. 2.35 SE detector images of a SAC 305 solder joint that was tested in impact at −70°C. **a** The fracture of the solder joint was clearly brittle, as indicated by the absence of plastic deformation. **b** Closer view of the area outlined by the small *frame* in Fig. 2.35a: the flat areas marked by *crosses* were nearly pure Sn, a fact that demonstrates the bulk embrittlement of the Sn-based solder at −70°C. The precipitate indicated by the *arrow* is a Cu_6Sn_5 IMC particle. Data from imec

when the 'mission profile' of the joint foresees combinations of shocks/intense vibrations with thermal excursions to low temperatures. The fracture behaviour of SAC 305 and other Sn-based commercial solders at room temperature was comparable to that of SAC 405, i.e. failure in the bulk of the joint occurred in a ductile manner [15, 16].

Figure 2.35 shows SEM images of the fracture surface of a SAC 305 solder joint that was tested in impact at −70°C. As may be seen, the joint failed in a brittle manner, i.e. failure was not preceded by plastic deformation, and the resulting fracture surface was macroscopically flat (Fig. 2.35a). On the microscopic level, the fracture behaviour of the nearly pure Sn solder matrix was also brittle, exhibiting a very flat fracture surface (Fig. 2.35b). The fracture behaviour of SAC 405 and the other Sn-based solders was similar to that of SAC 305 at low temperatures, i.e. brittle [15, 16]. In general, it was observed that the IMC particles in the tested Sn-based solders failed in a brittle manner over the whole test temperature range, i.e. between room temperature and −107°C, while the fracture of the Sn-rich solder matrix changed progressively from ductile to brittle as the test temperature decreased.

2.4.2 Microstructural Aspects of the Ductile-to-Brittle Transition

2.4.2.1 IMC Size Distribution and Spacing on DBTT

As discussed in Sect. 2.3.1, controlling the fracture behaviour of Pb-free solders requires the control of the shape, size distribution, and spacing of the IMCs. The results of an attempt to do just that are shown in Fig. 2.36, which presents the microstructural evolution of a SAC 405 solder caused by post-processing annealing. This alloy has been cast in a small mould, in order to produce a fine dispersion of Ag_3Sn IMCs in the as-cast bulk solder (Fig. 2.36a). Specimens of that casting were annealed for 100 and 1,000 h at 150°C (Fig. 2.36b and c, respectively), and 1,000 h at 175°C (Fig. 2.36d). The comparison of the micrographs in Fig. 2.36 reveals that the size and spacing of the Ag_3Sn IMC particles increases with the temperature and duration of annealing. On the other hand, the sharpness of these IMCs (as expressed by the radius of curvature of the particle edges) decreases with the intensification of the post-casting annealing. As-cast and annealed specimens were tested using a standard CVN impact test set-up, and the change in the DBTT of the alloy was related to changes in its microstructure [17]. Table 2.2 gives an overview of the change in the DBTT with the annealing conditions.

As may be seen for the results of this study, the DBTT decreases as both the temperature and the duration of the post-casting annealing increase. This may surely be related to the change in the size, sharpness, and spatial distribution of the IMC particles in the SAC 405 solder. It is thought that the finer IMCs in the as-cast solder pin the dislocations more effectively than the coarser ones after annealing. Dislocation pinning restricts the capability of the alloy for plastic deformation, facilitating embrittlement. Moreover, the sharper the IMC particles, the higher the intensity of stress concentration in their neighbourhood; making these particles more round is one way to lower the probability that brittle in-service failure will be triggered by their presence in the solder bulk.

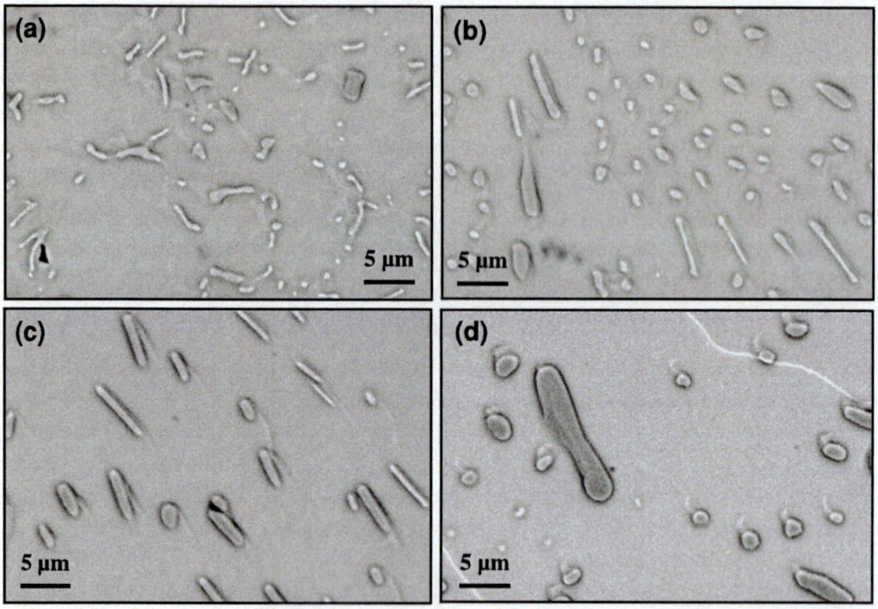

Fig. 2.36 BSE images of the changes in the microstructure of a SAC 405 solder alloy with the temperature and duration of post-processing annealing [17]. The alloy is studied in **a** the as-cast condition, **b** after 100 h at 150°C, **c** after 1,000 h at 150°C, and **d** after 1,000 h at 175°C. Data from imec

Table 2.2 DBTT of SAC 405 versus post-casting annealing conditions [17]

Material condition	DBTT (°C)
As-cast	−28 ± 6
100 h at 150°C	−42 ± 5
1,000 h at 150°C	−40 ± 5
1,000 h at 175°C	−48 ± 5

2.4.2.2 IMC Interaction with Crack Path During Thermal Cycling

Figure 2.37 shows SEM images of a SAC 405 solder alloy that has failed after 175 cycles of thermal cycling between 0 and 100°C. The images of Fig. 2.37 provide evidence of crack nucleation at IMC particles in the bulk of the solder joint. As already mentioned, sharp precipitates like the IMCs in Fig. 2.37 may be considered as internal 'notches' in Pb-free solders. Since Sn-based solders, like the SAC 405 studied here, are notch sensitive (see Fig. 2.30), one must be aware of the effect of the presence, sharpness, and size distribution of IMCs on the bulk embrittlement of such Pb-free solders.

At this stage of the research, it is thought that IMCs, like the Ag_3Sn particle in Fig. 2.37b, tend to pin dislocations, which in turn coalesce and result in crack

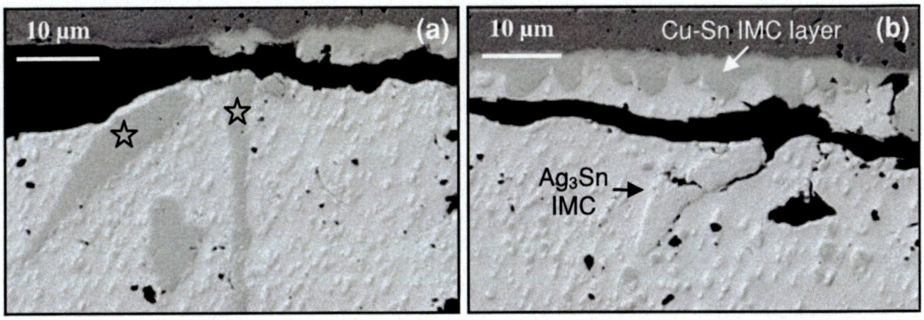

Fig. 2.37 BSE images of cross-sections through a solder joint made of SAC 405 that failed after 175 cycles of thermal cycling between 0 and 100°C. **a** Sharp Cu-Sn IMC particles *(stars)* act as sites of crack nucleation close to the joint surface. **b** Sharp Ag_3Sn IMC particle acts as site of crack nucleation far away from the joint surface. Data from imec

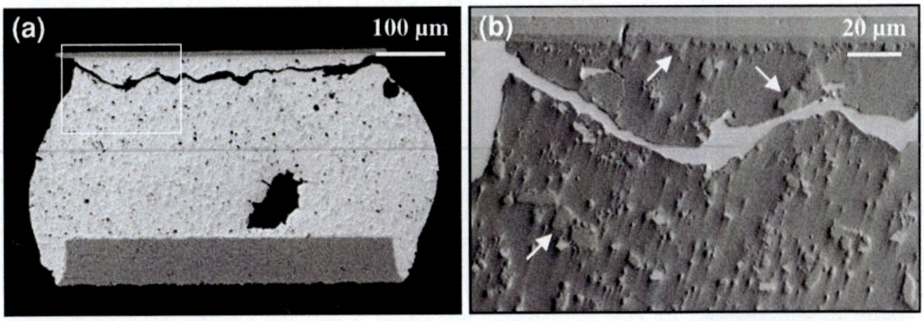

Fig. 2.38 BSE images of a cross-section through a Sn-0.7%Cu-0.1%Ni solder joint that failed after 434 cycles of thermal cycling between −40°C and 125°C. **a** Overview of the solder joint, which has failed due to the propagation of a fatigue type of crack. **b** Detail of the area inside the white *frame* in Fig. 2.38a. The *arrows* indicate IMC particles, which have mainly the $(Ni, Cu)_3Sn_4$ composition. Data from imec

nucleation. Once nucleated, the cracks may propagate either in a catastrophic (i.e. brittle) way through the Pb-free solder joint or in a more energy-absorbing (i.e. ductile) manner, like in the case of the thermo-mechanical fatigue cracks shown in Figs. 2.38 and 2.39. The way the crack will propagate through the solder joint will primarily depend on the service conditions, i.e. temperature, strain rate, and degree of material constraint for a specific structural application. Figures 2.38 and 2.39 show the propagation of fatigue-like cracks in two different joints made of the Sn-0.7%Cu-0.1%Ni solder. These solder joints belonged to BGAs subjected to thermal cycling between −40 and 125°C. The solder joint in Fig. 2.38 is a corner bump, while the joint in Fig. 2.39 is the bump next to the corner one in the array. Due to the fact that the corner bumps suffer the most during thermal cycling, the solder joint in Fig. 2.38 has failed completely, while the degree of damage in the joint of Fig. 2.39 is smaller. Since the corner joints are situated furthest away

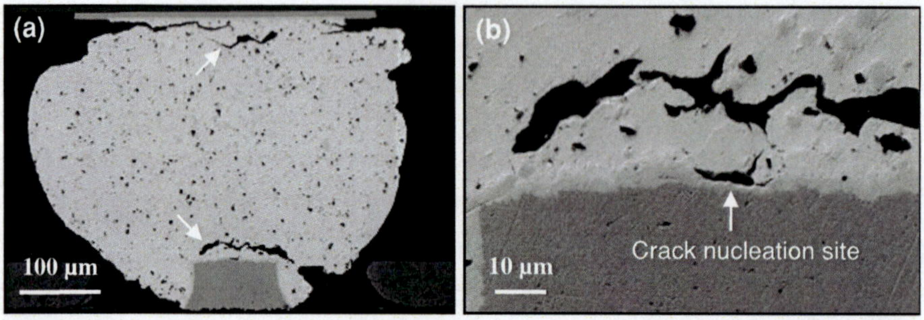

Fig. 2.39 BSE images of a cross-section through a Sn-0.7%Cu-0.1%Ni solder joint that failed after 577 cycles of thermal cycling between −40°C and 125°C. **a** Overview of the solder joint: fatigue-like cracks (indicated by *arrows*) propagate close to both upper and lower bond pads. **b** Crack nucleation between the solder bulk and the interfacial IMC layer. Data from imec

from the neutral point of the whole array, they are subjected to very high thermal stresses during thermal cycling. Thermal stresses are created due to the thermal expansion coefficient mismatch between the joint and its immediate surroundings, e.g. bond pad and PCB.

At this point, one may argue that the failure of the solder joints in Figs. 2.38 and 2.39 is thermo-mechanical fatigue and, as such, it falls outside the scope of this chapter. This is partly true, because the zigzagging of the cracks responsible for the failure of the two Sn-0.7%Cu-0.1%Ni joints suggest an energy-absorbing, rather ductile failure. However, a closer inspection of the sites of crack nucleation reveals that they are mostly related to the IMCs in the solder bulk or the IMC layers formed at the joint/bond pad interfaces. As emphasised throughout the development of this text, IMCs may be compared to 'internal notches' in the solder bulk. Notches are stress concentrators on their own account; if they are also subjected to external stresses (like the IMCs close to the two bond pads), they might be responsible for the initiation of failure (Fig. 2.6). Moreover, IMCs are believed to be responsible for dislocation coalescence and void formation leading to crack nucleation, as shown in Figs. 2.37b and 2.39b. Since ductile failures are mostly the result of microvoid coalescence [2], the propagation of fatigue cracks like those shown in Figs. 2.38 and 2.39 requires essentially the linking of these microvoids. In that sense, even the thermo-mechanical fatigue of Pb-free solders may be correlated with one of the main causes of bulk embrittlement: IMCs. Obviously, more research is required to validate the above hypothesis.

2.4.3 "Tin Pest"

Another case of possible in-service embrittlement of Sn-based Pb-free solders is the formation of "tin pest". As mentioned earlier, tin pest is a common name for α-Sn (also known as 'grey' Sn), which is the low-temperature allotropic form of β-Sn (also known as 'white' Sn). The β-Sn-to-α-Sn transformation occurs at 13.2°C and is accompanied by a substantial volume increase of ~26% [13, 18], since the density of β-Sn is 7.31 g/cm^3 and that of α-Sn is 5.77 g/cm^3. Once started, this transformation proceeds very fast, compromising the performance of the component. Interestingly, α-Sn is more brittle than β-Sn, since it is a semiconductor and not a metal [18]; hence, the formation of tin pest qualifies as a case of embrittlement. Figure 2.40a shows samples of Sn-0.5%Cu solder, which were aged at −18°C for up to 2 years. Forty percent of the specimen surface transformed into α-Sn after 1.5 years, and this percentage increased to 70% after 1.8 years of exposure [13, 18]. Embrittlement due to tin pest might be of concern for Sn-based Pb-free solders, since these solders may essentially be considered as a dispersion of IMC particles in a nearly pure Sn matrix. The risk of embrittlement

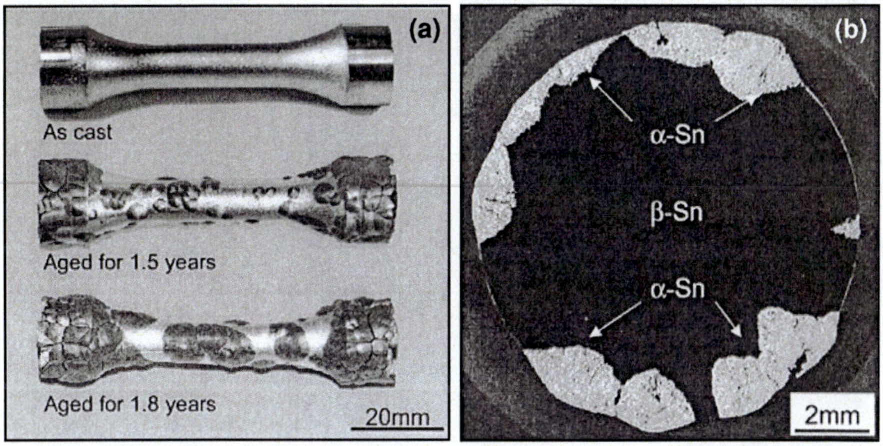

Fig. 2.40 **a** Appearance of Sn-0.5%Cu ingots aged at −18°C [13]$^\infty$. **b** Cross-section at the grip of a Sn-0.5%Cu sample after exposure for 1.5 years at −18°C. The tin pest formation reaction started at the surface of the sample and proceeded inwards [13]$^\infty$. $^\infty$ Reproduced with permission of Emerald Group Publishing Ltd. © 2000 by MCB University Press

Table 2.3 Tin pest-retarding effect of various alloy additions to pure Sn [18]

Alloy	Time to tin pest formation
99.998% Sn	2 h
Sn + 0.024 wt% Pb	12 days
Sn + 0.48 wt% Pb	29 days
Sn + 50 wt% Pb	190 days
Sn + 0.1 wt% Sb	14 h
Sn + 0.5 wt% Sb	Nothing visible after 5 years
Sn + 0.3 wt% Bi	330 days
Sn + 0.1 wt% Cu	4 h
Sn + 0.5 wt% Cu	7 h

becomes severe for these solders under the following two conditions: (a) the pure β-Sn grains reach the surface of the solder joint, since β-Sn transforms easier to α-Sn at free surfaces (Fig. 2.40b), and (b) the solder joint is exposed for long periods to low service temperatures. The good news is that avoiding tin pest seems to be feasible with the addition of small quantities of *tin pest retardants*, like bismuth (Bi) and antimony (Sb). Table 2.3 shows the tin pest-retarding action of various alloy additions to β-Sn specimens that have been inoculated (i.e. 'seeded') with 0.5 wt% α-Sn [18].

Acknowledgments The author would like to acknowledge certain persons, whose comments improved the quality of this text. These persons are listed in alphabetical order: Dr. Dag Andersson (IVF, Sweden), Prof. Ingrid De Wolf (imec, Belgium), Dr. Paul Michelis (IMMG, Greece), Prof. Marc Seefeldt (Department MTM, K. U. Leuven, Belgium), Dr. Geert Willems (imec, Belgium). The author also wishes to acknowledge the financial support provided by IWT (Flemish Government) in the framework of the ALSHIRA (Aspects of Lead-Free Soldering for High-Reliability Applications) Project. Last but not least, the author would like to thank Dr. B. Vandevelde, Mr. P. Limaye, and Mr. F. Duflos from imec, as well as Prof. B. Verlinden and Mr. W. Maurissen from the Department MTM of the Katholieke Universiteit Leuven (K. U. Leuven), for their collaboration on the research related with the embrittlement of Pb-free solder alloys.

References

1. Barsom JM, Rolfe ST (1999) Fracture and fatigue control in structures: applications of fracture mechanics, 3rd edn. ASTM Man Ser: MNL41, ASTM Int, West Conshohocken, USA
2. Hertzberg RW (1996) Deformation and fracture mechanics of engineering materials, 4th edn. Wiley, New York, USA
3. Hosford WF (2005) Mechanical behavior of materials, 1st edn. Cambridge University Press, New York, USA
4. Hurlich A (1968) Low temperature metals. In: Prodell AG (ed) Proc 1968 summer study supercond devices accel, pp 311–325
5. Callister WD Jr (2000) Materials science and engineering: an introduction, 5th edn. Wiley, New York, USA
6. Reed-Hill RE, Abbaschian R (1994) Physical metallurgy principles, 3rd edn. PWS Publishing Company, Boston, USA
7. Askeland DR, Phulé PP (2006) The science and engineering of materials, 5th edn. Nelson, Toronto, Canada
8. http://www.key-to-steel.com. Accessed 28 July 2008
9. Askeland DR (1984) The science and engineering of materials, 1st edn. Brooks/Cole, Monterey, USA
10. ASTM E 23-06 (2006) Standard test methods for notched bar impact testing of metallic materials. ASTM Int
11. Ratchev P, Loccufier T, Vandevelde B, Verlinden B, Teliszewski S, Werkhoven D, Allaert B (2005) A study of brittle to ductile fracture transition temperatures in bulk Pb-free solders. Proc EMPC 2005:248–252
12. Ratchev P, Vandevelde B, Verlinden B, Allaert B, Werkhoven D (2007) Brittle to ductile fracture transition in bulk Pb-free solders. IEEE Trans Compon Packag Technol 30:416–423

13. Kariya Y, Gagg C, Plumbridge WJ (2000) Tin pest in lead-free solders. Solder Surf Mount Technol 13:39–40
14. Metallography and Microstructures of Tin and Tin Alloys (2004) In: ASM Handb, vol 9: Metallography and Microstructures. ASM Int
15. Limaye P, Maurissen W, Lambrinou K, Duflos F, Vandevelde B, Allaert B, Hillaert J, Vandepitte D, Verlinden B (2007) Low-temperature embrittlement of lead-free solders in joint level impact testing. Proc EPTC 2007:140–151
16. Lambrinou K, Maurissen W, Limaye P, Vandevelde B, Verlinden B, De Wolf I (2009) A novel mechanism of embrittlement affecting the impact reliability of tin-based lead-free solder joints. J Electron Mater 38:1881–1895
17. Ratchev P, Vandevelde B, Verlinden B (2005) Effect of the intermetallics particle size on the brittle to ductile fracture transition in a bulk Sn-4 wt%Ag-0.5 wt%Cu solder. CD-ROM Proc IPC/JEDEC 10th Int Conf Lead-Free Electron Compon Assem
18. Ganesan S, Pecht M (eds) (2006) Lead-free electronics. IEEE Press, Wiley, Hoboken, USA

Chapter 3
Thermal Fatigue Analysis

Rainer Dudek and Ellen Auerswald

3.1 Introduction

One main reliability concern of the solder joints is related to thermal cycling loading, induced by either environmental temperature changes or active power on/off cycling. Low-cycle fatigue damage can be caused by this type of loading. The main source for thermal fatigue failure is the thermal expansion mismatch of the different materials soldered together, when the assemblies are subjected to a thermal cyclic environment already discussed previously. The most damaging mismatch occurs usually between the component and the printed wiring board (PWB) the component is soldered to, also called "global mismatch". Another kind of mismatch is that between the component and the solder material itself, the so-called "local mismatch". Its importance depends on the size of the solder joints; for current joint sizes, it is gaining importance only for cycle numbers greater than approximately 1000 of the characteristic temperature range −40 to 125°C. The so-called "inner mismatch", occurring between different solder phases and grains, causes solder fatigue also within free-standing solder layers in the absence of any other material. In tin-based lead-free solders, the anisotropic nature of the tin grains has turned out to be the major source for inner mismatch effects [1, 2]. Its effects on the long-term stability of the joints are still an open question.

R. Dudek (✉)
Fraunhofer ENAS, Micro Materials Center Berlin and Chemnitz,
Technologie-Campus 3, 09126 Chemnitz, Germany
e-mail: Rainer.Dudek@enas.fraunhofer.de

E. Auerswald (✉)
Fraunhofer IZM, Micro Materials Center Berlin and Chemnitz,
Volmerstr. 9B, 12489 Berlin, Germany
e-mail: Ellen.Auerswald@izm.fraunhofer.de

G. Grossmann and C. Zardini (eds.), *The ELFNET Book on Failure Mechanisms, Testing Methods, and Quality Issues of Lead-Free Solder Interconnects*,
DOI: 10.1007/978-0-85729-236-0_3,

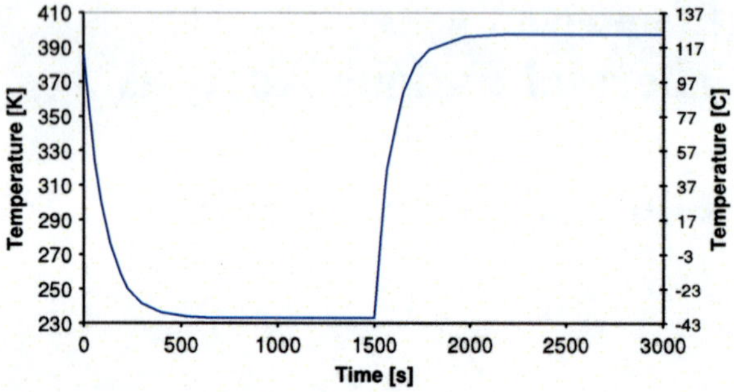

Fig. 3.1 Characteristic test temperature cycle

Two fundamental approaches are available to characterize the solder joint fatigue resistance. The first, which is applied to the majority of electronic products, is thermal test cycling. To achieve comparable statements in a reasonable time, various different test cycles have been defined. They operate usually with higher-temperature cyclic amplitudes and shorter cyclic times compared to the use conditions and are therefore called accelerated tests. Figure 3.1 shows a characteristic temperature versus time function as it occurs at a solder joint during air to air thermal cycling.

Several types of thermal test cycles are available, see Table 3.1. It is important to note that cycling out of use temperature range or with inappropriate high-temperature ramp rates, e.g. liquid to liquid, can engage inappropriate damage mechanisms. Thermal loading also can be dominated by active heating of the devices, i.e. by power cycling. The most complicated temperature-loading conditions occur in those cases when a cyclic temperature environment is overlaid by active heating.

The second approach for characterizing solder fatigue resistance is based on a theoretical fatigue model. Two basic tasks have to be solved by the theoretical model: the first is to define a physical measure to indicate failure, e.g. inelastic strain, or dissipated energy for appropriate strength theories, a damage parameter used in damage theories, or fracture toughness if a fracture mechanical approach is selected. The second task is to link this failure indicator to a critical cycle number.

3.2 Fatigue Failure Mechanism

Because the use temperatures of lead-free solders are also above 0.5 of their melting points (in K), as it was the case for Sn–Pbn solders, creep is a deformation mechanism which is important for materials degradation. Combined with slow loading during the different kinds of thermal cycles, the creep deformation kinetics determines the thermal fatigue behaviour of a solder to a high degree.

Failure of solder joints is a complex sequence of possible failure mechanisms. For conventional Sn–Pb solders, the sequence involves grain/phase coarsening, grain boundary sliding, matrix creep, micro-void formation and linking, and results in macro-crack initiation and crack propagation.

In the case of tin-based lead-free solder joints, the damage accumulation process leads to less well-known mechanisms. It includes thermo-mechanically

Table 3.1 Overview on thermal testing

Thermal testing	Influencing parameters	Methods
Thermal cycling (including thermal shock air to air)	Temperature rate	One-chamber cycling
	Dwell time	Two-chamber cycling
	Temperature extremes	
	Mean temperature	
Thermal shock (liquid/liquid)	Dwell time	Dip in hot and cold liquid
	Temperature extremes	
	Mean temperature	
Power cycling	Inhomogeneous temperature distribution	Power up/down
	Temperature rate	Ambient temperature
	Dwell time	Elevated temperature
	Temperature extremes	
	Mean temperature	

induced recrystallization, voiding and micro-cracking at regions where high induced strains cause local deformations in zones of local strain incompatibilities, e.g. colony or grain boundaries of different nature [3–5]. The micro-cracks tend to nucleate to form a macro-crack, also in dependence on the outer stress field. This behaviour can be related to the strengthening mechanism of Sn-based alloys, which can be seen in the tiny dispersed intermetallic particles embedded in the inherently soft ß-Sn matrix, mainly composed of $AgSn_3$.

During transition, there is also likely to be mixed assembly of both tin–lead and lead-free systems on the same board. The issue was studied by several authors, see e.g. [6, 7].

3.3 Thermal Test Cycling Standards

Because of the high importance of thermal loading in various electronic applications, thermal testing of lead-free solder joints has been studied extensively, see e.g. [8–12]. The accelerated thermal cycling test is a standard procedure to evaluate product reliability.

Currently available standards:

- IPC-9701A, "Performance Test Methods and Qualification Requirements for Surface Mount Solder Attachments"
- IEC/EN 60068 Part 2–14; "Environmental testing-Temperature Change"
- JESD 22-A104, "Temperature Cycling"
- various industry/company internal standards

Table 3.2 Product categories and worst case use environments for surface-mounted electronics (for reference only), cited from IPC-9701 A

Product category (typical application)	Temperature (°C)[a]	Worst case use environment							
	Storage	Operations	T_{min} (°C)[b]	T_{max} (°C)[b]	ΔT (°C)[c]	T (h)[d]	Cycles (year)	Typical years of service	Approx. accept. failure risk (%)
Telecom	–40	–40	–40	85	35	12	365	7–20	0,01
Computer/peripherals	–40	0	0	60	20	2	1460	5	0,1
Commercial aircraft	–40	–40	–55	95	20	12	365	20	0,001
Industrial arid automotive passenger compartment	–55	–40	–55	95	20 &	For all 12	185	10–15	0,1
					40 &		100		
					60 &		60		
					80		20		
Automotive (underhood)	–55	–40	–55	125	60 &	1	1000	10–15	0,1
					100 &	1	300		
					140	2	40		
Consumer	–40	0	0	60	35	12	365	1–3	1

& in addition

[a] All categories may be exposed to a process temperature range of 18–260°C

[b] T_{min} and T_{max} are the operational (test) minimum and maximum temperatures, respectively, and do not determine the maximum ΔT

[c] ΔT represents the maximum temperature swing, but does not include power dissipation effects

[d] The dwell tune to is the time available for the creep of the solder joints during each temperature half-cycle

Table 3.3 Temperature cycling requirements, mandated and preferred test parameters within mandated conditions, cited from IPC-9701A

Test condition	Mandated condition
Cycle condition	
TC1	0°C... +100°C (Preferred Reference)
TC2	−25°C... +100°C
TC3	−40°C... +125°C
TC4	−55°C... +125°C
TC5	−55°C... +100°C
Test duration	Preferred test duration: Until 50% (Preferred 63.2%) cumulative failure or
Number of thermal cycles (NTC)	requirement
NTC A	200 cycles
NTC B	500 cycles
NTC C	1,000 cycles (preferred for TC2, TC3. TC4)
NTC D	3,000 cycles
NTC E	6,000 cycles (preferred for TC1)
Low-temperature dwell	10 min
Temperature tolerance	0°C/−10°C [0°C/+5°C]
High-temperature dwell	10 min
Temperature tolerance	+10°C/0°C [+5°C/0°C]
Temperature rate	≤20°C/min
Full production sample size	>33
Printed wiring (circuit) board thickness	2.35 mm
Test monitoring	Continuous monitoring (event detector preferred)

Since the service conditions differ strongly, no general standard test can be defined for electronic products. A classification for various industries is given in IPC-9701A, see Table 3.2. The variety of thermal test cycles as defined by IPC-9701A is given in Table 3.3.

An overview on the different thermal fatigue testing strategies is given in Tables 3.4 and 3.5.

3.3.1 Fatigue Failure Detection

The occurrence of fatigue failure is detected either by electrical resistance measurements or by visual inspection of the solder joints' cross sections with regard to solder cracking. A third methodology to detect fatigue failure is the shear test, which is limited to simple component geometries, e.g. leadless ceramic components, and becomes very insensitive at states of progressed damage.

The electrical resistance measurement is the simplest technique and allows non-destructive failure detection for a statistically relevant number of components. However, often electrical opens resulting from solder joint failures are intermittent and may be difficult to detect accurately. This issue has turned out to be even more

Table 3.4 Overview on thermal cycling testing

	Slow thermal cycling (air/air)	Thermal shock (air/air)	Thermal shock (liquid/liquid)	Field cycling
Standards	IPC-970A JESD 22-A104	IPC-9710A JESD 22-A104	IEC/EN 60068 Part 2–14	None, company specific
Methods/ Equipment	Place in temperature chamber at varying temperature. One-chamber oven. Hot/cold with air circulation	Move (automatically or manually) between hot and cold chamber/two temperature chambers or a two-chamber oven with air circulation	Dip in hot/cold liquid bathes with heating/ cooling equipment	Power up/down, optionally environmental temp. change/ electrical loading equipment, optionally temperature chamber
Usage	Frequently	Most frequently	Not recommended for solder fatigue, tends to induce failure mechanisms different from the field	Rare
Advantages	Relatively slow, induces best field-related creep-dominated failure. Can be combined with other tests (Vibration, moisture)	Relatively fast, induces field-related creep-dominant failure	Fast. Cheap equipment	Shows field behaviour
Drawbacks	Long testing times, expensive equipment	Expensive equipment	Failure mechanisms different from filed failures	Slow, results device dependent
Influencing parameters	Dwell time, temperature extremes, mean temperature, ramp rates	Dwell time, temperature extremes, mean temperature	Temperature cycling parameters	Temperature cycling parameters, ramp rates, dwell times, gradients in power up/ down

(continued)

Table 3.4 (continued)

	Slow thermal cycling (air/air)	Thermal shock (air/air)	Thermal shock (liquid/liquid)	Field cycling
Coupling to theoretical analysis prediction	Coffin-Manson type empirical laws, FEA (based on secondary creep law and creep strain/ energy dissipation)	Coffin-Manson type empirical laws, FEA (based on secondary creep law and creep strain/ energy dissipation)	FEA (should include coupling of temperature gradients and primary/ secondary creep)	Empirical acceleration factors partially available, FEA (may include coupling of temperature gradients, secondary creep mostly sufficient)

important for lead-free solders than for the tin–lead materials. Whenever possible, online measurements should be preferred to avoid misleading results from the electrical measurements. Moreover, the latter method usually has to be combined with cross sectioning of the joints of interest. An example is shown in Fig. 3.2 for $SnAg_{3.5}$ solder, where in contrast to the totally broken joint, only a slight resistance increase was detected. A second example is shown in Fig. 3.3 for $SnAg_{3.5}Cu_{0.5}$ solder. It can be observed from the figure that cracking of large parts of the joints is not detectable by offline resistance measurement at room temperature. The resistance increases to infinity only if the cracked surfaces really move apart from each other, as shown in Fig. 3.4. Thus, a strict failure criterion is recommended for electrical resistance change, since it is not sensitive to cracking. Ideally, event detectors have to be preferred, as is recommended by IPC-9701A. According to this standard, offline measurement results are not precise enough to detect fatigue failure.

3.4 Theoretical Modelling of Thermal Fatigue

During the last decade, the rapid development of cheap and powerful computer technology has shifted the focus of reliability analysis of solder joints towards the use of finite element analysis (FEA), see e.g. [13–16]. However, since creep is the dominant deformation mechanism of solder, high effort is required to determine the creep properties. Several experimental techniques are in use to determine these creep properties of solders, which are summarized in Table 3.6. Because of the effects of microstructure on the creep behaviour, no unique creep properties seem to exist for both leaded and lead-free solders, i.e., the published properties are sometimes conflicting [14, 17, 18] and the ones used therefore approximate in character. Additionally, the properties-metallurgy dependencies of lead-free solders are still a subject of research [19, 20].

Table 3.5 Overview on thermal test cycling related to different standards

Thermal cycling/shock	IPC-970A, Feb. 06 Test methods and qualification requirements for SM solder attachments	IEN/EN 60068 part 2–14, 2000 Environmental testing: change of temperature	JEDEC standard JESD 22-A104-A, Dec 1989 temperature cycling	Company-project internal data related to solder fatigue
Temperature rate	Air to air, ≤20°C/min measured at samples (no liquid to liquid)	(a) Fast air to air, defined ramps (b) Slow air to an, defined ramps (c) fast liquid to liquid	Air to air, ≤1 min transfer time, heating of samples within 15 min	<10–30°C/min
Equipment	One (preferred) or two chambers	(a) Two chambers (b) One chamber (c) Two liquid bathes	One (preferred) or two chambers	One of two chambers (preferred), liquid to liquid rarely applied due to the risk to induce irrelevant failure mechanisms
Dwell time	10 min (SnPb) 10 min (SAC) for "stand alone" life assessments, not to compare to SnPb due to slower creep) ≥30 min (creep damage somewhat comparable to SnPb)	(a) Ramp rates 1, 3 or 5°C/min, dwell times 3 h to 10 min (b) Ramp times 2–3 min or 20–30 min or <10 s, dwell times 3 h to 10 min (c) Ramp times 8 s or 2 s, dwell times 20 min to 5 min or 15 s to 5 min	≥10 min	10 min to 2 h

(continued)

Table 3.5 (continued)

Thermal cycling/ shock	IPC-970A, Feb. 06 Test methods and qualification requirements for SM solder attachments	IEN/EN 60068 part 2–14, 2000 Environmental testing: change of temperature	JEDEC standard JESD 22-A104-A, Dec 1989 temperature cycling	Company-project internal data related to solder fatigue
Temperature extremes	−55 to 125°C different combinations, 0–100°C preferred (SnPb, SAC, SnBi), −25 to 75°C SnBi on SnPb plating	(a), (b) Worst case different combinations (c) Characteristic from 0 to 100°C	−65 to 200°C, worst case, different combinations	−55 to 150°C, characteristic −40 to 125°C
Definition of failure	Continuous intermittent event monitoring, ≥1,000 Ω, >1 μs, 10 events, report first event (preferred) and/or continuous intermittent event monitoring, ≤20% increase, max. 5 readings, report average	Loss of device electrical functionality under nominal and worst case conditions, mechanical damage	Loss of device electrical functionality under nominal and worst case conditions, mechanical damage (cracking, chipping, breaking, loss of hemieticity)	Offline measurement (char. ≥50% increase), on line resistance monitoring (char. ≥20% increase), online event monitoring (char. ≥1,000 Ω, >1 μs), cross sectioning (crack progress), shear testing (char. ≥50% load drop)
Recommendation for "pass condition"	Product category dependent (table)	Characteristic (a) 5 cycles (b) 2 cycles (c) 10 cycles	1,000 cycles	Product or test dependent, characteristic 1,000 cycles −40 to 125°C

Fig. 3.2 Chip resistors 0805 with SnAg solder on FR-4 PCB after 2,000 cycles –40…125°C, 60 min: resistance increased only by 164 mΩ at room temperature (H. Berek, FnE GmbH, Freiberg)

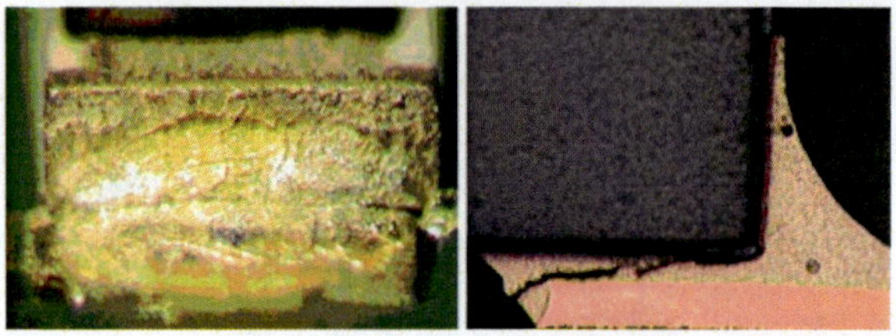

Fig. 3.3 Chip resistors 1206 with SAC solder on FR-4 PCB after 1,000 cycles –40…125°C, 60 min: resistance increased from 26 to 28 mΩ at room temperature (Z. Drozd, Warsaw University of Technology [10])

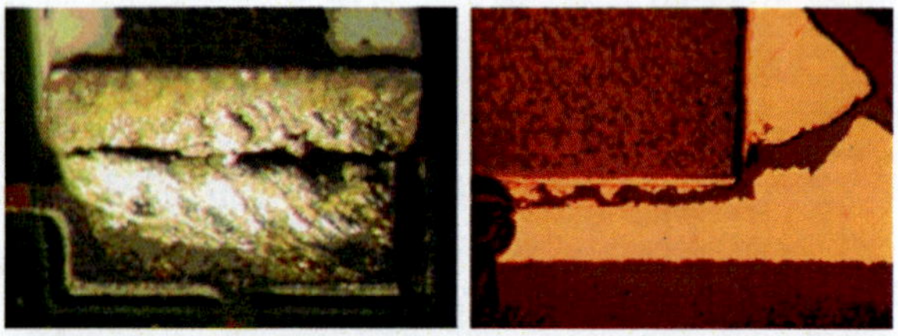

Fig. 3.4 Chip resistors 1206 with SAC solder on FR-4 PCB after 2,000 cycles –40…125°C, 60 min: resistance increased from 26 mΩ to ∞ at room temperature (Z. Drozd, Warsaw University of Technology)

It was already mentioned that the thermal cyclic environment can vary, and that the relation between test cycles and field cycles, the so-called acceleration factors, are of special interest. Simplified analytical models can be very useful in cases to avoid the high effort of numerical modelling. An experimentally supported model for determining the acceleration factor $N_{\text{test}}/N_{\text{field}}$ for lead-containing solders was developed by Norris and Landzberg [21]. This model was adopted by Pan et al. [22] for SAC387 solders as shown below.

SnPb

$$N_{\text{field}(x)} = N_{\text{test}(x)} \left(\frac{\Delta T_{\text{test}}}{\Delta T_{\text{field}}}\right)^{c1} \left(\frac{f_{\text{field}}}{f_{\text{test}}}\right)^{\frac{1}{3}} \exp\left[1414\left(\frac{1}{T_{\text{field}}} - \frac{1}{T_{\text{test}}}\right)\right] \tag{3.1}$$

Table 3.6 Overview on solder creep and fatigue characterisation

	Tensile test	Lap shear test with large joint area	Component type laps shear test	Ring- and plug test
Specimens	Macro sized dog bone; miniaturized dog bone	Single/double lap shear with large joint area (Cu on Cu)	Single/double lap shear with component type joints (FC, BGA, LGA)	Copper plug soldered in copper ring subjected to torsional ortensional load; copper wire soldered in via subjected to tensional load
Measuring apparatus	Standard/miniaturized tensile testers; commercially available	Standard/miniaturized tensile testers; commercially available	Special purpose test equipment	Special miniaturized torsion/ tensile testers; weights
Usage	Frequently; creep characterization; fatigue characterisation; includes a high risk of buckling effects	Most frequently; creep characterization; fatigue characterisation	Rare; creep characterization; fatigue characterisation	Rare; creep characterization; fatigue characterisation; creep characterization only
Advantages	Simple specimens with high precision; microstructure can be controlled by temperature regime; easy to measure/high precision; one-dimensional stress state	Interaction with pad finishes, alloying content and intermetallics match real joints; high joint size precision difficult; volume comes close to real joints;shear loading is dominant joint loading	Interaction with pad finishes, alloying content and intermetallics match real joints; volume matches real joints; shear loading is strongly overlaid by three-dimensional stress state	Volume comes close to real joints/matches real joints; shear loading is dominant joint loading (both wire and plug); simple measurement
Drawbacks	Missing interaction with pads (alloying effects, intermetallics); volume large related to actual joints, even for miniaturized specimens	Almost pure shear for large joints, but loss of joint matchdifficult to process with high precision; cooling regime/microstructure can differ from joint	Multi-dimensional stress state for small joints (FEA required); difficult to measure with high precision	Strain gradient and multidimensional stress state can beimportant (FEA required, both wire and plug); difficult to process with high precision

(continued)

Table 3.6 (continued)

	Tensile test	Lap shear test with large joint area	Component type laps shear test	Ring- and plug test
Major sources of errors	Strain measurement; microstructure is not characteristicfor joint	Dwell time, temperature extremes, mean temperature	Temperature cycling parameters	Temperature cycling parameters, ramp rates, dwell times,gradients in power up/down
Specimens	Macro sized dog bone; miniaturized dog bone	Coffin-Manson type empirical laws, FEA (based on secondarycreep law and creep strain/energy dissipation)	FEA (should include coupling of temperature gradients and primary/secondary creep)	Empirical acceleration factors partially available, FEA (mayinclude coupling of temperature gradients, secondary creep mostly sufficient)

SAC 387

$$N_{\text{field}(x)} = N_{\text{test}(x)} \left(\frac{\Delta T_{\text{test}}}{\Delta T_{\text{field}}}\right)^{2.65} \left(\frac{t_{\text{test}}}{t_{\text{field}}}\right)^{0.136} \exp\left[2185\left(\frac{1}{T_{\text{field}}} - \frac{1}{T_{\text{test}}}\right)\right] \quad (3.2)$$

$N_{\text{field}}(x)$, number of field cycles at which a percentage x of connections has failed, for example, $N_{\text{field}}(50)$ = mean cycles to failure at which 50% of connections have failed; $N_{\text{test}}(x)$ analogous number of test cycles; c_1, empirical constant, applicable is $c_1 = 1.9$ for SnPb; f cycling frequency expressed in number of cycles/day; t dwell times (equal dwell times for hot and cold dwell preferred); T maximum cycling temperature in K.

However, because of the complicated dependencies of SAC solder on various effects like type of components, metallization and joint size, the derived equation seems questionable [23]. The SnPb constants for the Norris–Landsberg model seem to be a better fit to the existing Pb-free data than the revised constants provided in the paper by Pan et al. Further work is required with respect to the eutectic SAC solder compositions and those improved by spurious alloys.

3.5 Open Questions

- Long-term fatigue performance of different lead-free solders in field use in dependence on the component type and adequate formulation of acceleration factors;
- Reliability of different lead-free solders under the combined action of different leadings, e.g. low-cycle fatigue overlaid by vibration;
- Effects of spurious alloys on the creep and thermal fatigue performance of second-generation lead-free solders;
- Development of virtual reliability design methodologies, aiming at relations between test cycling and mission profiles for use in high-reliability applications for low-cycle fatigue loadings and for combined loadings;
- Materials characterization and modelling of second-generation lead-free solders including SnZn, Bi-containing solders, under-eutectic SAC solders with Ni and other spurious alloys, and nanoparticle-containing solders or adhesives. Consideration of miniaturization-related effects on materials properties and modelling, respectively, e.g. the shift of plastic-creep properties of tiny interconnects that approach the meso-structure size (e.g. grain size, primary intermetallics size). Application of nano-indentation techniques towards elastic–plastic, elastic-creep or viscoelastic strongly localized materials characterization, also at thin-film materials or intermetallics;
- Development of sophisticated modelling techniques including viable ones for the meso-structure of materials and damage progress.

References

1. Bieler TR, Jiang H, Lehman LP, Kirkpatrick T, Cotts EJ (2006) Influence of Sn grain size and orientation on the thermomechanical response and reliability of Pb-free solder joints. In: Proceedings, 56th electronic components & technology conference, San Diego, pp 1462–1469
2. Gong J, Liu Ch, Conway PP, Silberschmidt VV (2006) Grain features of SnAgCu solder and their effect on mechanical behavior of micro-joints. In: Proceedings of 56th electronic components & technology conference, San Diego, pp 250–257
3. Dunford S, Canumalla S, Viswanadham P (2004) Intermetallic morphology and damage evolution under thermomechanical fatigue of Pb-free solder interconnections. In: Proceedings of 54th ECTC, Las Vegas, pp 726–735
4. Grossmann G, Jud PP, Sennhauser U (2005) Microstructure and deformation behavior of SnAg3.8Cu0.7 lead-free solder. In: The world of electronic packaging and system integration. Fraunhofer Institute Reliability and Microintegration, Berlin
5. Dudek R, Faust W, Gollhardt A, Michel B (2006) A FE study of solder fatigue compared to microstructural damage evaluation by in situ laser scanning and FIB microscopy. In: Proceedings ITherm 2006, San Diego, USA
6. Dušek M, Nottay J, Hunt C (2001) Compatibility of lead-free solders with PCB materials. National Physical Laboratory, UK, NPL Report MATC (A) 89
7. Clech J-P (2004) Lead-free and mixed assembly solder joint reliability trends. IPC Printed Circuits Expo®, SMEMA Council APEX, Designers Summit 04
8. Andersson et al (2004) Effect of different temperature cycle profiles on the crack propagation and microstructural evolution of lead free solder joints of different electronic components. In: Proceedings of Eurosime, Brussels
9. Schubert A, Dudek R, Auerswald E, Gollhardt A, Michel B, Reichl H (2003) Fatigue life models for SnAgCu and SnPb solder joints evaluated by experiments and simulation. In: Proceedings of 53rd electronic components & technology conference, pp 603–610
10. Drozd Z (2006) Testing methodology and results from thermal shock tests in the project GreenRoSE. ELFNET Project meeting, Athens
11. Vandevelde B, Ratchev P, Limaye P (2006) Latest results from ALSHIRA. ELFNET Project meeting, Athens
12. Dušek M (2006) NPL—lead-free projects of electronics interconnection group. ELFNET Project meeting, Dubendorf
13. Vandevelde B, Gonzalez M, Limaye P, Labie R, Ratchev P, Vanfleteren J, Beyne E (2004) Leadfree solder joint reliability estimation by finite element modelling: advantages, challenges and limitations. In: IPC-/JEDEC 7th international conference on leadfree electronic components and assemblies, Frankfurt
14. Dudek R (2006) Characterization and modelling of solder joint reliability. In: Zhang GQ, van Driel WD, Fan XJ (eds) Mechanics of microelectronics. Springer Science+Business Media, Dordrecht, pp 377–466
15. Lau J, Dauksher W, Vianco P (2003) Acceleration models, constitutive equations, and reliability of lead-free solders and joints. In: Proceedings, 53rd electronic components & technology conference, pp 229–236
16. Zhang Q, Dasgupta A, Haswell P (2003) Viscoplastic constitutive properties and energy-partitioning model of lead-free Sn3.9Ag0.6Cu solder alloy. In: Proceedings of 53rd electronic components & technology conference, pp 1862–1868
17. Clech JP, Handwerker C (2002) Review and analysis of lead-free solder material properties BGA, Flip-Chip. ESPI Inc., NIST
18. Pang JHL, Xiong BS, Neo CC, Zhang XR, Low TH (2003) Bulk solder and solder joint properties for lead free SnAgCu solder alloy. In: Proceedings of 53rd electronic components & technology conference, pp 673–679

19. Wiese S, Meusel E, Wolter K-J (2003) Microstructural dependence of constitutive properties of eutectic SnAg and SnAgCu solders. In: Proceedings of 53rd electronic components & technology conference, pp 197–206
20. Wiese S, Roellig M, Mueller M, Rzepka S, Nocke K, Luhmann C, Kraemer F, Meier K, Wolter K-J (2006) The influence of size and composition on the creep of SnAgCu solder joints. In: Proceedings ESTC, Dresden, Germany, pp 912–926
21. Norris KC, Landzberg AH (1969) Reliability of controlled collapse interconnections. IBM J Res Dev, pp 266–271
22. Pan N, Henshall GA, Billaut F, Dai S, Strum MJ, Benedetto E, Rayner J (2005) An acceleration model for Sn-Ag-Cu solder joint reliability under various thermal cycle conditions. In: Proceedings of SMTA international conference, Chicago
23. Hillman C (2006) Assessment of Pb-Free Norris-Landzberg model to JG-PP test data. Presentation DfR Solutions

Chapter 4
Electrochemical Behaviour of Solder Alloys

C. Zou and C. Hunt

4.1 Introduction

Achieving high reliability is a key issue in today's electronics assemblies, with a loss of continuity or short circuits failures, caused by metal corrosion an important failure mode. The corrosion source can originate from the manufacturing process, flux residues that may contain ionic or hydrophilic materials. Also operating in harsh environments where other aggressive species maybe encountered are another source of contaminants. Contaminants when combined with moisture result in a lowering of Surface Insulation Resistance (SIR) [1, 2] between conductors on the circuit and can accelerate corrosion. On a circuit assembly, corrosion may take place with and without the assistance of an applied voltage. Without the voltage bias, corrosion is driven by the electrochemical force, which depends thermodynamically on the anodic and cathodic processes. In general, the metals used for electronic circuit have a high corrosion resistance, without bias metal corrosion should not be a big concern. With applied bias, the voltage can overcome the electrochemical force driving further corrosion reactions to take place. In addition, the electric field on an insulating surface can also provide a driving force for metal ion migration and result in dendrite formation [3, 4]. Catastrophic failure occurs when dendrite growth between oppositely charged metal conductors bridge the conductor gap and cause short circuits, and is a very important failure mechanism in electronic assemblies.

A fundamental study of the electrochemical process that to dendrite formation is difficult, since the adsorbed atmospheric moisture layer is only approximately 100 nm thick. Typically, electrochemical studies use electrode systems that are on

C. Zou (✉) and C. Hunt
National Physical Laboratory, Hampton Road, Teddington,
Middlesex, TW11 0LW, UK
e-mail: ling.zou@npl.co.uk

G. Grossmann and C. Zardini (eds.), *The ELFNET Book on Failure Mechanisms, Testing Methods, and Quality Issues of Lead-Free Solder Interconnects*,
DOI: 10.1007/978-0-85729-236-0_4, © Springer-Verlag London Limited 2011

a significantly greater dimensional scale than 100 nm. Dendrite formation is a three-step process. Firstly, metal is oxidized to metal ions into electrolyte, then metal ions migrate from the anode to cathode under an electric field, and finally metal ions are reduced to metal and deposited at the cathode. Clearly, all these processes are necessary for dendrite formation, and there are many factors that can influence these processes. *Metal oxidation at the anode*: each metal will have a different corrosion rate [5], and a metal with a fast corrosion rate will clearly form higher metal ion concentrations in the thin water film, hence enhancing the formation of dendrites. *Metal ion transportation*: if the metal ion forms an insoluble compound, it will not migrate from the anode to cathode, and therefore, solubility of the metal hydroxide is key to dendrite formation [6]. Finally, *metal deposition* at the cathode, Zamanzadeh [7] found that there is limiting over-potential for dendrite formation, below which dendrite formation can never occur. Most investigations have avoided such difficulties by using statistical evaluation of failures from SIR measurement, without addressing the electrochemical nature of potential failure processes. However, some researches on electrochemical corrosion of electronic metals are usually carried out in bulk solution without consideration of high ionic resistance between conductor lines on electronic circuits [8–11].

The conductor–insulator structure of a comb pattern with an adsorbed moisture layer is shown schematically in Fig. 4.1 and when an AC signal is applied an equivalent circuit representing each significant conduction process is also shown [12, 13]. $\boldsymbol{R_C}$ is the ionic resistance of the thin water layer with contaminants across the inter-electrode gap of the comb, and $\boldsymbol{C_C}$ is the capacitance of the comb pattern. The circuit elements in the dotted box $\boldsymbol{R_p}$ are a representation of electrochemical processes at the interface of the contaminant layer and electrode. These reactions represent the transition from electronic conduction in the electrode to ionic conduction in the contaminant layer. The equivalent circuit for these electrochemical processes includes the charge-transfer resistance $\boldsymbol{R_{CT}}$ associated with a faradic reaction at the electrode surface, $\boldsymbol{C_{DL}}$ a double layer or blocking capacitance at the interface, which may allow ionic current to flow in the absence of a faradic reaction, and $\mathbf{Z_W}$ the Warburg impedance, detectable when a faradic reaction occurs under diffusion control near the interface. The different conduction processes can be separated by electrochemical impedance (EI) measurements due to the different frequency dependences. EI results can separate ionic resistance $\boldsymbol{R_c}$ from overall impedance of the whole system ($\boldsymbol{R_T}$). This can provide the information on where the controlling step in the overall conduction process lies. In this conduction system, $\boldsymbol{R_C}$ is an important parameter, dominating metal corrosion and dendrite formation. In contrast, the SIR measurement can only measure the total resistance of the system.

The aim of this work was to investigate the corrosion of three solders and two PCB finish materials using EI. The two-electrode system of a comb test pattern, manufactured from the studied materials, was used and tested at an elevated temperature and humidity. The inter-electrode gap of 200 μm of the comb patterns reflected current high-density electronic circuit design. Dendrite formation was also assessed using SIR measurements, and these measurements were correlated

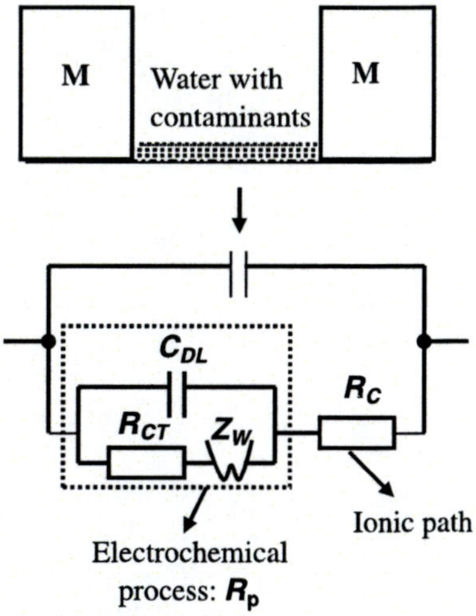

Fig. 4.1 Conduction path for conductor–insulator structure

with the ionic resistance R_C. The applicability of using simple and quick non-destructive EI measurements to predict reliability was evaluated.

4.2 Experimental

4.2.1 Test Board

A test board was designed consisting of four identical comb patterns, as shown in Fig. 4.2. The size of each comb pattern was 25 × 25 mm with 0.60 mm pitch and 0.20 mm gap. Boards were prepared with five finish materials: SAC, Sn, SnPb, ENIG and copper. Sn, ENIG and Cu-finished boards were obtained from the PCB supplier directly. SAC and SnPb finish boards were manufactured from Sn boards printed with proprietary no-clean solder pastes. The pastes used were $Sn_{63}Pb_{37}$ for the SnPb-finished boards and $Sn_{96.5}Ag_{3.0}Cu_{0.5}$ for the SAC-finished boards. Sn finish boards were used here to manufacture SAC and SnPb finish boards, the effect of underlying metal on the studied metal finish was therefore minimized, as solder may not completely wet the track surface particularly for SAC alloy. The wetting problem with the lead-free soldering process was found in a previous study [14].

4.2.2 Test Board Cleaning

The aim of this work was to investigate the corrosion behaviour, and propensity of solder alloys and PCB finish materials to form dendrites. In order to eliminate any effects of contamination associated with the manufacture processes, the cleanliness

Fig. 4.2 Test board with five finishes

of test boards with different finishes had not only to be assured but also to achieve similar high levels of SIR. The SAC and SnPb boards were prepared by printing onto the Sn finished as received boards with solder paste and reflowed before the following cleaning procedure. The cleaning procedures were as follows:

For the ENIG- and copper-finished boards:

Ionograph cleaning in 75% IPA + 25% DI water mixture at 45°C for 15 min

For Sn-finished board:

A single pass through a reflow process with a maximum temperature of 250°C
Ionograph cleaning in 75% IPA + 25% DI water mixture at 45°C for 15 min

For SAC-board and SnPb-finished board:

Ultrasonic cleaning in 100% Zestron FA + at 50°C for 5 min
DI water rinsing at 50°C for 5 min
Ionograph cleaning in 75% IPA + 25% DI water mixture at 45°C for 15 min

4.2.3 Test Flux

A commercial flux was used for the study. The flux was recommended by a flux supplier and was a no-clean flux suitable for lead-free soldering. The flux was a low resin content, halide-free and solvent based.

4.2.4 Test Board Contamination

The cleaned boards were fluxed using 50 μl of the test flux to cover each comb pattern. Fluxed boards were dried in a 100°C dry oven for 5 min for testing.

4.2.5 Flux Concentration

4.2.5.1 Electrochemical Impedance (EI) Measurements

For EI measurements, the flux was diluted to a certain level to achieve pronounced effects in the EI results from the electrochemical process. A dilution of 25% flux in IPA was used for all board finishes defined from the preliminary testing.

4.2.5.2 Surface Insulation Resistance (SIR) Measurement

For this work, it was necessary to determine the minimum flux concentrations that caused dendrite formation on the five board finishes. While this appears to be an artificial manipulation of the flux residue level, it does allow characterization of the flux residues in the critical regime for each board finish and is relevant since process residues vary over a wide range. Therefore, the flux concentrations were reduced as necessary from the as-received flux concentration by dilution with IPA. Three different concentrations (low, middle and high) were found for the different metal finishes from preliminary testing, and these are listed in Table 4.1.

4.2.6 Electrochemical Impedance (EI) Measurement

The EI was measured for each board finish using an ACM instruments GillAC. A two-electrode system was used for the test, as shown in Fig. 4.3. EI measurements were performed on three test patterns for ENIG board finish to check the repeatability of the test, and only one test pattern for the rest board finishes. As soon as the chamber reached the desired temperature and humidity, 40°C/93%RH, the open circuit potential (corrosion potential) was monitored for 60 minutes, then sample impedance were measured with the application of a 50mV peak-peak AC single with frequencies ranging from 3000 to 0.01 Hz.

4.2.7 Surface Insulation Resistance (SIR) Measurement

A single board, four test patterns, for each board finish was SIR tested. The SIR measurements were performed under the same condition as the EI measurements, 40°C/93%RH. A bias voltage of 5 V DC was applied during the test period of

Table 4.1 Test flux concentration for different board finish

Board finish	ENIG	Cu	SnPb	Sn	SAC
Flux concentration (% in IPA)	5	50	15	15	15
	10	65	20	25	25
	15	80	25	35	35

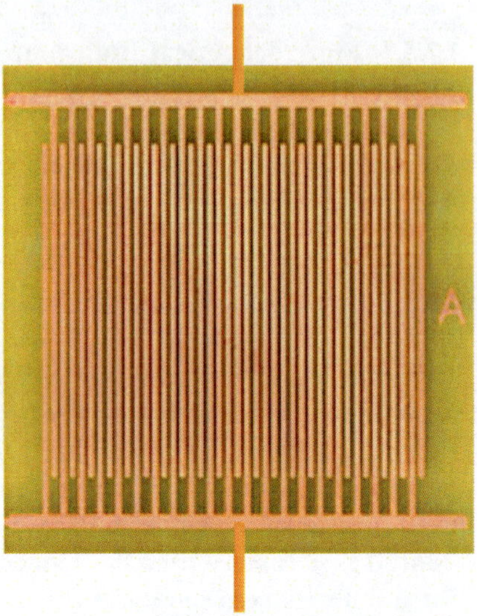

Fig. 4.3 Two-electrode system for EI measurements

72 h, and the SIR was measured every 15 min. The test equipment had a 10^6 Ω resistor incorporated into each test channel to preserve dendrite formation, so the minimum SIR measurement value of each channel was 10^6 Ω.

4.2.8 EDX Analysis of Dendrites

The dendrites formed during the SIR measurement on different board finish were identified and located using an optical microscope with back lighting, and the dendrite composition was assessed using an energy-dispersive X-ray (EDX) analyser attached to a SEM.

4.3 Results

4.3.1 Electrochemical Impedance Spectra on Five Board Finishes

EI results, Bode, Nyquist and phase angle plots, for five board finishes are plotted in Figs. 4.4, 4.5, 4.6, 4.7 and 4.8. Three tests, A, B and C, were performed for the ENIG finish to check the repeatability of the testing. The results for three tests, as shown in Fig. 4.4, showed a reasonable good repeatability; hence, only one test was conducted for the rest board finishes.

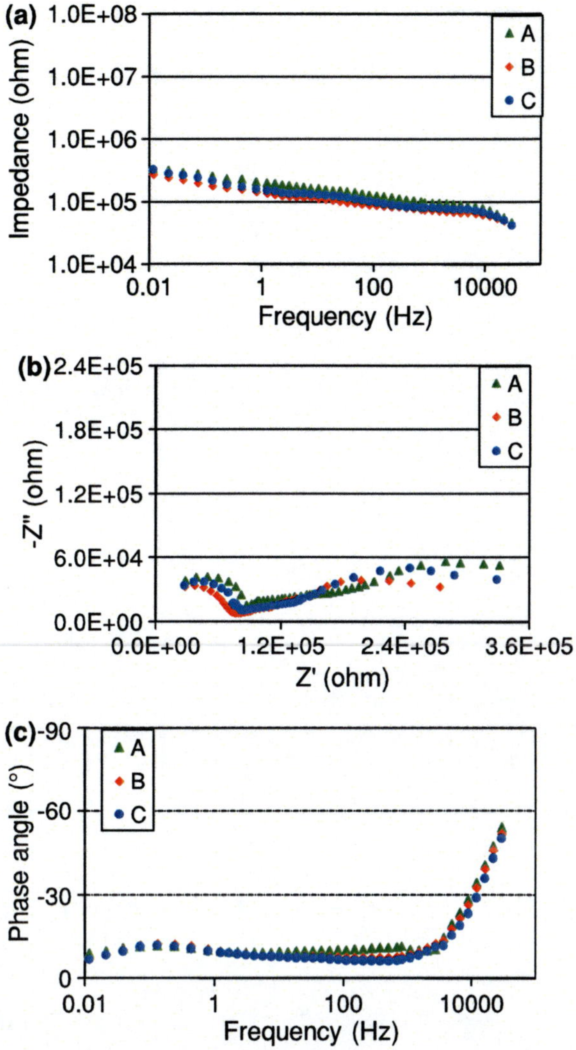

Fig. 4.4 Bode, Nyquist and phase angle plots for ENIG board

The EI results for the different board finishes revealed a range of responses. For Sn, SAC and Cu board finishes, a single arc was observed in the Nyquist plot, demonstrating that the comb behaved as a simple parallel combination of R_c and C_c [12]. The components corresponding to electrochemical process were not pronounced in the results. Therefore, the R_c can be easily estimated as the diameter of the arc. In contrast, the electrochemical processes were clearly apparent in the Nyquist plot for ENIG and SnPb boards, as shown in Figs. 4.4 and 4.5. The first arc in the Nyquist plot was due to a parallel combination of R_c and C_c, and the following behaviour in the plots represented mixed electrode reactions and diffusion processes, and correspond to the impedance components R_{ct}, C_{dl} and Z_w. A commonly observed second arc was not always seen in the plots due to a high R_c, which

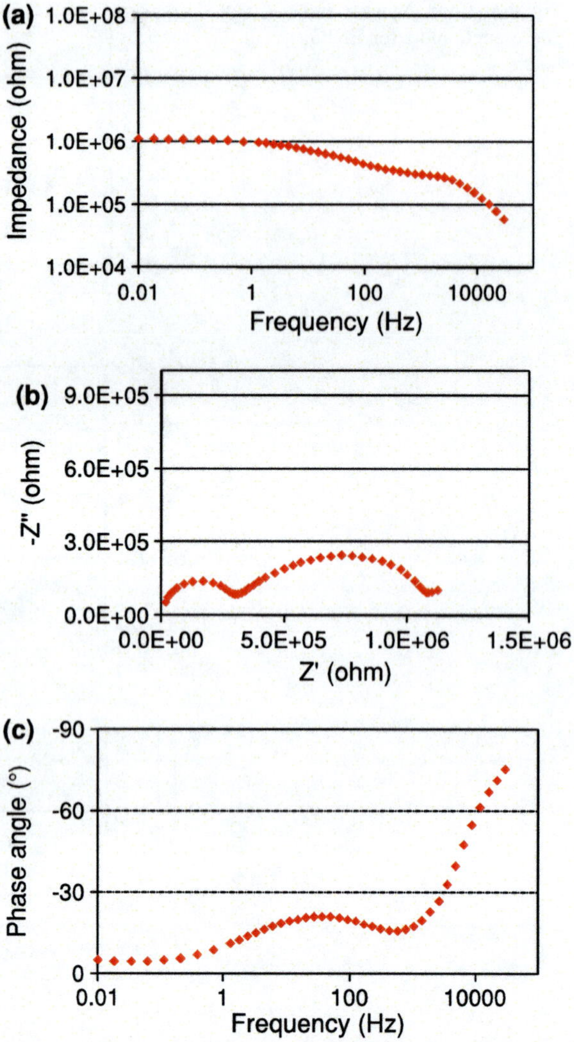

Fig. 4.5 Bode, Nyquist and phase angle plots for SnPb board

also can be distorted due to the complication of the above processes. Therefore, these components are not always easy to interpret from the results. However, $\boldsymbol{R_c}$ always can be evaluated as the diameter of the first arc.

Another important parameter can be interpreted from the results is total impedance at low frequency (at 0.01 Hz). At the low frequency, the impedance measured from Bode plot is the overall resistance for all conduction processes. The impedance from capacitive components become negligible, and this is evidenced by the phase angle. The phase angle tends to 0° for all board finishes at 0.01 Hz.

The ionic resistances $\boldsymbol{R_c}$ evaluated from Nyquist plots and the total impedance at 0.01 Hz ($\boldsymbol{R_{p\ total\ (0.01)}}$) from Bode plots for the five studied metals are listed in

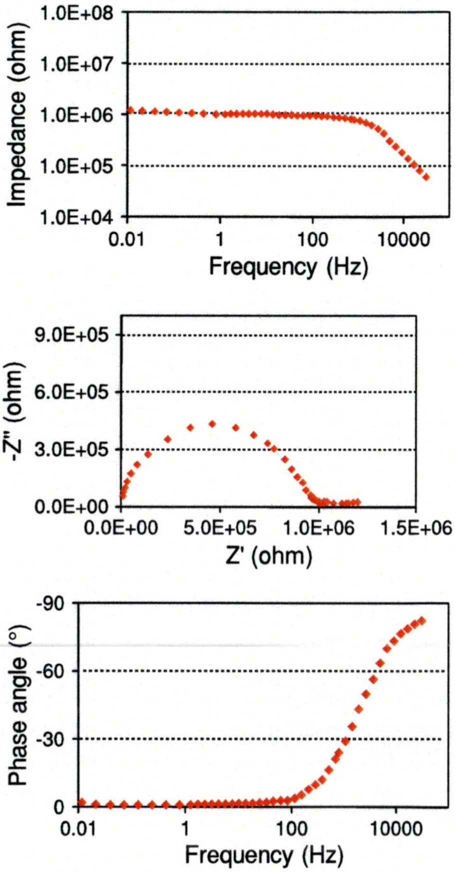

Fig. 4.6 Bode, Nyquist and phase angle plots for SAC board

Table 4.2 and plotted in Fig. 4.9. These results show that the ionic resistance $\boldsymbol{R_c}$ dominates the overall impedance $\boldsymbol{R_{p\ total(0.01)}}$, as seen for Sn, SAC and Cu board finishes. At low ionic resistance, the impedance due to corrosion processes $\boldsymbol{R_p}$ becomes significant, as seen for SnPb and ENIG board finishes. Therefore, the lower ionic resistance of the system, the more pronounced the corrosion processes. The ranking from the highest to the lowest in terms of the ionic resistance is Cu > Sn $\approx$ SAC > SnPb > ENIG.

4.3.2 Surface Insulation Resistance Results

The SIR values from four test patterns contaminated at three concentrations of flux for ENIG-finished boards are presented in Fig. 4.10 and those for SnPb, SAC, Sn and Cu-finished boards in Figs. 4.11, 4.12, 4.13 and 4.14, respectively.

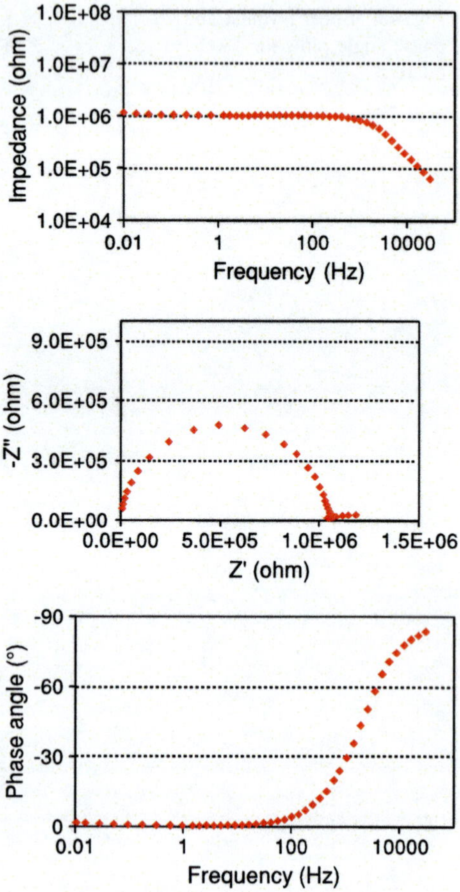

Fig. 4.7 Bode, Nyquist and phase angle plots for Sn board

Some important findings should be noted from these results. For all board finishes at low flux concentrations, the SIR values slowly increased with time and stabilized within the test period. This increasing SIR was typical of the situation in which the ionic contaminants were depleted from the insulating surface due to mobile ions migrating towards the electrodes and then undergoing reduction and oxidation processes at both electrodes, and therefore no longer contributing to the conduction process. In the case of the intermediate flux concentration levels, occasional rapid changes in SIR values were noted, resulting from dendrite formation. Hence, this flux level was defined as the minimum flux concentration to promote dendrite (FCPD) growth. When the flux concentration was increased still further, dendrite formation was more prevalent, as shown by the third plot in each figure.

The FCPD are listed in Table 4.3 and plotted in Fig. 4.15 and the ranking for the resistance to dendrite formation is Cu > Sn ≈ SAC > SnPb > ENIG, similar to that of the ionic resistance $\boldsymbol{R_c}$ described earlier.

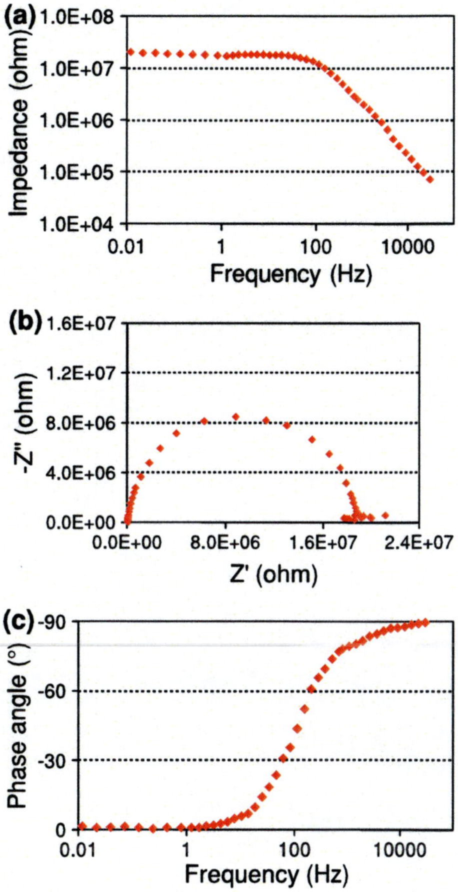

Fig. 4.8 Bode, Nyquist and phase angle plots for Cu board

Table 4.2 $R_{p\ total}$ and R_c for five board finishes

Metal	Cu	Sn	SAC	SnPb	ENIG
$R_{T(0.01)l}$ (MΩ)	21.22	1.28	1.20	1.12	0.27
R_c (MΩ)	20.11	0.98	0.92	0.34	0.07

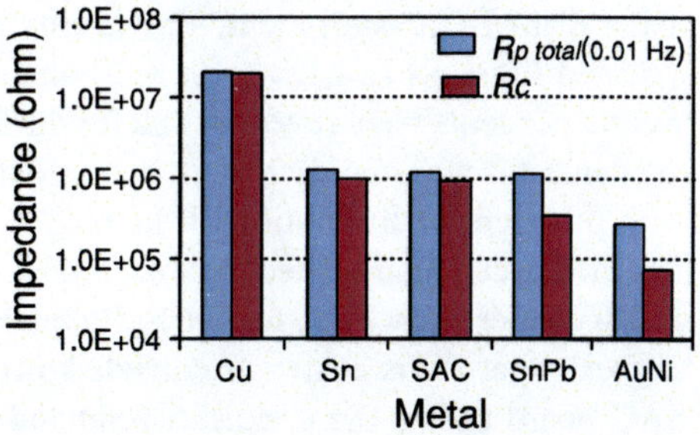

Fig. 4.9 $R_{p\ total}$ and R_c for five board finish boards

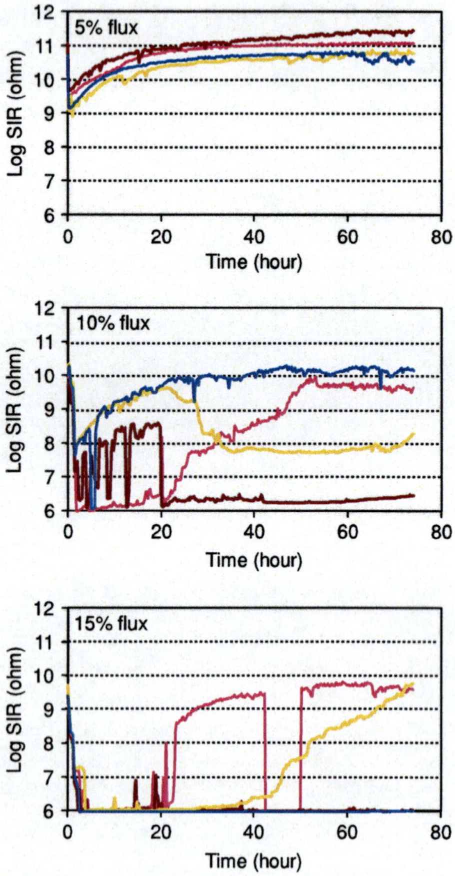

Fig. 4.10 SIR results from ENIG boards with three concentrations of flux

4.3.3 SEM-EDX Analysis of Dendrites on SIR Boards

The dendrites formed on different board finish during the SIR measurements were photographed using an optical microscope, as shown in Fig. 4.16, and analysed using SEM-EDX equipment. The results are listed in Tables 4.4, 4.5, 4.6, 4.7 and 4.8, and Figs. 4.17, 4.18, 4.19, 4.20 and 4.21. They are normalized and given in weight per cent. They revealed that for the ENIG-finished boards, the composition of the dendrites was mainly nickel, consistent (from an electrochemical point of view) with preferential dissolution of the nickel, compared with the noble metal gold.

On the copper-finished boards, the dendrites' composition was, as expected, 100% copper. However, for the SnPb board finish, the high Pb content in dendrites suggests that Pb is easier to corrode to form dendrite than Sn in SnPb alloy. On SAC board finish, the dendrites reflected the alloy composition and were mainly

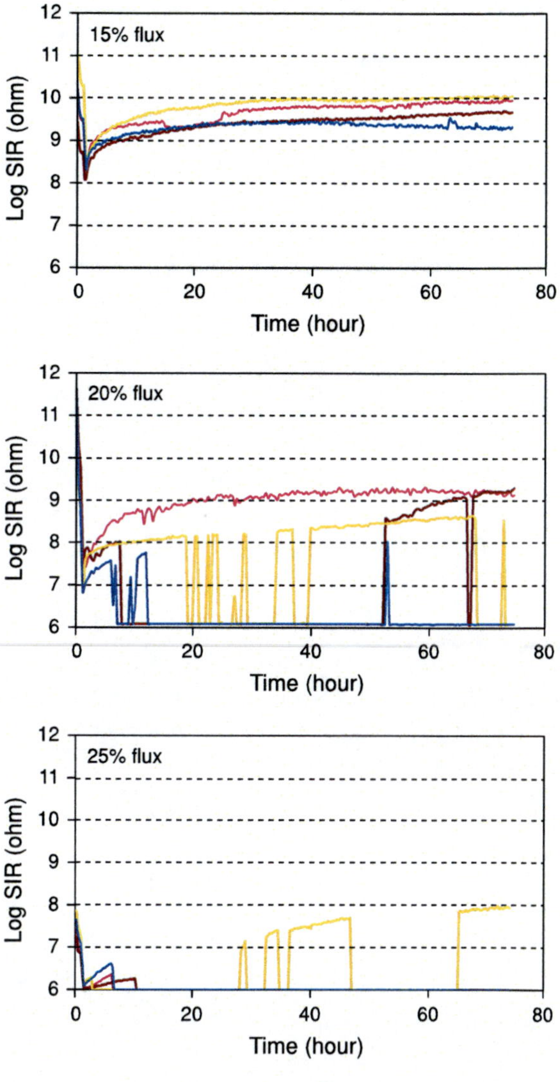

Fig. 4.11 SIR results from SnPb boards with three concentrations of flux

Sn, for some dendrites there was higher Ag and Cu. These results do not suggest a particular sensitivity to Ag, further work is need here to substantiate earlier work at NPL [14] that Ag migration is an issue when Ni is present. A contributory factor suggested by Yu [15] is the formation of inter-metallic compounds in the solder alloy that prevent Ag migration. For Sn finish, dendrites were 100% Sn.

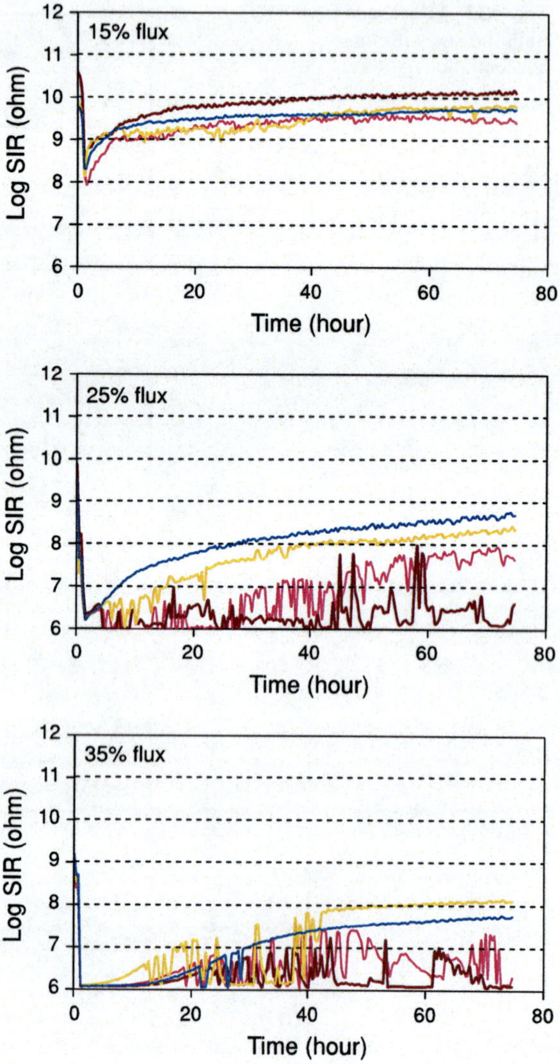

Fig. 4.12 SIR results from SAC boards with three concentrations of flux

4.4 Discussion

4.4.1 Ionic Resistance for Different Board Finishes

An aim of EI measurements is to evaluate the metal corrosion rate at the anode, but for our test system, the impedance attributed to corrosion process $\boldsymbol{R_p}$ is not easy to

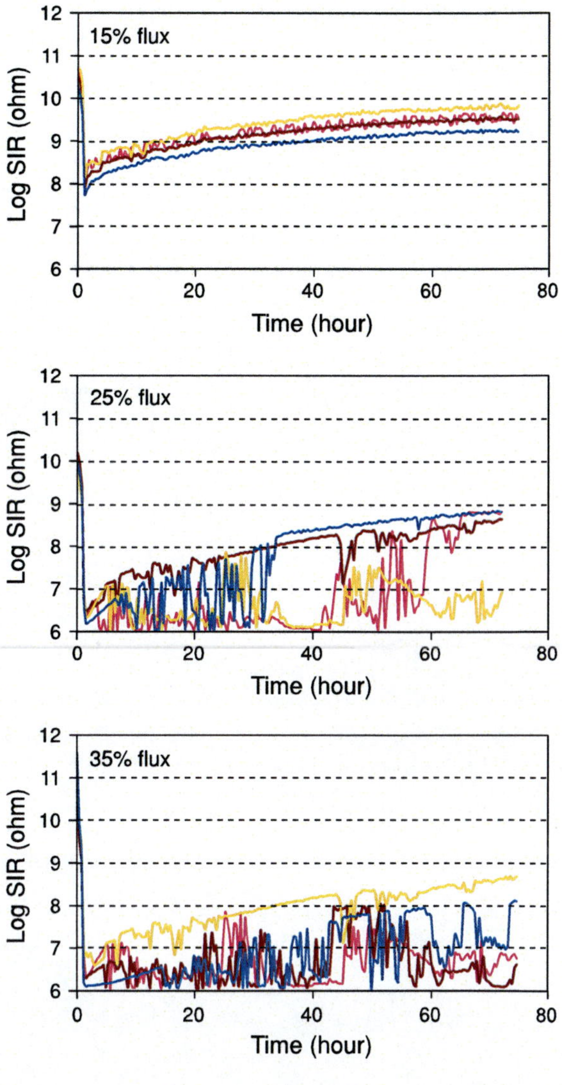

Fig. 4.13 SIR results from Sn boards with three concentrations of flux

interpret, due to the high ionic resistance between the two electrodes. The different finished boards were similarly contaminated, if metal corrosion was not significant, the ionic resistance between two electrodes will be the most significant part of the overall impedance. Therefore, under these circumstances, the $\boldsymbol{R_c}$ should be the same for different board finishes. The difference on ionic resistance $\boldsymbol{R_c}$ on different board finishes must be due to the metal corrosion. Metal corrosion can affect $\boldsymbol{R_c}$ in two ways: Firstly, if the metal has high corrosion rate, more metal ions

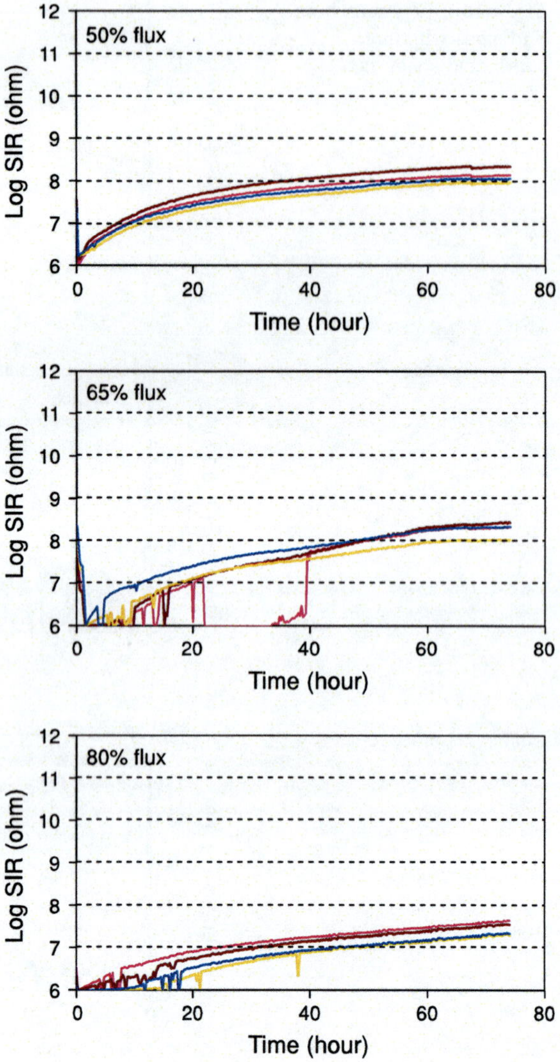

Fig. 4.14 SIR results from Cu boards with three concentrations of flux

Table 4.3 The FCPD on different finish boards

Board finish	FCPD (% in IPA)
ENIG	10
SnPb	20
SAC	25
Sn	25
Cu	65

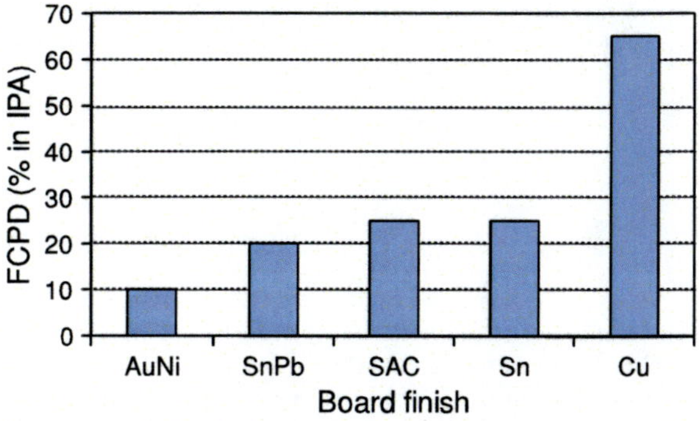

Fig. 4.15 THE FCPD on different finish boards

Fig. 4.16 Optical images for dendrite formed on different board finishes

can be produced from anode into the electrolyte lowing ionic resistance $\boldsymbol{R_c}$ of the system. Secondly, if the corrosion products metal hydroxide has high solubility, corroded metal ions will stay in electrolyte and contribute to conduction process lowing $\boldsymbol{R_c}$. Therefore, although $\boldsymbol{R_c}$ is not a direct measure of metal corrosion rate, it is a function of the metal corrosion rate and metal hydroxide solubility.

4.4.2 Effect of Ionic Resistance on Dendrite Formation

As discussed in the previous section, ionic resistance reflects a combination of the metal corrosion rate and solubility of the metal hydroxide. In this section, the

Table 4.4 Dendrite formed on an ENIG-finished board

Spectrum	Ni	Au
1	100.0	
2	92.8	7.1
3	100.0	
4	100.0	
5	100.0	

Table 4.5 Dendrite formed on a copper-finished board

Spectrum	Cu
1	100.0
2	100.0
3	100.0
4	100.0
5	100.0

Table 4.6 Dendrite formed on a SnPb-finished board

Spectrum	Sn	Pb
1	17.1	82.9
2	56.2	43.8
3	26.2	73.8
4	16.3	83.7
5	32.3	67.7

Table 4.7 Dendrite formed on a SAC-finished board

Spectrum	Cu	Ag	Sn
1	0.5	3.7	95.8
2		7.9	92.1
3	5.6		93.4
4	0.9	2.3	96.8
5		4.5	95.5
6	2.5	3.1	94.4

Table 4.8 Dendrite formed on a Sn-finished board

Spectrum	Sn
1	100
2	100
3	100
4	100
5	100
6	100

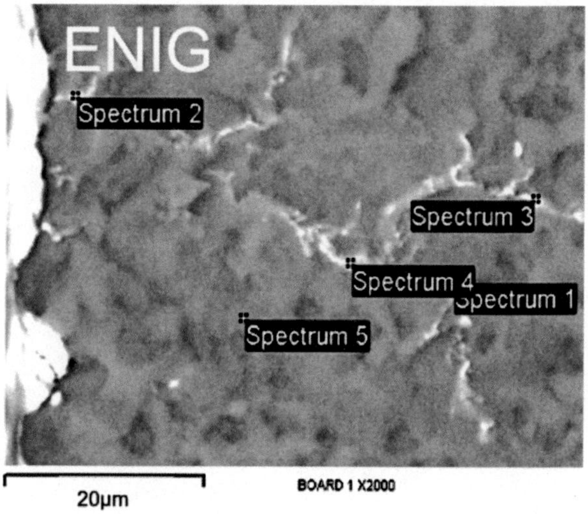

Fig. 4.17 Dendrite formed on an ENIG-finished board

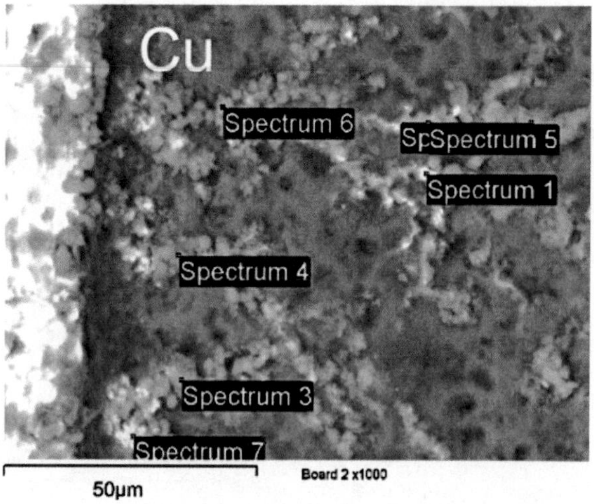

Fig. 4.18 Dendrite formed on a Cu-finished board

effect of the metal corrosion rate and solubility of the metal hydroxide on dendrite formation are discussed with the three steps involved for dendrite formation.

Dendrite formation can be considered as three-step process as mentioned in the introduction, the most important step is metal ion dissolution at the anode, and a higher corrosion rate will produce more metal ions to migrate from the anode to form dendrites. The metal ion produced at the anode must be able to migrate to the cathode without being precipitated as insoluble compounds; hence, the solubility of metal hydroxide will also affect the metal migration.

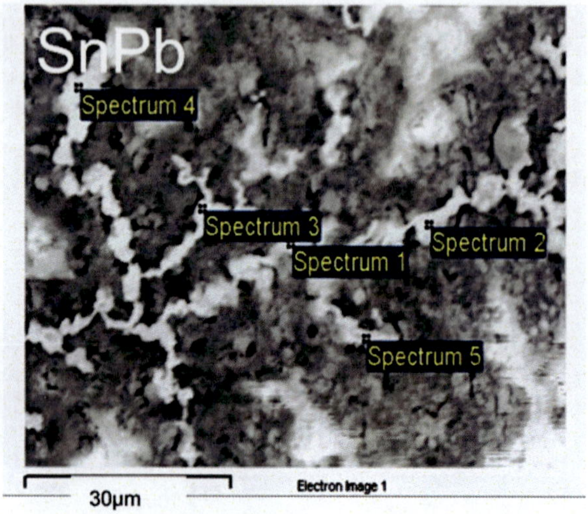

Fig. 4.19 Dendrite formed on a SnPb-finished board

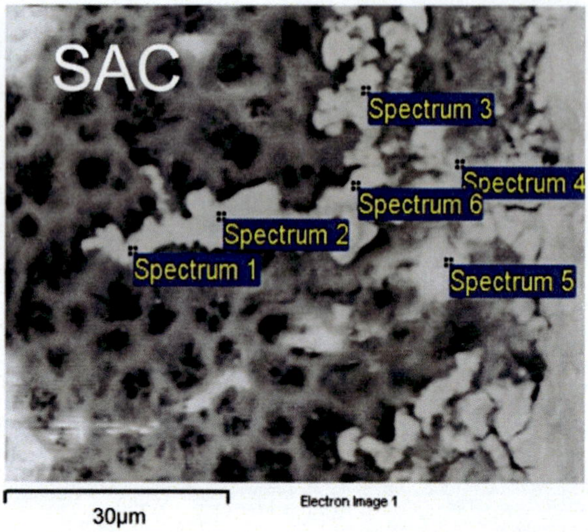

Fig. 4.20 Dendrite formed on a SAC-finished board

With SIR testing when 5 V DC is applied to the test comb, thermodynamically metal dissolution at the anode will dominate the anodic reactions. At the cathode, hydrogen and oxygen reduction will be preferable reactions, as there is sufficient H^+ from flux residues and O_2 in the thin water film. Metal ion deposition may not occur unless the over-potential is sufficiently high since the metal ion concentration may be low due to limited dissolution at the anode. Previous work also found that a minimum cathodic over-potential was required for the appearance and propagation of dendrite [7]. The voltage value, 5 V, between the anode and cathode would

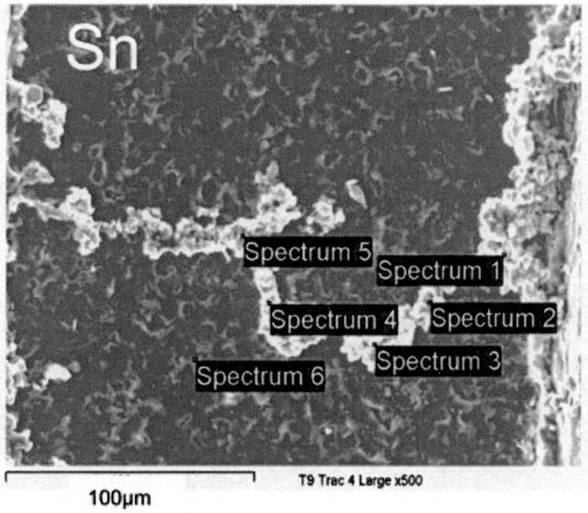

Fig. 4.21 Dendrite formed on a Sn-finished board

normally be considered to be high for an electrochemical process in solution. However, the electrolyte on the comb is a thin water film, and hence, there is a significant voltage drop across the inter-electrode gap $\boldsymbol{R}_{\mathrm{C}}$. Therefore, the activation over-potential at the electrodes is very small and depends on the $\boldsymbol{R}_{\mathrm{C}}$ value. If system has low $\boldsymbol{R}_{\mathrm{C}}$, then more voltage is dropped at the electrodes and a higher over-potential is achieved. With a higher over-potential at the cathode, there is an increased the ability for metal ion deposition.

Furthermore, a high over-potential at the anode accelerates metal dissolution, and in turn, more metal ions will be produced in the electrolyte. Thus, a system with low $\boldsymbol{R}_{\mathrm{C}}$, both the metal ion concentration and the cathodic over-potential will be high and help to promote metal ion deposition at the cathode. Therefore, ionic resistance $\boldsymbol{R}_{\mathrm{C}}$ is a very important indicator on dendrite formation for the system.

4.4.3 EI Measurement to Predict Dendrite Formation

The ranking on propensity to form dendrite for five studied metals from SIR measurements demonstrates an inverse correlation with ionic resistance ($\boldsymbol{R}_{\mathrm{C}}$) interpreted from the EI results. Therefore, $\boldsymbol{R}_{\mathrm{C}}$ can be used to predict the propensity of the studied metals to form dendrite. The EI measurements use a small amplitude AC single, and the technique itself does not actually promote dendrites. Hence, this predictive capability could be developed into a non-destructive fast test method.

The higher propensity to form dendrite on SnPb-finished board compared with SAC- and Sn-finished board is due to its low ionic resistance, which was probably caused by high corrosion rate of Pb in SnPb solder alloy, or, and high solubility of $Pb(OH)_2$. This was supported by the EDX analysis of the dendrites showing high

Pb content. Low propensity of lead-free solders implies that the reliability of the circuit will be improved using lead-free solders over a SnPb solder. This will not always be necessarily true, because high temperature soldering process and different metallurgy require more active fluxes for lead-free solder; hence, there is the potential of leaving more flux residues on the circuit. The dendrite formation strongly depends on the level of flux residues, as the SIR results in 3.3, and discussed in previous report [14]. The propensities to form dendrites in lead-free solders, SAC and Sn, are likely to be similar and dominated by the corrosion of Sn, the main constituent in these alloys.

Comparing two PCB finishes of Cu and ENIG, the low propensity to form dendrite for the Cu finish, compared to the ENIG finish, can be attributed to the low corrosion rate and solubility of $Cu(OH)_2$. The low solubility of $Cu(OH)_2$ was evidenced by some green corrosion products observed around the anode in Fig. 4.16.

4.5 Conclusions

Electrochemical Impedance measurements have been made of three solders and two PCB finish materials. Effect of metal corrosion on reliability of circuit (dendrite formation) has been investigated using SIR measurement. In developing the impedance technique, results were compared with those taken with SIR. The corrosion conditions were manipulated so that dendrite formation occurred, and the dendrite composition was confirmed with EDX. The dominant conduction process for dendrite formation has been identified.

EI results were recorded over a wide range frequency on comb test pattern, as typically used in the SIR technique. This allowed the ionic resistance $\boldsymbol{R}_\mathrm{C}$ between two electrodes to be separated from the overall impedance of the system $\boldsymbol{R}_{T(0.01)}$. $\boldsymbol{R}_\mathrm{C}$ was found to be dependent on the metal corrosion rate and solubility of the metal hydroxide. The ranking from the highest to the lowest in terms of the ionic resistance is Cu $>$ Sn $\approx$ SAC $>$ SnPb $>$ ENIG.

The relative propensity for dendrite formation assessed using SIR measurement demonstrates an inverse correlation with ionic resistance, $\boldsymbol{R}_\mathrm{C}$, interpreted from the EI results. In a system with low $\boldsymbol{R}_\mathrm{C}$, a significant amount of metal ion migration occurs, resulting in dendrite formation and probable short circuit. Hence, from these results, the $\boldsymbol{R}_\mathrm{C}$ value could be interpreted in the relative likelihood for dendrites to form and so the reliability of electronic assemblies. Since the EI technique only applies a 50 mV AC signal, and a scan takes less than one hour, the $\boldsymbol{R}_\mathrm{C}$ measurement can be considered to be a fast and method non-destructive.

Dendrite formation is a complex electrochemical processes, dependent on both the metal corrosion rate and the solubility of metal hydroxide. The high propensity to form dendrite of a SnPb solder compared with a SAC solder and Sn is due to high corrosion rate of Pb in SnPb solder alloy, or, and high solubility of $Pb(OH)_2$. Therefore, under the same contamination conditions, an electronic circuit will be more reliable if soldered with lead-free solders. Furthermore a Cu-finished PCB will be less likely to form dendrites compared to the ENIG finish due to its low corrosion rate and low solubility of $Cu(OH)_2$.

Acknowledgments The work was carried out as part of a project, Measure Electrochemical Corrosion of Lead-Free Process Residues in Electronic Assemblies, in the Processing Programme of the UK Department of Trade and Industry.

References

1. Chan HA (1996) Surface insulation resistance methodology for today's manufacturing technology. IEEE Trans Compon Packaging Manuf Technol Part C 19(4):300–307
2. Hunt C, Zou CL (1999) The impact of temperature and humidity condition on surface insulation resistance values for various fluxes. Soldering Surf Mt Technol 11(1):21–24
3. Zou CL, Hunt C (1999) The effect of test voltage, test pattern and board finish on surface insulation resistance (SIR) measurements for various fluxes. NPL Report CMMT(A)222
4. Harsanyi G (1995) Electrochemical process resulting in migrated short failures in microcircuit. IEEE Trans Compon Packaging Manuf Technol Part A 18(3):602–610
5. Zou CL, Hunt C (2006) Electrochemical behavior of solder alloy. NPL Report DEPC-MPR (PAPER) 040 September
6. Steppan JJ, Roth JA, Hall LC, Jeanotte DA, Carbone SP (1987) A review of corrosion failure mechanisms during accelerated test. J Electrochem Soc 134:175–190
7. Zamanzadeh M (1990) Electrochemical examination of dendrite growth on electronic devices in HCl electrolytes. Corrosion 46(8):665–671
8. Yang S (2007) Failure model for silver electrochemical migration. IEEE Trans Divice Mater Reliab 7(1):188–196
9. Brusic V, Dimilia DD, Macinnes R (1991) Corrosion of led, tin, and their alloys. Corrosion 47(7):509–518
10. Mohanty US, Lin KL (2007) Electrochemical corrosion study on Sn-XAg-0.5Cu alloys in 3.5% NaCl solution. J Mater Res 22(9):2573–2581
11. Beccaria AM, Mor ED, Bruno G, Poggi G (1982) Corrosion of lead in sea water. Br Corros J 17(2):87–91
12. Zou CL, Hunt C (2009) Characterization of the conduction mechanisms in adsorbed electrolyte layers on printed circuit boards using AC impedance. J Electrochem Soc 156(1):C5–C15
13. Takahashi KM (1991) Conduction paths and mechanisms in FR-4 epoxy/glass composite printed wiring boards. J Electrochem Soc 138(6):1587–1593
14. Zou CL, Hunt C (2007) Susceptibility of lead-free system to electrochemical migration. NPL Report MAT1
15. Yu DQ (2006) Electrochemical migration of SnPb and Lead free solder alloys under distilled water. J Mater Sci Mater Electron 17:219–227. doi:10.1007/s10854-006-6764-0

Chapter 5
Void Formation by Kirkendall Effect in Solder Joints

M. J. M. Hermans and M. H. Biglari

5.1 Diffusion

With diffusion, the atomic movement within a solution is meant. Diffusion can be treated either as an atomistic or as a continuum approach. In the former, the nature of the diffusing species is considered on an atomic level, whereas the latter the system is treated as a continuous medium on a more micro- and macroscopic level.

Diffusion may occur by migrating of interstitial or substitutional atoms, depending on the sites the atoms occupy in the lattice.

Usually the concentration of interstitial atoms is small and only a fraction of the available sites is occupied. This means that there are always neighboring site where the interstitial atom can jump to. For substitutional atoms, vacancies must be present in the lattice.

Assume an ideal solid solution with A, the solute component, and B, the solvent component. Due to vacancy motion, atoms can move through the lattice and the probability for jumping into the vacancy is the same for all the atoms surrounding the vacancy. This implies that the jump rate does not depend on the concentration.

When a concentration gradient exists within the solution, there will be a net flux of atoms down the concentration gradient. For a one-dimensional system, this flux can be described by Fick's fist law of diffusion:

$$J_A = -D_A \frac{\mathrm{d}C_A}{\mathrm{d}x} \tag{5.1}$$

M. J. M. Hermans (✉)
Delft University of Technology, Materials Science and Engineering Joining and Mechanical Behaviour, Mekelweg 2, Delft, The Netherlands
e-mail: m.j.m.hermans@tudelft.nl

M. H. Biglari
Mat-Tech BV, Ekkersrijt 4605, Eindhoven, The Netherlands
e-mail: m.biglari@mat-tech.com

G. Grossmann and C. Zardini (eds.), *The ELFNET Book on Failure Mechanisms, Testing Methods, and Quality Issues of Lead-Free Solder Interconnects*,
DOI: 10.1007/978-0-85729-236-0_5, © Springer-Verlag London Limited 2011

with D_A the diffusion coefficient or diffusivity and $\frac{dC_A}{dx}$ the concentration gradient.

The diffusion coefficient depends on the activation energy for the migration and can be expressed as:

$$D_A = \left[\frac{1}{6}\alpha^2 zv \exp\frac{\Delta S_m}{RT}\right] \exp\frac{-\Delta H_m}{RT} = D_{A0} \exp\frac{-Q}{RT} \tag{5.2}$$

with α the jump distance (Å), z the number of nearest neighbors, v the lattice vibration frequency (s^{-1}), ΔS_m the activation entropy (J mol^{-1} K^{-1}) and ΔH_m the activation enthalpy (J mol^{-1}).

Although derived for interstitial diffusion, the equations are also applicable for any diffusing specie in a cubic lattice.

For non-steady conditions, where the concentration varies with distance and time, Fick's second law of diffusion should be used.

$$\frac{\partial C_A}{\partial t} = \frac{\partial}{\partial x}\left(D_A \frac{\partial C_A}{\partial x}\right) \tag{5.3}$$

This equation relates the rate of change of composition with time to the concentration profile. When the variation of D_A with concentration can be ignored, the equation can be simplified.

$$\frac{\partial C_A}{\partial t} = D_A \frac{\partial^2 C_A}{\partial x^2} \tag{5.4}$$

It should be kept in mind that D is in solid solutions often dependent on the composition as can be seen in Fig. 5.1 for three gold alloys.

In binary substitutional alloys, the situation is more complex. The rate at which the solvent and solute atoms can move to a vacancy is not equal and each atomic species must be given its own intrinsic diffusion coefficient.

$$J_A = -D_A \frac{A\partial C_A}{\partial x} \text{ and } J_B = -D_B \frac{A\partial C_B}{\partial x} \tag{5.5}$$

J_A and J_B are the fluxes of A and B atoms across a given lattice plane (cross-sectional area A). If these fluxes are in opposite directions and when they are not equal, a net flux exists which should be matched by a flux of vacancies in the opposite direction of the net flux of atoms, see Fig. 5.2. When the vacancy concentration should be maintained near the equilibrium, vacancy concentration on one side of the interface vacancies should be created, while on the other side they should be annihilated. Jogged edge locations can provide a source and sink of vacancies. This means that extra atomic planes are introduced on one side while planes are annihilated on the other side of the interface, see Fig. 5.3. The velocity of any given plane can be related to the flux of vacancies crossing it:

$$v = (D_A - D_B)\frac{\partial X_A}{\partial x} \tag{5.6}$$

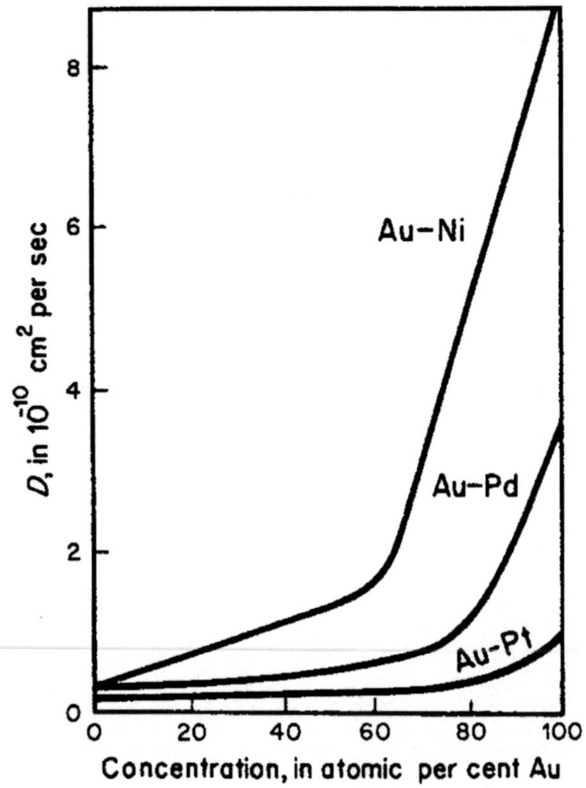

Fig. 5.1 Variation in diffusion coefficient (D) with composition for three Au alloys [2]

It can be derived that Fick's second law for diffusion in substitutional alloys is:

$$\frac{\partial C_A}{\partial t} = \frac{\partial}{\partial x}\left(\tilde{D}\frac{\partial C_A}{\partial x}\right) \tag{5.7}$$

in which $\tilde{D}$ is the interdiffusion coefficient, depending on D_A and D_B.

$$\tilde{D} = X_B D_A + X_A D_B \tag{5.8}$$

with X_A and X_B the mole fractions of A and B, respectively.

It should be mentioned that the atomic mobility along defects (grain boundaries, surfaces, dislocations) can be considerable different from bulk diffusion. A shift in diffusion path mechanism can for instance occur with varying temperature. Atomic mobility is also influenced by electromigration.

Finally, when additional diffusing species are present in the system, the diffusivities are altered as the probability of a jump into a vacancy will change.

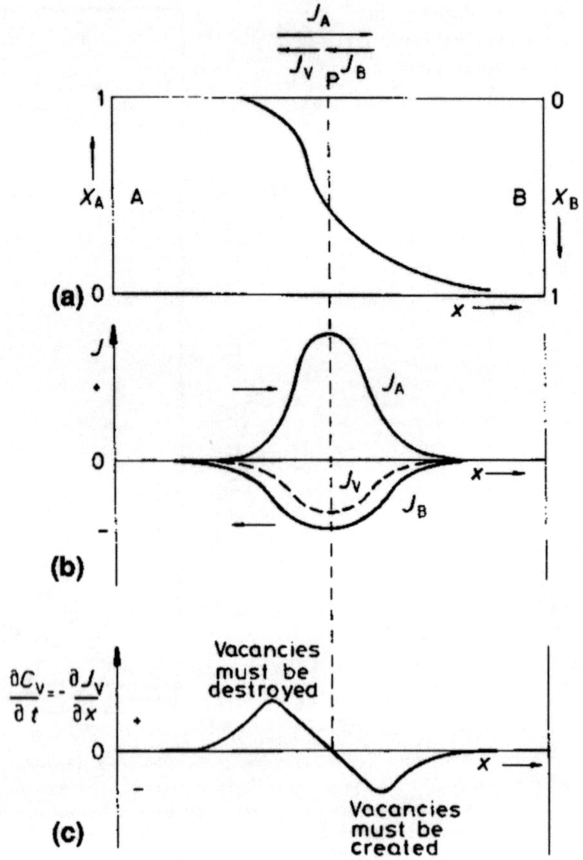

Fig. 5.2 Interdiffusion and vacancy flow [3]

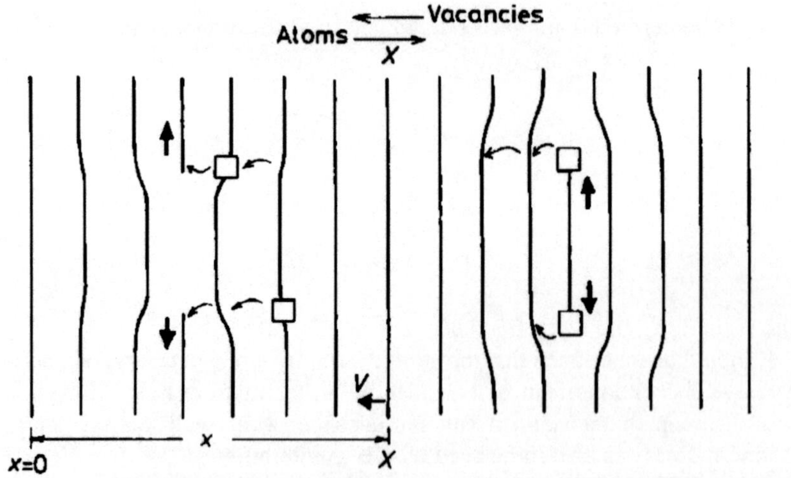

Fig. 5.3 The flux of vacancies causes the atomic planes to move through the specimen [3]

5.2 Kirkendall Effect

The experiment by Smigelskas and Kirkendall [1] studied the diffusion couple copper–zinc. At the original interface of the two pure metals, fine marker wires were incorporated. After annealing, the concentration profiles were determined across the interface. The interesting result of their study was that the marker wires had moved during the diffusion process. This is shown schematically in Fig. 5.4, where the upper figure represents the situation before the heat treatment, while the lower figure shows the position after diffusion had occurred. The distance of the marker movement was found to vary with the square root of time the specimen was kept at the diffusion temperature. The moving plane in which the markers are situated is called the Kirkendall plane.

This marker movement could only be explained by a different speed of diffusion for the different types of atoms. In this way, the effect confirms the vacancy mechanism of diffusion, with different rates of jumping into a vacancy for both types of atoms. Cooperative movement (direct interchange and Zener-ring mechanism) of atoms can be discarded in the case the Kirkendall effect is observed.

Darken's equations (Eqs. 5.6 and 5.8) makes it possible to determine the intrinsic diffusivities experimentally. The following assumptions are required for this derivation.

- volume expansion/contraction due to unequal mass flow only takes place in the direction perpendicular to the interface
- the total number of atoms per unit volume is a constant ($n_A + n_B$ = constant)
- porosity does not occur in the specimen during the diffusion process.

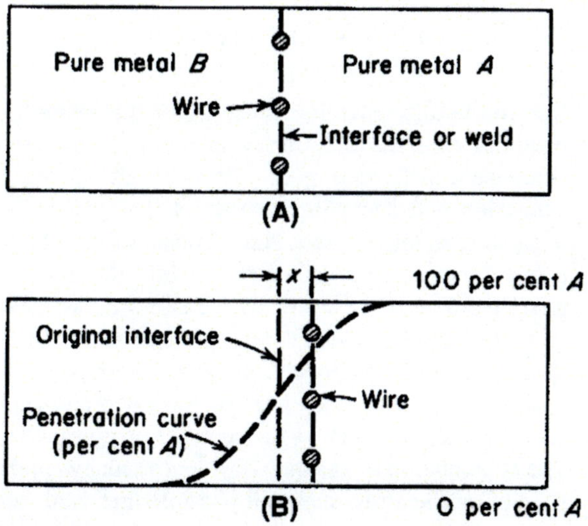

Fig. 5.4 Marker movement in a Kirkendall diffusion couple [2]

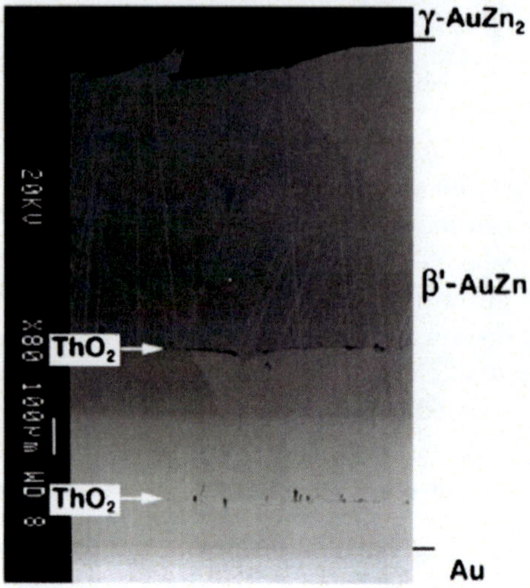

Fig. 5.5 Back-scattered electron image (BEI) of an Au_$Au_{36}Zn_{64}$ ("g-$AuZn_2$") diffusion couple annealed at 500°C for 17.25 h under flowing argon. After interdiffusion, the ThO_2 markers introduced between the couple halves are clearly visible as two distinct straight rows of inclusions [4]

Two standard methods for measuring the diffusion coefficient are:

- Diffusivity is assumed constant (Grube method)
- Diffusivity is a function of the composition (Matano method).

Van Dal et al. [4] and Paul et al. [5, 6] studied the Kirkendall effect for various diffusion couples and showed that multiple Kirkendall planes can develop. This can be seen in Fig. 5.5. The Kirkendall planes can be either stable, unstable or virtual. By using a Kirkendall velocity plot, it is possible to explain and predict the Kirkendall plane formation.

5.3 Kirkendall Void Formation

Due to the difference in diffusivity of the atoms in a binary solution, one of the components in the diffusion couple will experience a loss of mass while the other component will gain mass. As a result of the mass transfer, shrinkage and expansion will occur in the parts of the system. In this way, a state of stress is introduced in the diffusion zone. The part that suffers a loss of mass is placed under a two-dimensional tensile stress, while the side that gains mass will be placed under a compressive stress. These stress fields may bring about plastic flow.

Furthermore, if one of the components in a binary diffusion couple diffuses faster than the second component, a vacancy flux passes in the direction of the slowest component. The vacancies are both created and annihilated in the metal couple at sources and sinks such as dislocations or internal interfaces, as mentioned before. The combination of vacancy flow and vacancy condensation in combination with a state of tensile stress makes it possible that voids are formed, see Fig. 5.6.

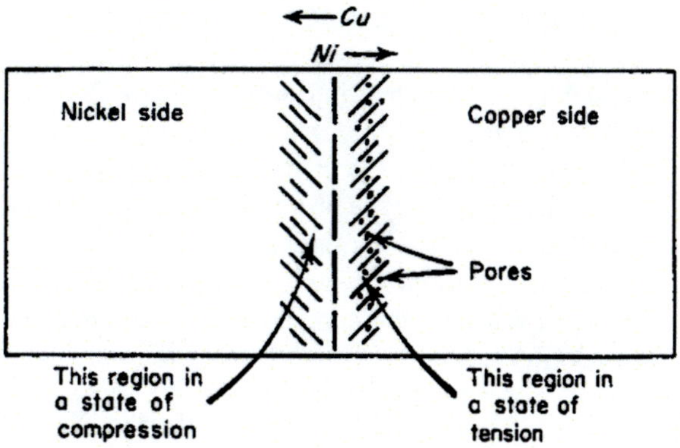

Fig. 5.6 Regions of compression and tension in a Ni-Cu diffusion couple. The formation of pores in the region in a state of tension [2]

5.4 Kirkendall Voids in Solder Joints

The evolution of the microstructure of a soldered joint is governed by the phenomena that take place during the soldering stage, where dissolution of components in the liquid metal occurs, and the subsequent solid state diffusion during its life time.

Over the last decade, effort has been undertaken to model the microstructural evolution and the occurrence of Kirkendall planes [7]. The phases can be predicted from thermodynamics and kinetics. Also the morphology of the phases plays an important role as it may influence diffusion.

It should be noted that in the case of one of the diffusion elements in the diffusion couple is deposited as a thin film, the layer may finally be completely consumed during the process, and underlying species may start to participate in the diffusion process.

A comprehensive overview of interfacial reactions between lead-free solders and common base materials is given by Laurila et al. [8]. In this review for a number of systems, values for the interdiffusion coefficient are given.

5.4.1 Diffusion Couple Sn/Cu

The phase diagram of the binary Sn/Cu system shows a series of peritectic reactions and several intermetallic compounds. In the lower temperature range (<415°C), the interfacial reactions of Cu with molten Sn-based solder result in the formation of Cu_3Sn (ε) and Cu_6Sn_5 (η) layers. This later IMC has a stable form η' at room temperature, but as available time for the transformation is short, the high temperature η remains as a meta-stable phase. The stable η' may form when the system operates at elevated temperatures. Apart from this ordening, the thickness of the layers grows during operation.

5.4.1.1 Reactions During Soldering: Cu Reactions with Liquid Sn

When liquid Sn comes into contact with Cu, Cu will dissolve until the solder becomes supersaturated. Locally, high concentrations of Cu can be realized at the vicinity of the liquid–Cu interface and Cu_6Sn_5 crystallites can form very fast in a more or less scallop-type of uniphase layer. In addition, a $Cu_6Sn_5 + Sn$ two-phase layer may form. The formation of Cu_3Sn (ε) requires long contact times and the thickness of the layer, when observed is much smaller. The morphology of the phases depends on the concentration gradients, distribution of alloying elements and cooling rates.

5.4.1.2 Reactions in the Solid State: Cu Reactions with Solid Sn

In the solid state, several temperature regimes can be addressed. Up to 60°C, only the Cu_6Sn_5 (η') phase will grow with an observable rate. The reaction is controlled by the release rate of Cu from the lattice. At room temperature, the main diffusing species is Cu. Above 60°C also Cu_3Sn (ε) will start to grow and the fraction increases with annealing time. The growth of the Cu_6Sn_5 (η) phase is controlled by the diffusion of Sn in the temperature range from 60 to 200°C. At the increased temperatures, the volume diffusion of Sn dominates over the grain boundary and interstitial diffusion of Cu. In this temperature regime, the ε phase continues to grow. The initially formed morphology of the phases during soldering will have an effect on the resulting morphology during aging.

Paul studied the intermetallic growth and the Kirkendall effect in the Cu/Sn diffusion couple [6]. Cu_6Sn_5 and Cu_3Sn layers were observed after annealing 215°C, 225 h. On the basis of the Kirkendall velocity diagram, a stable Kirkendall plane was predicted in the Cu_6Sn_5 layer. This was experimentally verified.

5.4.1.3 PbSn Solder/Electrodeposited Cu

Zeng et al. [9] studied the void formation during the reactions at the interface between eutectic PbSn solder and electrodeposited Cu. The study was initiated to the search for replacements for electroless Ni(P)/immersion gold (ENIG), which has a potential reliability issue due to 'black pad' formation. Five candidates were selected as alternative to ENIG plating: bare Cu, organic solderability preservative (OSP) on Cu, direct immersion gold (DIG) on Cu, immersion Sn on Cu and solder on Cu. They all have in common that solder will be in direct contact with Cu. The OSP will evaporate, while DIG will quickly dissolve in the solder. Sn-Cu intermetallic compounds will form during the soaking time and during aging. In this study, the microstructural development at the interface was investigated during reflow soldering (220°C) and solid state aging (up to 80 days at 150, 125 and 100°C). A large number of Kirkendall voids were observed at the interface between

Cu_3Sn and Cu. Mechanical testing (ball shear and pull) showed brittle fracture at the interface.

The microstructural development starts during the reflow process. After the flux has reduced the oxide layer, the Cu starts to dissolve in the molten solder. The layer adjacent to the interface becomes saturated with Cu. From the Sn–Pb–Cu phase diagram, the first IMC formed is Cu_6Sn_5, which forms scallop-like at the interface. The formation of the IMC takes Cu out of the saturated solder and further dissolution will take place. The phase diagram indicates that an interface between Cu and Cu_6Sn_5 is not stable and Cu_3Sn may form in between; initially, this is a very thin layer. As the Cu_6Sn_5 scallops join to become a continuous layer, the fastest diffusion routes are the channels between the scallops. If the soldering is followed by aging, the Cu_6Sn_5 scallops will transform to a layer. It is observed that the Cu_3Sn will grow faster and thicker. Kirkendall voids are introduced at the interface Cu–Cu_3Sn, see Fig. 5.7. This is observed after 3 days aging at 150°C. When aging continues, the microvoids coalesced into larger voids and finally into disk-like gaps. The number of voids did not increase significantly. The disk-like voids block the diffusion path of Cu. Without the supply of Cu, Sn becomes the fastest diffusing species, slowing down the grows of Cu_3Sn and eventually convert

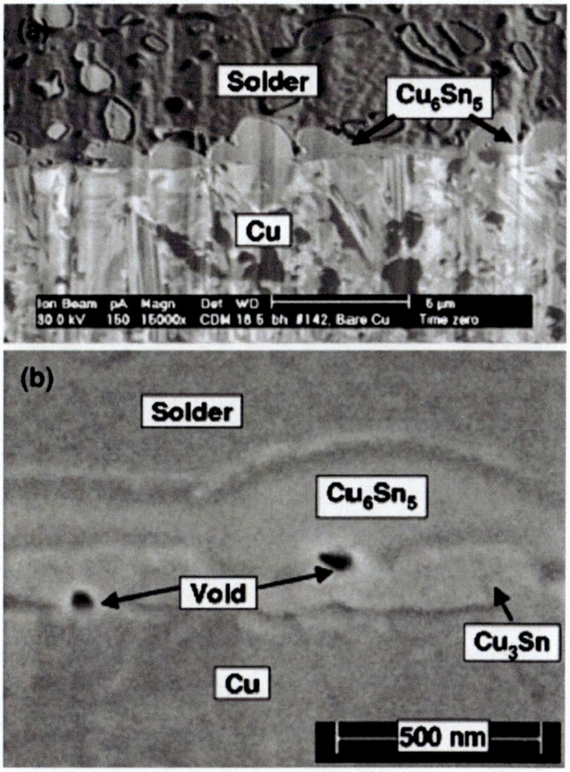

Fig. 5.7 Micrographs of cross sections through SnPb solder Cu interface after reflow of the ball attachment process (time-0), **a** ion beam image, **b** electron beam image [9]

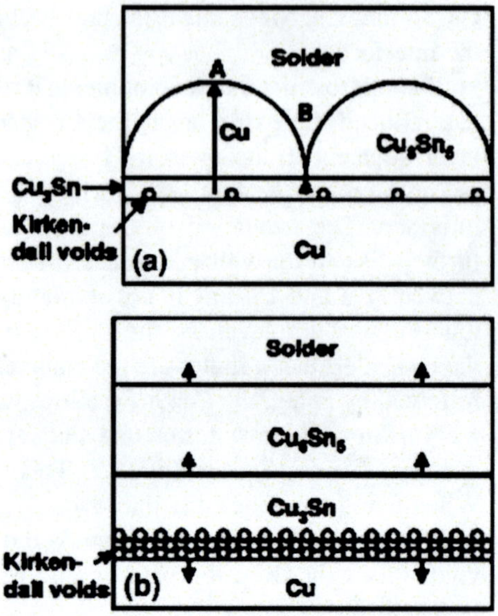

Fig. 5.8 Schematic illustration of microstructural evolution at the interface between solder and Cu: **a** after reflow; **b** after aging. The *vertical arrows* in (**a**) indicate the diffusion of Cu atoms, while those in (**b**) indicate the moving directions of the boundaries during aging [9]

it back to Cu_6Sn_5. At lower temperatures, the time required to observe the formation of voids increases.

The mechanism of the formation of the intermetallic compounds and the Kirkendall voids are schematically depicted in Fig. 5.8, in which a shows the situation after reflow and b after aging.

5.4.1.4 SnPb Solder/Sputtered Trilayer Cu/Ni(V)/Al Thin Film Metallization

Liu et al. [10] studied the reactions between eutectic SnPb solder and a sputtered trilayer Cu/Ni(V)/Al thin film metallization for UBM application, see Fig. 5.9. The initial reaction products were Cu_6Sn_5 and Cu_3Sn. The Cu layer was consumed by the Cu-Sn reaction after 1-min annealing. The scallop-type of Cu_6Sn_5 grains, in the as-received condition, grows during annealing in the direction normal to the UBM and gradually transform into a columnar morphology. At the interface Cu–Cu_3Sn, microvoids formation takes place after one reflow. The Cu_3Sn grains are grouped in clusters and Kirkendall voids are observed in the center of each cluster, see Fig. 5.10. The Cu_3Sn transforms in Cu_6Sn_5 after annealing for more than 1 min at 220°C. The Kirkendall voids that were observed to accompany the formation of Cu_3Sn disappeared when this layer transforms to Cu_6Sn_5. Cu is identified by marker movement as the dominant diffusing species and the out-diffusion of Cu is balanced by the in-diffusion of vacancies. The disappearance of the voids is explained by the diffusion of Sn through the Cu_6Sn_5 layer and the reaction with Cu_3Sn to form Cu_6Sn_5. The volume exchange between the Sn and the vacancies eventually leads to the disappearance of the voids.

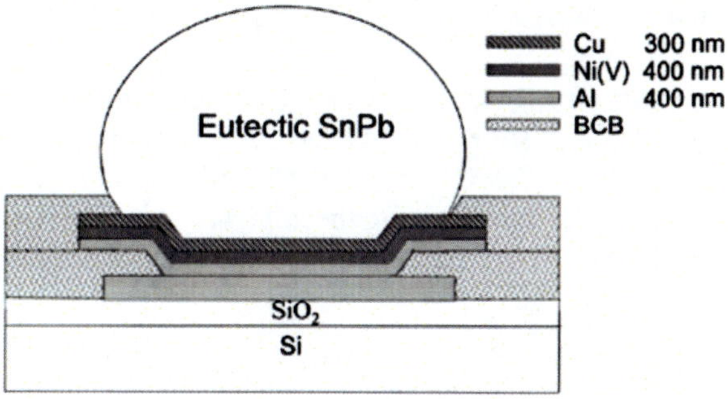

Fig. 5.9 Cross-sectional schematic representation of a SnPb solder ball on the trilayer UBM [10]

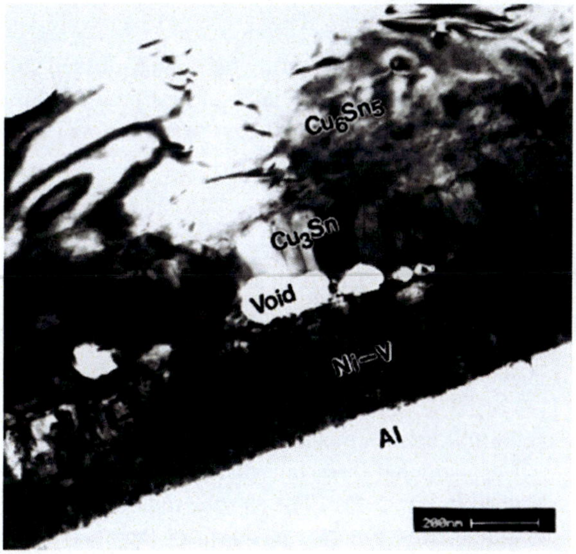

Fig. 5.10 TEM image of the Cu_3Sn grain and the Kirkendall voids [10]

The Ni(V) layer remains almost unchanged and there is no spalling of Cu_6Sn_5 and Ni(V). As the reaction between Ni(V) and solder is limited, it indicates that Cu_6Sn_5 acts as a diffusion barrier, preventing Sn to react with Ni(V).

5.4.1.5 Eutectic SnAgCu Solder/Sputtered Trilayer Cu/Ni(V)/Al Thin Film Metallization

A study was also conducted to the wetting reaction between eutectic SnAgCu and the Al/Ni(V)/Cu thin film UBM [11]. This solder shows a somewhat different behavior compared to SnPb solder as the Cu_6Sn_5 IMC does not form an optimal diffusion barrier. In fact due to the higher solubility of Cu in SnAgCu, the IMC layer dissolves. Super saturation of the solder with Cu may overcome this problem. No reference is made to the formation of Kirkendall voids.

5.4.1.6 $SnAg_{3.5}Cu_{0.7}$/CuOSP Board Metallization

Bennemann et al. [12] investigated microstructural development, IMC and defect formation in solder to package and solder to board metallization interfaces. After one reflow, the Cu_6Sn_5 and Cu_3Sn IMCs are formed. After high temperature storage, the thickness of these layers increase. Under the conditions mentioned no or only small pores are detected that were not considered to form reliability risks.

5.4.2 Diffusion Couple Sn/Ni(P)

In this section, the interaction between Sn-based solders and a Ni(P) substrate will be discussed. The interfacial reactions between Sn and Ni are described in detail by Laurila [8]. It should be noted that additional alloying components influence the evolution of the microstructure.

5.4.2.1 Sn-3.5Ag/Electroless Ni(P)

The microstructural evolution and mechanical properties were studied for soldered tensile testing specimens, as presented in Fig. 5.11a, and for UBM configurations [13, 14]. The reflow temperature was 251°C for 3 min, while the specimens were aged at different temperatures and times.

It appears that three interfacial layers are formed, Ni_3Sn_4, NiSnP and Ni_3P, as depicted in Fig. 5.11b. The as soldered specimens fracture during tensile testing in the bulk solder. Fracture position for the aged specimen gradually changes to the interface of the solder/Ni_3Sn_4, and finally to the Ni–P/Ni substrate interface, depending on temperature and time. During aging, the intermetallic compound layer thickness increases. At the later interface also, Kirkendall voids are observed inside the Ni_3P layer, see Figs. 5.12 and 5.13. Silver forms Ag_3Sn IMC particles randomly distributed inside the solder matrix. It is not likely that it has a major effect on Sn-Ni interfacial reactions.

The mechanism of the reactions of the solder with Ni–P UBM is sketched in Fig. 5.14. The formation of Ni_3Sn_4 caused a depletion of Ni in the Ni–P layer. This results in the formation of Ni_3P, which is crystalline. Growth of Ni_3Sn_4 is supposed to come from the net Ni out-flux through the Ni_3P layer. Diffusion through this layer is relatively easy due to the columnar grains. Sn does not diffuse up to the Ni_3P layer but remains in a ternary NiSnP layer.

Jeon et al. [15] found Kirkendall voids in the Ni_3Sn_4 layer close to the Ni_3P layer, both not inside this layer.

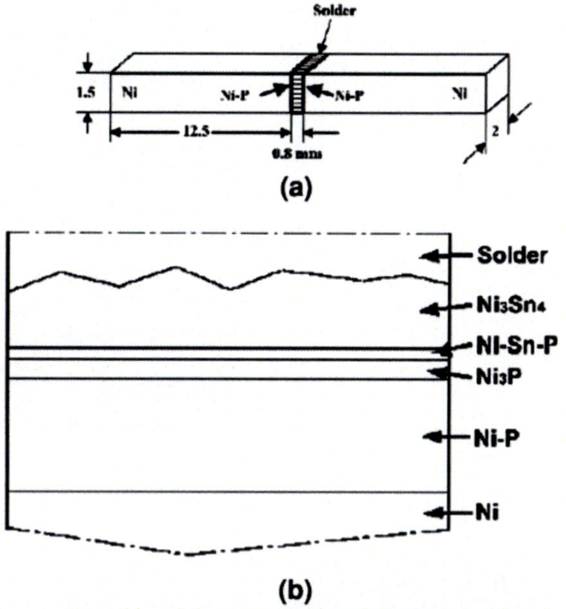

Fig. 5.11 **a** Dimensions of the tensile testing specimen used in this study. The dimensions are in millimeters, **b** Schematic diagram of the interfacial layer structure in a thermally aged Sn–3.5Ag/Ni–P solder joint [14]

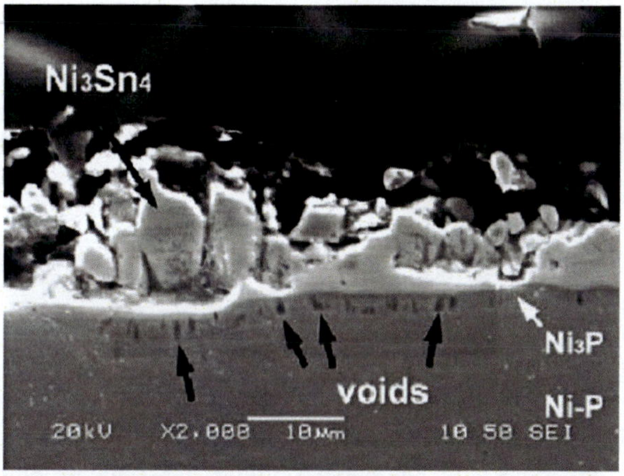

Fig. 5.12 Cross-sectional view of the magnified failure path, the failure between the solder and Ni_3Sn_4. Kirkendall voids are visible in the Ni_3P layer [14]

5.4.2.2 SnAgCu Eutectic/Ni(P)

Upon reflow at 240°C, apart from the Ni_3P layer, a $(Cu,Ni)_6Sn_5$ layer is formed rather than a Ni_3Sn_4 IMC [11]. During aging, the Ni atoms were diffusing back toward the NiSnP layer and Sn atoms were diffusing into the Ni_3P layer. Kirkendall voids were found in the NiSnP layer. Because Ni is coming from $(Cu,Ni)_6Sn_5$ to the NiSnP and no P is found in the $(Cu,Ni)_6Sn_5$, the voids should have been generated by the outward diffusion of Sn.

Research by Li et al. [16] states that $(Cu,Ni)_6Sn_4$ dominates at the interface. No Kirkendall voids are found, which can be due to the low aging temperature of 80°C.

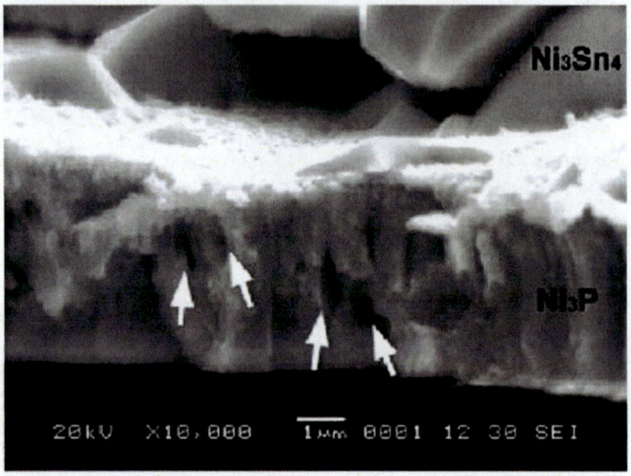

Fig. 5.13 Crack propagates through all interface layers, including the Ni_3P layer, which contains Kirkendall voids (shown by the *arrows*). The joint was aged at 190°C for 400 h [14]

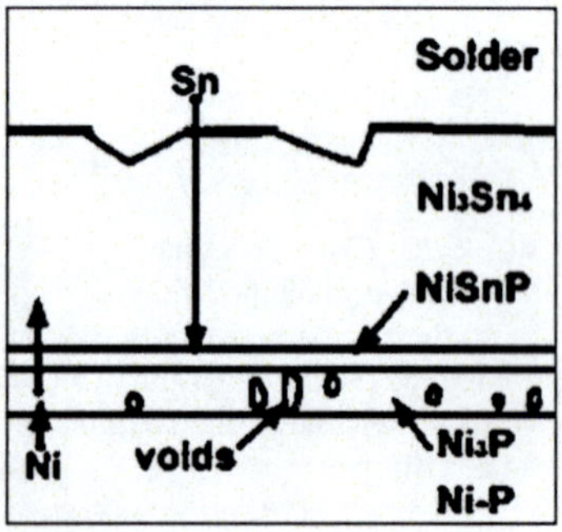

Fig. 5.14 Schematic representation of the mechanisms in Solder-Ni–P [14]

5.4.2.3 $SnAg_{3.5}Cu_{0.7}$/Electroless Ni(P) Metallization

Related work has been carried out by Benneman et al. on the microstructural development of second level interconnects using lead-free $SnAg_{3.5}Cu_{0.7}$ on an XFLGA package [12]. The interfaces solder/packaging metallization and the interface solder board metallization has been studied. The IMC morphology depends on the Ni deposition (electroless, electroplated) and the solidification. The results related to the NiAu board finish are described. Ni_3Sn_2 and Ni_3Sn_4 IMCs are observed after reflow. After high temperature storage, the Ni_3P layer shows small pores and microcracks. After 1000 temperature cycles, small defects are found in the P enriched zone.

5.4.2.4 Sn–Pb/Electroless Ni(P)

He et al. [13, 14] demonstrated that the lead in the solder decreases the activity of Sn during soldering, affecting the Sn/UBM reactions. When aging, the Pb accumulates at the interface creating a Pb-rich phase, reducing the Sn activity even more as Sn has to diffuse through this area. In other words, IMC growth is slower when a Pb containing solder is applied.

Jang et al. [17] did not report Kirkendall voids in his experiments. He assumes decomposition of the Ni_3P layer. The Ni diffuses through the Ni_3Sn_4 layer, causing it to grow, while the P returns to the Ni–P.

Zeng et al. [9] mention that Kirkendall voids observed in the Ni_3Sn_4 layer, near the Ni_3P for eutectic SnPb solder.

5.4.3 Diffusion Couple Sn–Zn/Cu

Islam et al. [18] investigated the formation of IMC in the system Sn-Zn solder/Cu substrate during reflow and extended reflow. The formation of IMCs γ-Cu_5Zn_8, β-CuZn and an unknown thin Cu–Zn layer is found. Kirkendall voids are observed in this layer.

5.4.4 Diffusion Couple Sn/Ag

Silver dissolves relatively quickly in liquid Sn and the formation of the IMC Ag_3Sn is mentioned [8]. When AgPd metallization is used to reduce the dissolution of Ag but mechanical problem may arise due to Kirkendall voiding. The number of investigations in this system is relatively small.

5.5 Conclusions

The Kirkendall effect manifests itself by the movement of marker planes in a diffusion couple system. It is the result of a net mass flow, accompanied by a vacancy flow in the opposite direction.

A large number of studies are dedicated to the microstructural evolution in solder-substrate systems. Apart from experimental work, models are becoming available to predict the microstructures.

No reference is found to predict the formation of Kirkendall voids. The voids are observed experimentally in many diffusion couple systems, but it receives not much attention.

IMC layers with Kirkendall voids are an easy path for crack propagation and determine the reliability of the solder joint.

5.6 Appendix

See Table 5.1.

Table 5.1

Solder	Substrate	Process	Aging	IMC	Kirkendall voids	References
SnPb	Cu, OSP on Cu, DIG on Cu, imm. Sn on Cu	Reflow 220°C,	100, 125, 150°C, up to 80 days	Cu_6Sn_5, Cu_3Sn	At interface Cu_3Sn and Cu	[9]
SnAgCu	Cu/Ni(V)/Al thin film metallization	Multiple reflow 1–20	260°C, 5, 10, 20 min	Cu_6Sn_5 dissolves, Ni consumed		[11]
SnAgCu	CuOSP board metallization	Reflow 1–10 260°C	150°C, 1000 h, TCoB 1000× −40 to 125°C,	Cu_6Sn_5, Cu_3Sn	Nanoscopic defects in Cu_3Sn	[12]
Sn-37 Pb	Ni–P UBM	Reflow 1 213°C, 120 s	130, 150, 170°C, 100, 225, 400, 625 h	After reflow: Ni_3Sn_4 (more needle type), NiSnP (thin layer) and Ni_3P Ag_3Sn randomly distributed After aging: Ni_3Sn_4 layer flatter and thicker	Not for 130°C For 150°C long aging time, for 170°C short aging time In: Ni_3P layer	[13]
Sn-3.5Ag	Ni–P UBM	Reflow 1 251°C, 180 s	130, 150, 170, 190°C, 100, 225, 400, 625 h Air 216°C	Ni_3Sn_4 (more chunky type), NiSnP and Ni_3P Ag_3Sn randomly distributed After aging: Ni_3Sn_4 layer flatter and thicker, faster than SnPb	Not for 130°C For 150°C long aging time, for 170°C short aging time In: Ni_3P layer	[13]
Sn-3.5Ag	Ni–P Ni	Reflow				[14]
Sn-37 Pb	Ni–P UBM	Reflow	200, 220, 240°C 1–40 min	Ni_3Sn_4, NiSnP and Ni_3P Ni_3P decomposes and P returns to Ni–P	No Aging time probably to short	[17]

Table 5.1 (continued)

Solder	Substrate	Process	Aging	IMC	Kirkendall voids	References
SnAgCu	Ni–P	Up to 5× Reflow 240°C	170°C, 64 h	$(Cu,Ni)_6Sn_5$ and Ni_3P (columnar) thin layer NiSnP	Voids in NiSnP layer	[11]
SnAgCu	Ni(P)	Reflow 1–10 260°C	150°C, 1000 h, TCoB 1000× −40°C to 125°C,	Ni_3P, Ni_3Sn_4	Non-critical voids	[12]
SnPb	Ni–P	Reflow 220°C		Ni_3Sn_4 (chunky and needle type) and Ni_3P	Yes, in Ni_3P layer after prolonged reflow	[11]
Sn_9Zn	Cu	Reflow 230°C 5, 10, 20 min		γ Cu_5Zn_8, β CuZn and unknown CuZn layer	In Cu–Zn layer	[18]
Sn_8Zn_3Bi	Cu	Reflow 230°C 5, 10, 20 min		Additional Bi precipitates	In Cu–Zn layer	[18]

References

1. Smigelskas AD, Kirkendall EO (1947) Trans AIME 171:130
2. Reed-Hill RE (1973) Physical metallurgical principles, 2nd edn. D. Van Nostrand Company, ISBN 0-442-06864-6, pp 378–406
3. Porter DA, Easterling, KE (1981) Phase transformations in metals and Alloys, 1st edn. Van Nostrand Reinhold, ISBN 0-442-30439-0
4. van Dal MJH, Gusak AM, Cserhati C, Kodentsov AA, van Loo FJJ (2001) Microstructural stability of the Kirkendall plane in solid state diffusion. Phys Rev Lett 86(15):3352–3355
5. Paul A, van Dal MJH, Kodentsov AA, van Loo FJJ (2004) The Kirkeldall effect in multiphase diffusion. Acta Mater 52:623–630
6. Paul A (2004) The Kirkendall effect in solid state diffusion, PhD thesis, Eindhoven University of Technology, ISBN 90-386-2646-0
7. Roönkä K, van Loo FJJ, Kivilahti JK (1998) A diffusion-kinetic model for predicting solder/ conductor interactions in high density interconnections. Metal Mater Trans 29A:2951
8. Laurila T, Vuorinen V, Kivilahti JK (2005) Interfacial reactions between lead-free solders and common base materials. Mater Sci Eng R 49:1–60
9. Zeng K, Stierman R, Chiu T, Edwards D, Ano K, Tu KN (2005) Kirkendall void formation in eutectic SnPb solder joints on bare Cu and its effect on joint reliability. J Appl Phys 97:024508
10. Liu CY, Tu KN, Sheng TT, Tung CH, Frear DR, Elenius P (2000) Electron microscopy study of interfacial reaction between eutectic SnPb and Cu/Ni(V)/Al thin film metallization. J Appl Phys 87(2):750–754
11. Zeng K, Tu KN (2002) Six cases of reliability study of Pb-free solder joints in electronic packaging technology. Mater Sci Eng R38:55–105
12. Benneman S, Graff A, Schischka J, Petzold M, Theuss H, Dangelmaier J, Pressel K (2006) A SEM and TEM study of the interconnect microstructure and reliability for a new XFLGA package, 1st Electronics System integration Technology Conference, Dresden Germany, 5th–7th September 2006, 26–34
13. He M, Chen Z, Qi G (2004) Solid state interfacial reaction of Sn–37Pb and Sn–3.5Ag solders with Ni–P under bump metallization. Acta Mater 52:2047–2056
14. He M, Chen Z, Qi GJ (2005) Mechanical strength of thermally aged Sn–3.5Ag/Ni-P Solder joints. Metal Mater Trans A 36A:65–75
15. Jeon YD, Paik KW, Bok KS, Choi WS, Cho CL (2001) Studies on Ni–Sn intermetallic compound and P-rich Ni layer at the electroless nickel UBM solder interface and their effects on flip chip solder reliability. In: Electronic components and technology conference, IEEE
16. Li D, Liu C, Conway PP (2005) Characteristics of intermetallics and micromechanical properties during thermal ageing of Sn–Ag–Cu flip-chip solder interconnects. Mater Sci Eng A 391:95–103
17. Jang JW, Kim PG, Tu KN, Frear DR, Thompson P (1999) Solder reaction-assisted crystallization of electroless Ni–P under bump metallization in low cost flip chip technology. J Appl Phys 85(12):8456–8463
18. Islam MN, Chan YC, Rizvi MJ, Jillek W (2005) Investigations of interfacial reactions of Sn–Zn based and Sn–Ag–Cu lead-free solder alloys as replacement for Sn–Pb solder. J Alloy Compd 400:136–144

Chapter 6
Tin Whiskers

Antonello Vicenzo

6.1 Introduction

Tin-based finishes are the dominant solution for replacing tin–lead alloys used in lead-frame plating due to economic and manufacturing reasons. Pure tin layers are also extensively used as contact surface and protective coatings for connectors, relays and other electronic parts, as well as solderable finish on printed circuit boards.

Outstanding characteristics of corrosion resistance, low melting point, high electrical conductivity, solderability, non-toxicity and large availability have motivated this choice.

However, tin-rich finishes pose a reliability risk due to potential growth of whiskers, i.e. filamentary crystals which are conductive and mechanically strong. In fact, pure tin and tin-rich finishes are all, in different degree, prone to whiskering.

The tin whisker is an old problem for the electronic industry and an enduring challenge for physical metallurgists. Following the approval of the RoHS directive by the European Parliament, the whisker problem was unexpectedly revived and continues to stand up as a major technical issue raised by the lead-free legislation.

It is about 60 years since the first time whiskers were recognized as a potential threat to electric and electronic equipment. During this relatively long lapse of time, indeed a long time from the perspective of technology development, innumerable experimental investigations and a few theoretical studies were conducted, thanks to which a significant progress in understanding was achieved. In more recent years, whisker studies have been strongly intensified because of the urgent

A. Vicenzo (✉)
Politecnico di Milano, Dipartimento di Chimica, Materiali,
Ingegneria Chimica "Giulio Natta", Via L. Mancinelli 7, 20131 Milano, Italy
e-mail: antonello.vicenzo@polimi.it

G. Grossmann and C. Zardini (eds.), *The ELFNET Book on Failure Mechanisms, Testing Methods, and Quality Issues of Lead-Free Solder Interconnects*,
DOI: 10.1007/978-0-85729-236-0_6,

need of mastering this insidious threat for the sake of industrial implementation. A review compiled by Galyon [1] is a detailed account of the history of this troublesome and fascinating phenomenon up to 2004.

The prevailing attitude towards the whisker issue is that the current state of knowledge is inadequate to provide a comprehensive picture of the phenomenon. In particular, the mechanistic details so far uncovered and the physical models developed are still unable to provide a solid base for the assessment of the kinetics of whisker growth. Moreover, the technique of mitigating whisker growth is today not much ahead of the level of efficacy and technological confidence of some decades ago. Nevertheless, tin whisker mitigation practices have seen some refinement and consolidation over the last 10 years, and, most importantly, new plating processes have been developed ensuring lower propensity to whiskering.

Therefore, despite a persistent uncertainty, the present level of understanding, the development of mitigation practices and, finally, the advances in plating processes, all these elements have contributed to give confidence to users in adopting tin-based finishes for electronic products with short service life and low-reliability requirements. Still, given the present state of knowledge, the use of tin finish in mission critical applications is unanimously considered as a high-risk scenario. However, both management and market factors may leave no choice but using tin base finish also in high-reliability segments of the microelectronic industry. That is the simple reason why continued research on tin whisker and mitigation is a must.

6.2 Failure Mechanisms

Tin whiskers are reputed responsible of failure events by a number of different mechanisms. However, it should be stressed that not all of these mechanisms are either documented or undoubtedly ascertained.

- *Electrical shorts at relays and connectors*. This is the best documented failure occurrence caused by whisker formation; a number of reported cases are included in the list of whisker failures compiled by Leidecker and Brusse [2].
- *Electrical shorting between neighbouring terminations of lead-frame*. In low-voltage, high-impedance circuits, the current may be insufficient to fuse a whisker open. According to Arnold, the fusing current for a 2 μm diameter, 2-mm-long whisker is in excess of 10 mA [3]. The fusing current of a whisker depends on its diameter and its structure, if hollow. According to Dunn [4], the short-circuit current ranges from 10 to 30 mA as the diameter increases from 1 to 3 μm. A current range of 7–15 mA to fuse open a tin whisker depending on its geometry was reported in a recent study [5]. A burnout maximum current of 75 mA was reported by Hada et al. [6]. An empirical model for quantifying the probability of failure by short circuit caused by tin whiskers as a function of voltage was developed by Courey et al. [7, 8].

- *Shorts caused by metal debris, bridging the gap between conductors remote from the whisker origin.* Such an event of failure requires a whisker to break loose from the surface because of mechanical shock or vibrations. This failure risk should be weighted in the light of the available information about whisker mechanical strength. Dunn performed mechanical shock and vibration testing on tin whiskers [9], concluding that, in his experimental conditions (vibration frequency in the range 10–2,000 Hz, shock loading in the range from 20 to 2,000 $\times g$), neither vibration nor mechanical shock caused any visible damage to whiskers. The same conclusion emerged from a study where similar conditions of vibration test were used [10]. In the later study, mechanical shock tests were also performed according to the drop impact test method for mobile phone giving as well a negative response.
- *Loose whiskers may interfere with sensitive optics functioning and actuators operation.* See comments above.
- *In the presence of high field, the risk exists of spark formation* at the tip, possibly initiating metal vapour arching in vacuum and combustion.
- *Interference in high-frequency devices.* According to a recent publication [11], the effect of whiskers on high-frequency circuits performance, in the range 1–20 GHz, and its interference with high-speed signals is deemed negligible. This conclusion is based on a modelling work. On the other hand, Galyon and Gedney [12] state that whiskers become an issue above 6 GHz and that whisker length needs to be less than 75 μm in order to avoid affecting high-speed circuits.

In summary, electrical shorting caused by whiskers bridging conductors could represent the highest risk event. This is a concern particularly for fine pitch components and for applications in static environments. In a dynamic mechanical environment, whiskers may be expected to break off if long enough, in which case the whisker risk would be that of a shorting event caused by loose filaments. However, this failure risk needs to be further assessed. Not only whiskers seem to have quite a strong endurance to vibrations and mechanical shocks but, in case of breaking, the risk of shorting may be significantly reduced if not nullified, because of the high contact resistance between the mating surfaces [13]. In fact, short circuits caused by whiskers bridging closely spaced circuit elements require a minimum relatively high contact pressure or, alternatively, local breaking of the surface oxide. According to Hilty and Corman [5], contact pressure in the range of 10^{-3}–10^{-2} N was necessary in order to overcome the contact resistance. In the same study, voltage breakdown was observed on applying voltage higher than 3 V to the contact interface.

6.3 Whisker Morphology

There are two main morphological types of spontaneous crystal growth, which are usually observed on tin-based finish and which can be classified as whiskers.

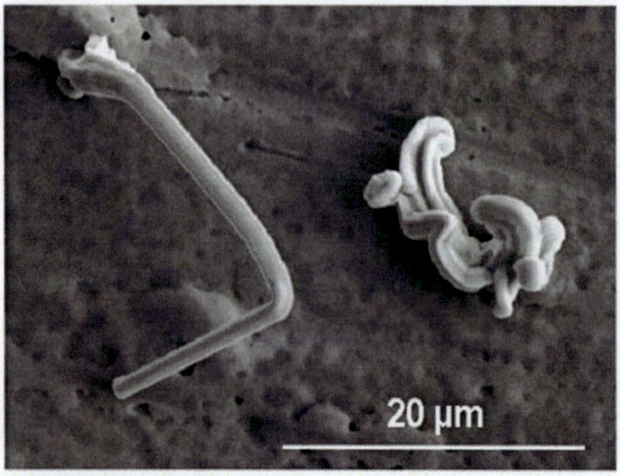

Fig. 6.1 Kinked whisker and eruption on a bright tin layer on CuFeP substrate with Ni barrier (2 μm nominal thickness). Ageing at 55°C/uncontrolled humidity/3,000 h

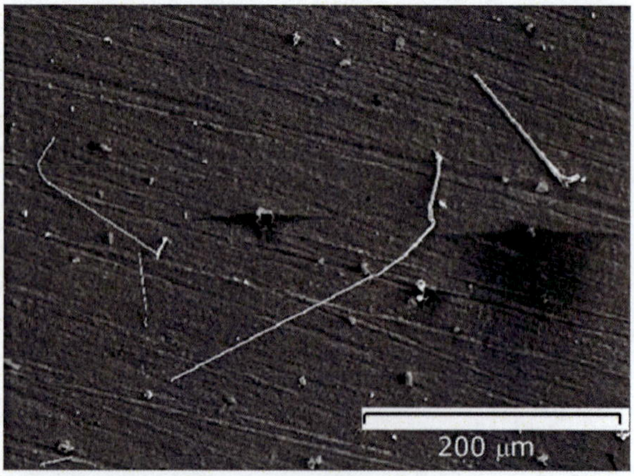

Fig. 6.2 Whiskers on bright Sn/Ni/CuFeP (Ni barrier nominal thickness: 2 μm). Ageing: 55°C/uncontrolled humidity storage/3,400 h after conditioning [17]

The first type is the hair-like or filamentary crystal, which is the archetype and the icon of the whisker growth phenomenon (see Figs. 6.1, 6.2). Filamentary crystals are characterized by a small diameter, in the range of 1–3 μm, length from a few tenths to several hundreds of micrometres and either a smooth or tapered lateral surface, e.g. whiskers on bright pure Sn and Sn(Cu) layers [14]. The second type is the columnar crystal, which is taking the place as the more familiar look of whiskers, being the most common whisker crystal shape observed on matt tin finishes (Fig. 6.3). Columnar whisker crystals, with diameter usually in the range of 2–6 μm, depending on the grain size of the tin layer, have length rarely exceeding 0.2 mm and a streaked surface. There are however reports of whisker growth on matt tin surfaces having length up to 0.45 mm under high temperature and humidity conditions, that is 60°C and 93% RH [15], or with a thin tin layer, namely 2.5 μm, on copper lead-frame under ambient storage conditions [16].

Fig. 6.3 Columnar whiskers, mostly curved or kinked, on matt tin layer on Cu lead-frame, with post-plating bake. Ageing according to JEDEC test conditions: 60°C/85% RH/4,000 h

A common and characteristic morphological feature of both filamentary and columnar whisker crystals is a constant cross section from the base to the tip.

A different class of morphological features, still referable to spontaneous crystal growth or recrystallization growth behaviour of a tin surface, engenders a number of different growth features, which are named by different locutions, such as nodule or bud and flower or simply eruption (see Fig. 6.4).

Nodules and, more generally, eruptions may be whisker growth warning signs. All these features may be seen on either matt or bright tin coatings, though odd-shape eruptions are more common on bright tin surfaces.[1]

Whiskers should be straightforwardly identified by their peculiar morphology, i.e. the hair-like or columnar shape of growth. However, from time to time, there are reports of abnormal crystal growth phenomena observed on tin layers as well as other metals, which are hastily classified as whisker. Unusual morphology includes conical, pyramidal whiskers and the like. A recent example of such odd growth morphology is the report of ribbon-like whiskers observed on matt tin layer on Cu–5%Sn–0.35%P substrate after exposure to 55°C/100% RH [18]. Unusual, rarely observed, growth morphologies are possibly relevant to mechanism, but cannot be deemed relevant to most in-use conditions.

The user should be aware that the plating process itself can produce a rather impressive gallery of peculiar growth features, including dendrites (i.e. branched crystals), needle-like crystals usually in thick bundles, conical or acicular protrusions and even ribbon-like crystals. These abnormal growth features can be produced during the plating process and are associated with conditions of either mass

[1] Tin is electrodeposited with either a matt or bright finish depending on the type and load of additives in the plating solution. The type of finish is basically characterized by the surface grain size, which is usually in the range of 3–5 µm for matt Sn (but may range from 1 to 10 µm scaling with layer thickness) and less than 1 µm for bright Sn deposits. The level of compressive residual stress (higher for bright, lower for matt tin) and the carbon content in the layer (less than 0.05% for matt Sn and in the range of 0.1–1% for bright tin) are also usually referred to as qualifying properties of the type of finish.

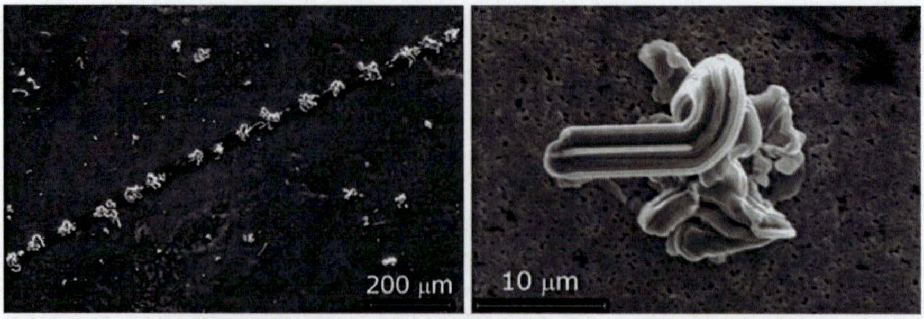

Fig. 6.4 *On the left*, a linear array of nodules along a surface scratch on a 5-μm bright tin layer on brass substrate, after 1,000-h ageing at room temperature and uncontrolled humidity. *On the right*, a tin nodule or flower from which a whisker is sprouting (10-μm-thick bright tin coating on CuFeP substrate)

transport polarization effects or out of control plating bath chemistry (a *library* of such abnormal morphological features was produced by dosing a tin electrolyte with a wetting agent in conjunction with lead impurity [19]).

Besides the consequences of process control deficiency, it should also be mentioned that corrosive environment may induce the growth of seemingly whisker crystals, likewise the special case of whisker growth on silver surface exposed to sulphide-containing environment.[2] Caution should be used in drawing conclusions from corrosion-induced tin whisker growth, meaning by this expression the formation of whisker crystals in the presence of a specific chemical participating in their formation, since field conditions are, for most applications, far less aggressive than those required for such an occurrence on tin finish.

6.4 Mechanisms

The detailed mechanism of whisker growth is still controversial and, as a result, the ultimate theory of whisker formation has yet to be written. Despite this state of affairs, there is a wide consensus that whisker formation is a stress relief phenomenon—a *discovery* now more than 50 years old [21, 22]. Namely, it is now generally recognized that this peculiar crystal growth phenomenon is driven by compressive mechanical stress in the tin deposit on which the whisker crystals grow. A well-informed account and an historical perspective of the tin whisker theory development are given in reference [1].

[2] Many other metal surfaces grow either metal or compound whiskers, in different and possibly quite special conditions [20] but, in most cases, with the notable exceptions of zinc, cadmium and silver whiskers, it is hard to say if this is of any practical relevance.

One of the arguments that is frequently put forward to justify the impossibility to define a solid theoretical framework and find a general agreement on the underlying mechanism of whisker formation is the inconsistency of the results reported in the plethora of whisker studies available, conducted during more than half a century. Indeed, it is common experience of those who have carried out whisker testing that the consistency of the results is by itself an achievement. Indeed, there are many different factors from which whisker growth may be influenced. Among these, the plating process itself is a critical and hard to control variable.

Thanks to the recent development of test methods, notwithstanding the fact that they are still under scrutiny, this seemingly inherent and frustrating characteristic of the whisker growth phenomenon seems to have been alleviated if not set aside.

However, the key reason why whisker formation is still not completely understood is its challenging inherent complexity. As any other crystal growth phenomenon, whisker formation is a two-stage process, involving nucleation and growth. In addition, whiskers usually grow only after a dormancy period, which can range from a few weeks to years.

Recently, there has been a strong resurgence of interest in the fundamental mechanism of whisker growth. The outcome of this intense activity has been twofold. On one hand, the concept that whisker growth may be associated with a multiplicity of mechanisms is gaining ground. On the other hand, the key mechanistic aspects have been identified and nowadays are also largely agreed on.

6.4.1 Stress Sources

From an industrial perspective, the most important aspect in whisker studies is the suppression or mitigation of whisker formation and growth. In this respect, as well as for the purpose of product qualification, the development of acceleration methods is an essential requisite. Obviously, a detailed knowledge of the parameters affecting whisker growth is key to advances in acceleration and mitigation.

It has been speculated that the driving force for whisker growth is a state of compressive strain within the Sn layer or a positive strain gradient in the direction from the base to the surface [23] and/or in the direction from the surroundings to the whisker root parallel to the surface of the layer [24]. Synchrotron radiation micro-diffraction analysis of the stress state in the region surrounding a whisker provided some piece of evidence to this principle (on Sn(Cu) deposit) [25]. The details of this mechanism are being developed so that it is still not possible to formulate a quantitative growth model effectively relating the growth kinetics to relevant physical and experimental factors, such as material properties, process parameters and part design.

The nature and origin of this strain should first be defined. Basically, the factors unanimously recognized as the most common determining conditions for whisker formation and growth are the following:

1. Formation of an intermetallic compound between the Sn-based finish and the substrate material, inducing compressive strain into the Sn or Sn alloy layer.
2. Temperature cycling generating stresses due to thermal expansion mismatch between coating and substrate materials.

Other sources of stress possibly having an impact on whisker growth, either as determining driving force or as an accelerating factor, were identified over the years, including the following:

- Mechanical stress and surface damage, an issue particularly for connectors [26].
- Surface chemical modification, primarily oxidation but also corrosion.
- Electric bias.

Some of these are discussed in more details in the following.

A mechanism of stress generation is a necessary condition for whisker growth but it is not sufficient to explain the formation of whisker. A further condition is the inhibition of a uniform strain relaxation in the tin layer or, analogously, the local relaxation of the strain resulting in the formation of a whisker.

6.4.1.1 Diffusion in the Copper–Tin System

The reference metallurgical system for the whisker issue is the metal couple Sn/Cu, the tin layer being a matt tin finish. The most widely accepted view is that chemical affinity between members of the couple triggers the room temperature inter-diffusion of the two metals, resulting in the formation of an irregular intermetallic compound layer and in the following build-up of compressive stress into the tin layer. The basic model of internal stress generation by a chemical reaction in the Sn/Cu couple is schematically illustrated in Fig. 6.5.

What is particularly relevant to the whisker problem is the identification of the dominant diffusing species at low temperature, the microstructural evolution of the interface interphases and the related stress evolution resulting from inter-diffusion and phase transformation.

A complete and unambiguous description of the behaviour of the Cu/Sn diffusion couple is still lacking. In particular, there are conflicting views on the kinetic law governing the growth of the intermetallic layer and confusing results concerning diffusion parameters. For a critical overview of the investigations into the Cu/Sn diffusion couple carried out during the last decades, the reader is referred to a paper of Laurila et al. [27].

The available experimental evidence indicates that copper is the fast diffusant of the couple at room temperature [27], though a conclusive marker study of the room temperature inter-diffusion in the Cu–Sn system, directly relevant to the whisker case study, has yet to be performed. A room temperature marker study of a thin-film Cu/Sn diffusion couple was conducted by Tu and Thompson [28]. The Cu/Sn couple was prepared by consecutive deposition of the metals by electron

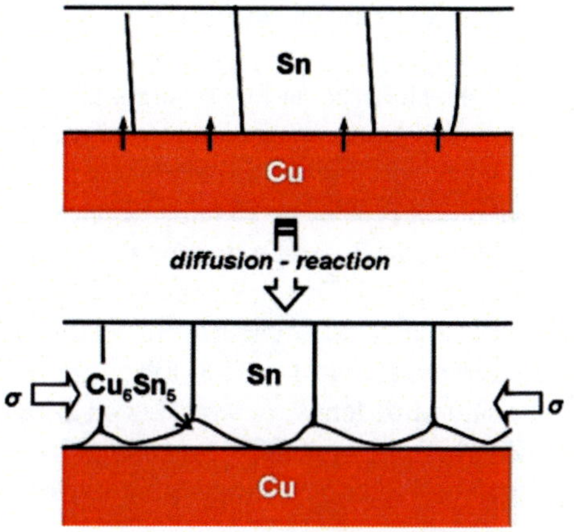

Fig. 6.5 Sn/Cu metal couple model system for whisker growth mechanism. From up down: Cu atoms can diffuse into Sn at room temperature, forming a wavy layer of intermetallic Cu_6Sn_5, and inducing compressive stress into the tin deposit

beam evaporation, using a discontinuous, island-like film of W as diffusion marker and measuring its depth position by Rutherford backscattering spectroscopy. According to this study, Cu atoms were the main diffusing species at room temperature. The other important finding of this work was that by isothermal annealing at T lower than 60°C only Cu_6Sn_5 formed and, since Cu was identified as the fast diffusant, the interface reaction was the Cu_6Sn_5/Sn. The formation of the intermetallic Cu_3Sn was observed together with the Cu_6Sn_5 phase following isothermal annealing at temperature in excess of 60°C.

According to the analysis of Laurila et al. [27], the low temperature behaviour of the Cu/Sn diffusion couple ($T < 60$°C) can be rationalized assuming that it is determined by the operation of short-circuit diffusion paths, namely interstitial diffusion of Cu atoms in the Sn and grain boundary diffusion of Cu atoms through the Cu_6Sn_5 reaction layer. This behaviour changes progressively with the increase in temperature above about 60°C, as a result of volume diffusion becoming predominant. In fact, different studies agree on the observation that Sn is the fast diffusant during formation of the Cu_6Sn_5 phase at high temperature [29–32] and that both Cu and Sn are mobile in Cu_3Sn, the fast diffusant being Cu [31, 32]. The Cu flux was found to be as much as three times greater than the Sn flux at 200°C. As pointed out by Ho et al. [33], the latter finding is in agreement with the observation that in the Cu/Sn diffusion couple Kirkendall voiding occurs at the Cu/Cu_3Sn interface or within the Cu_3Sn layer.

The room temperature diffusion behaviour and the reactivity of the Cu/Sn couple is particularly important to the whisker phenomenon: without the unbalance in the room temperature global diffusivity between Cu and Sn atoms and the simultaneous formation of an intermetallic layer—which takes place *within* the tin deposit, therefore inducing a compressive strain into the tin grains—a necessary condition for whisker growth would be lacking.

6.4.1.2 Thermal Cycling

Thermal stress arises as a result of CTE mismatch between Sn and substrate material (for stress being compressive on heating: $\alpha_{sub} < \alpha_{Sn}$—e.g. Alloy 42, ceramic substrate as for multi-chip module ceramic capacitors). This mechanism of stress generation is relevant to temperature cycling testing. In thermal cycling conditions of −55 to 85°C, whisker growth is reported to reach saturation after 1,500–2,000 cycles [34, 35]. The maximum tin whisker length observed on matt tin coatings over copper alloy lead-frame materials is about 50 μm at saturation. Similar observations are reported for test up to 8,000 cycles [36] and 10,000 cycles [37]. On the contrary, the saturation length of whiskers on alloy 42 lead-frame is in excess of 100 μm [38]. Tin whiskers of length exceeding 100 μm were also observed on ceramic chip capacitors after 500 temperature cycles (−40°C/90°C) [39].

6.4.1.3 Oxidation and Corrosion

The recognition of the accelerating influence on whisker growth by atmospheric O_2 and water vapour was an early observation of whisker studies [20]; the effect of air exposure on whisker growth on Cd was noted even earlier [40]. Kehrer and Kadereit [41] reported that 30–100 nm thick Sn films deposited on glass in a moist oxygen atmosphere of 10^{-4} Torr formed many whiskers after annealing at 60°C, while when the Sn films were deposited at 10^{-6} Torr, no whiskers were observed. Similarly, in ultra-high vacuum, no surface hillocks were found to grow on Al surfaces under compression [42]. Both these observations pointed to a specific mechanism of whiskering driven by oxidation.

The involvement of oxidation in the whisker mechanism of tin coatings on brass was also demonstrated, though in this case, the dominant interaction was that between oxygen and zinc out-diffusing from the base metal at the surface [43]. A similar behaviour, i.e. where a different element appears responsible for increased susceptibility to whisker growth induced or triggered by oxidation, emerges from the results of different studies, including Sn–Mn electrodeposits [44], Sn–3Ag–0.5Cu–1Ce [45] and Sn–Zn solder [46] and bulk Sn–Al alloy [47].

The accelerating effect of humidity on whisker growth was discussed in relation to oxidation/corrosion first by Oberndorff et al. [48, 49] and later by Osenbach et al. [50]. The interpretation emerging from these works was that the influence of oxidation on whisker formation is through the induction of compressive stress due to the volume expansion accompanying the formation of oxidation products.

The issue of corrosion during storage in high temperature and humidity environment was carefully considered and studied in the frame of the activities carried out within the iNEMI Tin Whisker Accelerated Test Project, due to its particular relevance to the development of test methods for evaluating tin whiskers. Corrosion of the tin, possibly enhanced by galvanic contact with the underlying copper [49] or alloy 42 [16] substrate, is now generally acknowledged as a promoting factor for whisker growth, though its relevance to in-use condition is still debated.

Fig. 6.6 Massive whisker growth in a corroded zone (punching edge) on a matt tin layer on Cu lead-frame. Ageing according to JEDEC test conditions: 60°C/85% RH/3000. Photo courtesy of P. Crema, STMicroelectronics

An example of intense whiskering in a corroded zone on a matt tin layer on Cu lead-frame is displayed in Fig. 6.6.

The relevance of corrosion as a potential synergetic factor in whiskering was raised in the PROTIN project, in the light of the results of whisker growth testing on tin-plated copper lead-frame at 60°C/93% RH storage conditions [51]. This phenomenon was later observed also by Osenbach et al. [52] and described as a "condensation-induced corrosion-assisted whisker growth". In this study, trim and formed components were found to be particularly vulnerable to corrosion-assisted whisker formation compared to non-trim and formed devices, which may be explained in terms of acceleration due to galvanic coupling between the Sn layer and the Cu substrate, according to previous studies by Oberndorff et al. [48]. In addition, it is worth noting that the morphology of the growth features, called by these authors "flower clusters of Sn whiskers", was quite different from the conventional tin whisker filamentary or columnar growth and that also tin-based materials usually considered practically immune from whiskering may grow whisker if exposed to corrosive environment or contamination during processing, e.g. by flux residues [53].

The occurrence of corrosion-assisted whisker formation may continue to be a concern for damp-heat testing of component finishes for whisker growth, where the temperature/humidity condition may incidentally get close to the condensation point or as a result of incidental contamination. Control over this phenomenon does not seem to be easily manageable, since it does not depend exclusively on storage conditions but also on surface contamination and other accidental factors. The influence of corrosion is currently explained according to the same interpretation scheme used to rationalize the influence of oxidation [49, 50].

6.4.1.4 Electric Bias

Another possible stressing/accelerating factor is electric bias; its evident connection to in-use conditions is the reason why the Phase 4 evaluation of the iNEMI

Tin Whisker Accelerated Test Project was designed to investigate the effects of electrical bias on whisker formation on matte tin-plated components, assembled with both lead-free and tin–lead solder.

Published information is relatively scarce and somewhat confusing, possibly not comparable due to different bias level or stress (current density). Early investigation performed by Arnold seemed to rule out the influence of bias on whisker growth [3]. These findings were confirmed in recent works. According to Osenbach et al., there is no effect of electric bias on the propensity for whisker growth [52]. Similarly, Hilty et al. [54] found no appreciable influence of applied bias on whisker growth. The latter authors followed the test conditions suggested by iNEMI (5 V bias, with variable electric field intensity). Tests performed on samples continuously exposed to an 50°C/50% RH environment showed that whiskers grow both at the anode and cathode end; the whisker density was observed to decrease in the presence of bias, while the length of whisker was observed to increase [55]. According to a report from Texas Instruments (TI) [56], based on results of combined environmental (51°C/85% RH) and bias tests, bias is a relevant additional stress condition, producing more consistent whisker growth results, though no evidence of acceleration was actually reported. The 5 V bias suggested by the iNEMI consortium for testing was later adopted by TI and Alcatel [55].

The electromigration behaviour and its relationship with whisker growth was studied in pure tin using Blech structure by Liu et al. [57]. Tests were performed at current density of 7.5×10^4 and 1.5×10^5 A cm^{-2}, i.e. two to three orders of magnitude higher than in standard electronic applications, at room temperature and 50°C. They found that whiskers grew on the anode, while voids were observed on the cathode, concluding that whisker growth occurred as a result of compressive stress induced by the Sn atoms electromigration flux.

The influence of bias on whisker growth is still to be completely clarified. There seems to be no conclusive evidence that bias or current flow has an influence on whiskering of matt tin; on the other hand, there is some evidence suggesting that whisker formation on bright tin finishes may be enhanced and/or accelerated by the presence of bias [58, 59].

6.5 Theory of Whisker Formation

Early theories of whisker growth were based on dislocation dynamics and reached their final stage of development with the models proposed by Lindborg [60] and Lee [61]. The mechanism of whisker growth based on dislocation theory was first questioned by Ellis [62] and has been the object of much criticism since then. It can be now considered superseded by the recent developments. The experimental evidence disproving dislocation-based theories is reviewed and discussed in references [1, 14, 63].

Briefly speaking, there are two main theoretical approaches to the whisker growth phenomenon, with significant overlapping [14]. The first has its origin in a reference theoretical study by Tu [64] and is generally known as the *Cracked Oxide Theory*. The second one, largely based on recrystallization concepts, can be traced back to early studies of the whisker phenomenon, stemming from the work of Ellis [62], Glazunova [65] and Furuta [66]. It was then revitalized and developed by the Tin Whisker Modelling Group of the iNEMI consortium and is currently being further refined by group members, e.g. [67, 68].

The key difference between the two approaches lies in the explanation of the initiation of whisker formation, namely while in the former, a key role is attributed to the surface oxide, in the latter, the mechanism relies completely on the recrystallization behaviour of the material, namely in the formation of *recrystallized* grains. Grains acting as initiation site for whiskers are expected to have peculiar characteristics, such as oblique angle grain boundaries [63] or yield stress lower than that of as-deposited grains [69]. The detailed mechanism of whisker development will change accordingly.

6.5.1 *Cracked Oxide Theory*

According to the analysis of Tu [64], the driving force for whisker formation is the compressive stress generated by the interface reaction between the tin film and the copper substrate to form the Cu_6Sn_5 intermetallic, as shown schematically in Fig. 6.7.

The mechanism proposed requires three conditions: (1) the room temperature interstitial diffusion of Cu atoms into the Sn lattice and the reaction between the two metals to form Cu_6Sn_5; (2) the room temperature self-grain boundary diffusion of Sn atoms, that is the diffusional process enabling stress relief; (3) a stable and protective surface oxide, which is the condition required to prevent a uniform relaxation of the compressive stress.

It is important to stress that chemical reactions provide a continuous driving force, as long as reactants are available, contrary to sources of thermal and mechanical stresses, which are not usually constantly active, though there are exceptions (e.g. connectors). The Cu_6Sn_5 intermetallic compound formed by the reaction between the solder finish and the substrate continues to grow overtime in the Sn layer grain boundaries, generating and maintaining, like a wedge [61], the compressive stress. Stress relieving is made possible by Sn grain boundary diffusion down the stress gradient, feeding the growth of a whisker. In order for this stress relief process to result in whisker formation, a further condition is necessary, the localized fracture of the oxide layer on the tin surface, enabling the extrusion of the needle like crystal. In the absence of a protective surface oxide, whisker growth would be impossible, since stress could be uniformly relieved. Similar to much older concepts theorizing the inhibition of general extrusion for whiskers to grow [21, 70], the cracked oxide theory presumes that free deformation in the

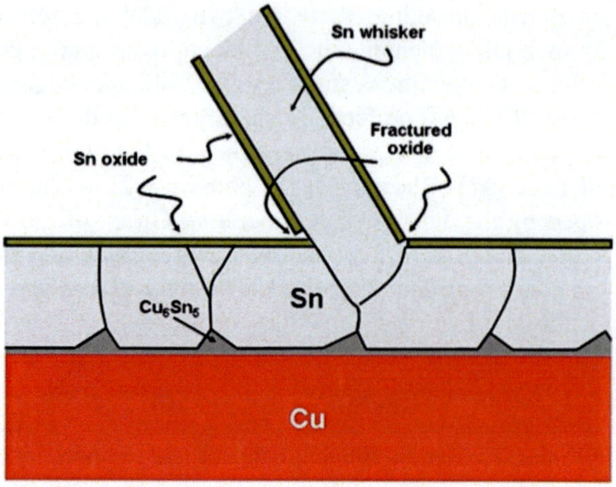

Fig. 6.7 A schematic diagram illustrating the Cracked Oxide Theory [64]. Copper migration is driven by chemical potential gradient to form the intermetallic layer Cu_6Sn_5 in grain boundaries. This induces a compressive stress in the Sn layer. To relieve the stress a flux of tin atoms driven by the stress gradient (creep) diffuses away and supplies Sn atoms to a growing whisker

direction normal to the film plane is restrained by the tin surface oxide. On the other hand, this surface oxide must not be thick so that it can be broken at certain weak spots on the surface.

This weakness of the oxide layer allowing local breaching is explained as a consequence of the tin lattice elastic anisotropy, resulting in strain in the direction normal to the film plane varying with the orientation of grains. It is therefore assumed that difference in strain along different crystallographic directions will give rise to shear cracking of the surface oxide [61]. A weak spot in the protective layer can be identified with a structural discontinuity, such as the misorientation of a grain to the adjoining grains (e.g. (210) grain in a (321) + (211) matrix) [71].

According to Galyon: « The weak oxide approach remains viable, but no direct proof has yet been published » [72]. In fact, recent results, though based on a small base of data, show that there is no apparent crystallographic relationship between whiskers and adjacent grains [73].

Experimental evidence against the involvement of oxidation in whisker formation was claimed by Moon et al. [74]. Sn(Cu) coatings were cleaned using an Ar^+ ion beam and aged in the 2×10^{-9} Pa Auger system chamber. Whiskers and other features present during Ar^+ ion cleaning left visible shadows on the surface. During ageing in the ultrahigh vacuum system, new whiskers, identified by the absence of the shadows, nucleated and grew. According to the authors, this was the demonstration that whiskers can be formed in the absence of a surface oxide. However, it is known from an early study that ion beam bombardment of a tin surface may induce whisker growth [75]. Therefore, the observation of Moon and colleagues does not seem to be conclusive in this respect.

On the other hand, there are several observations that do support the relevance of oxidation in whisker formation, as already noted previously. Moreover, if grain boundary diffusion and grain boundary pinning are both taken as necessary conditions for whisker growth, the role of the surface oxide would become decisive, for both the initiation (the latency period) and the nucleation of whiskers [76].

In fact, given that premise, the opening of local discontinuities (breaches) in the oxide layer is needed for allowing a vacancy flow into the material.

An alternative explanation for the so-called weakness of the oxide layer—not relying on either the elastic anisotropy influence [61, 71] nor the oxidation rate difference between adjoining grains with different orientation [50, 52]—was proposed on the basis of the electrochemical behaviour of the Sn metal, that is the fast ion-exchange kinetics and active–passive behaviour of the metal [17]. This concept could help in clarifying the accelerating effect of humidity on whisker growth and also the interplay between corrosion and whisker formation.

6.5.2 Recrystallization Theory

The existence of a latency period preceding the growth suggests that whisker formation may take place only after the achievement of a yet to define critical state, which most likely depends on the microstructure evolution of the material. The view of the dormancy period as an incubation time for a recovery and recrystallization stage [77] is currently an accepted interpretation [78]. As defects and impurities pile up at grain boundaries, the dormancy period goes towards its end, since recrystallization becomes inhibited. The time length of the dormancy period depends therefore on the initial level of the excess energy stored into the material in the as-plated state. As already noted earlier, these principles were resumed by the NEMI Tin Whisker Modelling Group and reawakened in the frame of the current recrystallization theory of whisker growth. A recent paper attempts to gain direct evidence to support this interpretation, presenting interesting results [79].

Within the iNEMI Modelling Group, the basic elements for a recrystallization theory were first discussed by Boguslavsky and Bush [80], in connection with a dislocation mechanism, later completely abandoned. The general framework of the so-called *Integrated Theory* was outlined in a paper by Galyon [72] and later complemented by a paper by Smetana [63].

Special *whisker grains* are assumed to form during the initiation stage of whisker growth. According to Smetana [63], whisker grains are recrystallized grains characterized by oblique grain boundaries, contrary to as-grown grains, which have vertical boundaries. This is the key feature of such whisker grains, since, while the stress state in columnar grains is biaxial—i.e. there is no stress component normal to the layer plane—on the contrary, the stress state in a grain with oblique boundaries is triaxial at the root—there is a stress component along the oblique boundary—and biaxial at the free surface. This shear stress at oblique grain boundaries is thought to be essential to whisker development. Based on a simple geometric model, the oblique grain boundary is described as a lower stress interface compared to vertical grain boundaries, thus accounting for a positive stress gradient towards the whisker root. The theory further assumes that Sn atoms diffuse towards this lower stress zone activating grain boundary sliding, the final stage leading to whisker growth.

Probably, the most appreciated result of the *Integrated Theory* has been the attempt to provide a simple and effective description of the metal couples of interest for lead-frame components (namely, the Sn/Cu, its variants Sn/Ni/Cu and the couple Sn/Fe–Ni42), which notoriously show different propensity to develop whisker as well as different sensitivity to stress factors in terms of the stress state induced into their layered structure.

The Sn/Cu couple is described by the zone model reproduced in Fig. 6.8. According to Gaylon's analysis [72], grain boundary diffusion of Cu atoms into the Sn creates a vacancy-rich zone in the Cu substrate, which may eventually cause the formation of microvoids in the reaction layer. Because of the excess volume made available, a tensile stress develops in the Cu-depleted zone; on the other hand, the formation of Cu_6Sn_5 at grain boundaries results in compressive stress within the tin layer. This is the driving force responsible for Sn diffusion towards the surface and, given the correct conditions, for whisker growth.

Similar zone models were also proposed for the other systems, providing a persuasive tool to explain the build-up of internal stress into the different systems, their sign—compressive/tensile—and their influence on the whisker formation [81]. In support to his thesis, Galyon published a series of SEM micrographs showing the presence of microvoids at the Cu_6Sn_5/Cu interface [72]. As to the stress state developing in the metal couple after formation of the Cu_6Sn_5 layer by room temperature annealing, Tu [82] reported that the strain in the remaining Cu and Sn layer was tensile and compressive, respectively, in agreement with the zonal concept proposed by Galyon.

Nevertheless, this pictorial view of the dynamics of whisker growth, though persuasive, is not actually based on strong experimental evidence. The zone structure of the Sn coating/Cu substrate has yet to be firmly supported by convincing evidence. According to the classical theory of diffusion in the solid state [83], vacancy concentration may not reach equilibrium in a reactive diffusion couple with a strong unbalance of the members flux, possibly resulting in the

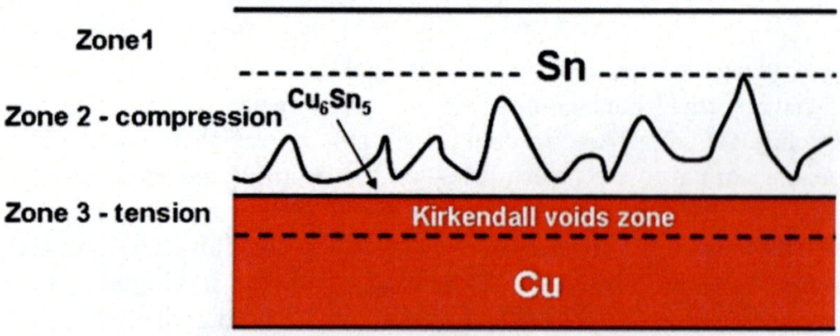

Fig. 6.8 Schematic diagram of the 4-zone structure of a Sn/Cu couple after intermetallic formation as depicted by Gaylon [72]

formation of voids either into the reaction layer or at the interface between the product phase and the fast diffusant member. In principle, the Sn/Cu couple could show a similar behaviour on ageing, in conditions leading to the formation of only the Cu_6Sn_5 phase, provided that Cu is the fast diffusant in these conditions. However, this room temperature morphological evolution of the Sn/Cu interface has not been demonstrated. Actually, Kirkendall void formation in the Sn/Cu diffusion couple was reported to occur only when the product phase was Cu_3Sn not with Cu_6Sn_5 [27, 33]. The kinetics of formation of the Cu_3Sn phase is diffusion controlled, and the Sn atoms are a much slower diffusant with respect to the Cu atoms in the Cu_3Sn compound. This behaviour may account for the absence of local equilibrium of point defects, the consequent rising of a supersaturation in vacancy concentration, and the actual observation of voiding.

Finally, the very possibility of Kirkendall void formation in the Sn/Cu couple, even for the case of Cu_3Sn formation, was also questioned based on the observation that the presence of impurities—such as into an electroplated Cu pad—seems to be a condition for its occurrence [27].

6.6 Mitigation

Mitigation strategies are processes performed during manufacturing to reduce the risk of whisker growth by minimizing compressive stress induced into tin-based surface finishes.

There is no mitigation practice that will guarantee a zero whisker risk; that is, there is no proof that any of the mitigation techniques so far developed can provide exemption from whiskers for tin-based finishes.

Mitigation strategies currently used in the industry are largely based on previous experimental and theoretical achievements, though they have been thoroughly re-examined and proven effective on a much larger and more reliable base of data only in recent years.

Mitigation practices are addressed in the JEDEC/IPC joint publication JP-002, "Current Tin Whisker Theory and Mitigation Practices Guideline" [78].

The final scope of a mitigation technique should be the suppression of the driving force for whisker growth. In the real world, the achievement of this objective is not straightforward. In fact, the whisker growth theory itself is not mature enough to provide the basic elements necessary for the confident design of effective measures of mitigation. According to another view, a further obstacle against the safe development of mitigation techniques is that there is not a single whisker mechanism. In other words, if the assumption is made that compressive internal stress or stress gradient is the root cause of whisker formation and that there are different stress-generating mechanisms, then mitigation techniques may be effective only with respect to one of these different mechanisms. That is the reason why mitigation techniques should not be regarded as prevention techniques, rather as whisker risk reduction methods.

There are basically two different strategies of mitigation, having the largest diffusion in the industry and the highest acceptance among users:

- Recourse to an underlayer as diffusion barrier, usually a nickel layer (silver was also shown to be an effective barrier [51, 84]).
- Annealing shortly after plating (so-called post-bake treatment).

Both these mitigation techniques address but one stress source, i.e. that associated with the otherwise uncontrolled interface reaction between the plated Sn and the substrate material. Indeed, as noted by Tu et al. [85], the use of a barrier layer to slow down or prevent the chemical reaction at the interface is inherently inadequate to completely suppress whisker growth, since it has no effect on the self-grain boundary diffusion of tin, which is the transport process responsible for feeding atoms to a growing whisker. The relevance of this remark is not diminished by recent reports questioning the key role of intermetallic formation at the deposit/substrate interface as the major stress source for whisker growth, based on the observation of the effects of co-deposited impurities [86]; neither could it be disputed were the interface reactivity disproved as a necessary condition for a tin layer/substrate system to initiate whisker growth, as recently reported [87].

Eventually, from the point of view of industrial applications, intermetallic formation in the Sn/Cu couple remains the most important stress-generating factor, and then the first to be targeted by any mitigation technique; and the more so, since it operates continuously over a long time span contrary to other possible stress sources.

6.6.1 Nickel Barrier

The use of nickel plate as a barrier layer has an established industrial tradition for both matt and bright Sn electrodeposits. With regard to the whisker risk, there are about 15 years of experience in surface mount ceramic capacitor. Interestingly, this mitigation practice was proven ineffective against thermal cycling in laboratory testing on ceramic chip capacitors [39]. This suggests that the results of whiskers acceleration tests should be carefully scrutinized since they may not be straightforwardly related to in-use conditions.

Mitigation is primarily achieved by eliminating compressive stress build-up due to Cu diffusion and the concurrent formation of the Cu_6Sn_5 intermetallic at the Sn/Cu interface. A further benefit is claimed in the frame of the integrated theory of Gaylon [72, 81], based on a four-zone model of the Sn/Ni/Cu system, where diffusion of Sn into Ni is thought to cause tensile stress build-up in the Sn at the interface with the Ni underlayer, counteracting the compressive growth stress in the Sn layer. Internal stress in Sn deposits over a Ni barrier on copper alloy substrate was actually reported to be tensile for thickness of the tin layer in the range up to 2 μm and after 40 days room temperature ageing [88]. Accordingly, it was also shown that a Sn/Ni finish on copper may also perform effectively against whisker growth in thermal cycling conditions, provided the thickness of the Sn layer is suitably tailored in order to maintain an average tensile stress in the Sn also during the heating cycle.

A few warning remarks should be noted in connection with the use of the nickel barrier: first, barrier films may fail if porous or thin, may be damaged or cracked; during reflow, solder may react with nickel and form Ni_3Sn_4 within the Sn, causing compressive stress. A minimum thickness of 1.27 μm is usually specified for the nickel barrier, following iNEMI recommendation.

6.6.2 Annealing or Post-Plate Bake

Tin whisker mitigation by annealing is a strategy that has a quite good reputation and has been in use as long as the nickel barrier solution. The post-plate treatment of matt tin-plated lead-frame (1 h at 150°C within 24 h of plating) has been adopted as mitigation technique by the major electronic components producers. This mitigation technique must be used in combination with a minimum tin layer thickness of 7 μm or 10 μm nominal [78].

Some 35 years ago, it was already established that a post-plate annealing is effective in increasing the incubation time and decreasing the maximum whisker length [89]. Recent results, see e.g. [52], corroborate this conclusion.

According to Oberndorff et al. [84, 90], the post-plating annealing (1 h at 150°C, directly after plating) promotes the formation of a more uniform and even intermetallic layer (see Fig. 6.9), consisting of Cu_3Sn and Cu_6Sn_5, inhibiting if not preventing the formation of the compound at grain boundaries. Such an intermetallic layer is believed to be a durable and effective diffusion barrier against grain boundary diffusion at ambient conditions, slowing down the otherwise continuing growth of a discontinuous layer of the Cu_6Sn_5 intermetallic. The improved planarity of the layer, possibly in connection with microstructural modifications of the tin coating, is believed to reduce compressive stresses and more particularly localized stresses in the interface region.[3] These effects of the post-bake treatment on the structure and morphology of the intermetallic layer and on the internal stress state of the tin deposit were fully confirmed in a recent work by Sobiech et al. [92].

The annealing treatment needs to be applied shortly after plating, because intermetallic particles may form under ambient conditions and penetrate deep into grain boundary, growing to the height of about 1 μm within a few weeks. The thickness of the intermetallic layer that is formed by the post-bake procedure is about 1 μm thick [78].

---

[3] An early view on the effect of post-plating annealing of electroplated tin, recently revived in a paper by Pinsky [91], attributes its mitigation efficacy to hydrogen relief. However, it is hard to say if there is any direct and uncontroversial evidence of the involvement of hydrogen codeposition in whiskering of tin finish.

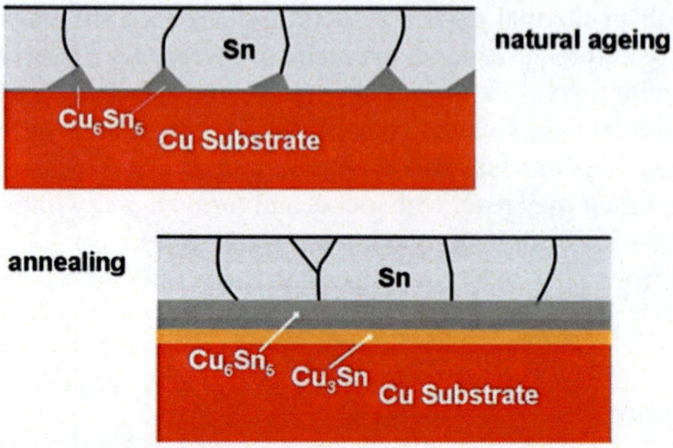

Fig. 6.9 A schematic diagram illustrating the structure of a post-plate annealed Sn/Cu couple (*below*) compared to the structure formed on ageing at room temperature (*above*). By annealing an uniform and even intermetallic layer can be formed, acting as an effective diffusion barrier. This reactive barrier has a two-layer structure, due to the high-temperature reaction of Cu and Sn resulting in the formation of both Cu_3Sn and Cu_6Sn_5

While the key feature of the post-plating bake treatment is believed to be the formation of an intermetallic two-layer Cu_6Sn_5/Cu_3Sn barrier at the Sn/Cu interface, further beneficial effects of an annealing treatment may be residual stress relief, grain growth and recrystallization [51, 61, 76, 93].

The modification of the stress state of the tin coating due to post-plating annealing was purposely investigated by Sobiech et al. [92] using a methodology to perform X-ray stress measurements with controlled depth of penetration. According to the results of this study, a residual tensile stress persists in the surface region of tin coatings subjected to post-plate annealing even after prolonged ageing, while compressive stresses appear concentrated in the interface region. This depth-dependent stress distribution is believed to result in the absence of a significant stress gradient in the near surface region of the tin deposit, which in turn is claimed as the reason for whisker suppression [92]. In this respect, it should be noted that, according to the analysis of Osenbach et al. [76], the improvement is one of degree not of kind that is the increasing of the incubation time not the elimination of the potential for whisker growth.

6.6.3 Chemical Etching or Roughening Copper Substrate

Recent work has shown that surface roughening of the copper-based alloys before plating could reduce tin whisker growth, when the etch depth is in the range of 3–4 μm. However, there is limited evidence so far. Deep etch ($\sim$3.5) of copper lead-frames was reported to result in even growth of intermetallic layer and whisker-free deposits [94]. The mitigating effect of surface roughening was reported also for fine pitch connectors, though whisker testing was not performed according to standard methods [95]. According to Zhang et al. [94], whisker growth is affected by texture and roughness/topography of copper substrate. The higher the surface roughness, the lower is the whisker density and the average length. A similar result was also reported for sputtered tin films on brass substrate of varying roughness conditions [96].

This influence is apparently related to the growth characteristics of the intermetallic layer. Reasonably, the rougher the surface, the higher the nucleation intensity for intermetallic formation; this leads to the growth of a fine grained and relatively more smooth and regular intermetallic layer. This, in turn, may favourably affect the stress state brought about by intermetallic growth.

6.6.4 Fusing and Hot Dipping

Fusing is performed by immersion in a hot oil bath, shortly after plating. In principle, it may eliminate any residual stress, cause grain growth and possibly thermal tensile stress build-up. For tin deposits on copper substrate, fusing causes the formation of an intermetallic layer of Cu_6Sn_5 and Cu_3Sn [97]. This technique has a quite good reputation, but is not used for components.

Hot dipping is a molten tin process, which is used for components. Hot dipping with pure tin is not effective as mitigation technique, based on both industry experience and testing results [37], though the latter refer to the environmental conditions of 60°C/93% RH, i.e. testing conditions that were later abandoned because of the risk of corrosion. Hot dip Sn–Ag and Sn–Ag–Cu are reputed an effective mitigation [78].

6.6.5 Conformal Coatings

The application of conformal coatings aims at retarding and/or containing whisker growth within the coat by sealing the plated surface. Creating a mechanical barrier against whisker growth should prevent whiskers from causing shorts between adjacent conductors.

Actually, whiskers may penetrate also relatively thick conformal coating, but the risk of a whisker penetrating the coating on a facing surface or the event of two whiskers growing from near surfaces and making contact in between to create a short can be reasonably deemed negligible. Other issues with conformal coatings may be coverage, depending on the material of choice and the method of application, cost and reworkability.

Papers focused on the application of conformal coatings as a mitigation practice against tin whisker growth are relatively few in the literature [98–100] and all from the aerospace industry. A paper from Boeing presents a detailed investigation into different coating materials and changing thickness [101]. Among the different materials tested, Parylene C and silicone were found to be the most effective. Test conditions were as follows: 278 days at ambient +336 days at 50°C/50% RH. No obvious correlation was found between mechanical properties of the coatings and ability to suppress whisker growth; no obvious correlation between oxygen and water vapour permeability and ability to suppress whisker growth. Generally, thin coatings, with thickness below about 60–80 μm, are penetrated by whiskers. Acrylic coatings were penetrated at thickness up to about 100 μm. Thicker coatings, with thickness in the range 100–150 μm, were not penetrated.

According to Livingston [102], the use of conformal coatings or foam encapsulation over tin finish encounters some limitations. First, there may be variability in the quality and thickness of the coating coverage. Provision should be made to avoid that the coating does not bridge the gap from one surface to another, since this could provide a direct path for whisker. The application of the coating under mounted components may be difficult. If the coating is applied by a spray process, care should be taken to avoid shadowing effects due to high-profile components. Livingston also reports results concerning the evaluation of conformal coatings (Uralane 5750), according to which whiskers grew through a 6.3-μm-thick coating after 2.5 years of storage at room temperature. On the other hand, no whisker growth occurred through a 50-μm-thick coating over the time span of 3 years.

Other characteristics, such as modulus and coefficient of thermal expansion, should be considered to prevent cracking. Finally, conformal coatings permeability to moisture poses an additional reliability concern, i.e. electrolytic migration that may be trigged by surface contamination (e.g. flux residue [103]).

6.6.6 Mitigation by Alloying with Lead

The use of Sn–Pb solder alloy finish, with Pb content usually in the range 3–10%, was the industry standard mitigation measure against the whisker risk since the 1960. This peculiar virtue of Pb alloyed Sn finish was first discussed by Arnold [104] and is still the subject of investigation and speculation. In fact, the reasons why Sn–Pb is whisker free or, if not exactly so, less prone to whiskering are still not well understood.

The Sn–Pb binary system stands by itself among the phase diagrams of the binary Sn alloys, also when compared to other Sn eutectic binary systems. The room temperature mutual solubility is limited, about 2% Sn in Pb and practically nil for Pb in Sn[4]; however, at the eutectic temperature, the solubility of Sn into Pb is quite remarkable, about 19% (29% at) [105]. This means that some degree of

[4] The opposite occurs in the Sn-Bi binary system which is practically the only other simple eutectic system of Sn relevant to application as solder finish. The Sn–Zn system is also a simple eutectic system with essentially no mutual solubility [105].

supersaturation may be obtained in the electrodeposited alloy [106] and in fact Raub reported evidence of formation by electrodeposition of metastable Pb–Sn solid solution with Sn content up to 8% [107]. Since the solubility of Pb in Sn is very limited, the electrodeposited alloy shows a two-phase structure with an almost pure Sn matrix and a Pb(Sn) solution. By room temperature recrystallization or after heat treatment, the near room temperature equilibrium concentration of Sn in the Pb-rich phase, about 2%, is re-established [108], resulting in Pb segregation at grain boundaries. Grain boundary segregation of the Pb-rich phase is apparently a characteristic of deposit growth and not only a result of transformation taking place after deposition. In other words, the electrodeposition of Sn–Pb alloy is likely to proceed by a process of simultaneous nucleation of the two-phase alloy [109], so that the observed two-phase structure is the intrinsic growth structure of the alloy and not only the result of the structural rearrangement of a supersaturated solid solution taking place during or shortly after the deposition.

The peculiarity of the electrocrystallization process of Sn–Pb electrodeposits is also responsible for the microstructure of the layer (in particular of matt Sn–Pb coatings), which is not columnar or field oriented, as invariably is for either matt or bright Sn deposits. In fact, because of the simultaneous deposition of the two phases, grain growth of the Sn matrix may be prevented by the formation of the Pb reach phase, and new grains must be nucleated and grow so that the final microstructure shows an equiaxed rather than columnar type of structure. It should be noted that the formation of a well-developed equiaxed (not columnar) microstructure depends on both the Pb content in the alloy, the type of bath and the deposition conditions, as may be inferred from results reported in [109], [110] and [111].

The deformation behaviour of the Pb reach intergranular phase should play an important role in determining the stress relaxation of the deposit. First, contrary to the precipitation of Cu_6Sn_5 particles, Pb segregation causes tensile stress build-up. Lead has in fact higher density than tin and, given that the alloy is formed as a two-phase supersaturated solid solution, the following separation of almost pure Pb and its segregation at grain boundary causes the build-up of tensile residual stress in the Sn–Pb coatings. Over longer time, the evolution of internal stress may also be influenced by the formation of Cu_6Sn_5 at the coating/substrate interface. On the other hand, the almost equiaxed grain structure of some Sn–Pb deposits permits a uniform creep deformation contrary to the case of columnar microstructure such as those of pure matt Sn and Sn–Cu, where creep is localized, giving rise to hillocks and whiskers. Hillocks may occur when grain boundary is relatively mobile (pure Sn), and whiskers when grain boundary motion is impeded (Sn–Cu) [111].

Lead does not seem to be replaceable by any other metal as an alloying element of Sn in solder finishes. First, the systems that can be considered viable for solder finish are not many. Several requirements must be simultaneously satisfied: the alloying element must be electrochemically co-deposited; it should not impair solderability to any appreciable extent; it should not be a toxic or hazardous material. Based on such requirements, it is easily seen that a number of possible alloying elements are ruled out and that there are but a few options. Not surprisingly, these have all already been tested and/or are in use: Ag–Sn, Bi–Sn, Cu–Sn, Sn–Zn. A further option could be Sn–Sb; however, antimony raises concerns as to its toxicity.

6.7 Test Methods and Specification Standards

Standardized test methods and specification standards enabling the assessment of susceptibility and defining acceptance requirements for whisker growth are key to a save implementation of tin-based lead-free finishes in components manufacturing and electronic assembly.

Standard test methods are essential for providing the industry with a methodological framework and a set of operating procedures for process and technology qualification; standard tests are also vital to the ongoing activities in the development and qualification of tin plating processes with low whisker propensity.

The final goal is to develop an accelerated test enabling the qualification of products to serve in a given application.

Industry consortiums concerned with the tin whisker issue have been working intensely to identify and evaluate environmental test conditions for the assessment of whisker growth propensity of tin coating: Soldertec Global and the E4 group in Europe,[5] the iNEMI consortium in the US[6] and the JEITA whisker test method subcommittee in Japan. In addition, a Tin Whisker Team established in the CALCE consortium conducted research in this field, operating in test plans development and mitigation practices qualification.[7]

Test methods for whisker growth currently in use are still under examination and liable to modification, in particular, the high temperature/high relative humidity test. The following remarks should be taken into account as a common sense guidance for any further development.

- Not a single test is capable of simulating the action of all—so far known and agreed on—driving factors for whisker growth. This is the reason why different tests are needed to assess susceptibility. In this respect, the final application may

[5] Formerly E3: Philips Semiconductors, ST Microelectronics, and Infineon Technologies; Freescale Semiconductor joined in 2004.

[6] Formerly NEMI, National Electronics Manufacturing Initiative; the *i* prefix was inserted when becoming *International*, from January 1, 2005.

[7] Computer Aided Life Cycle Engineering (CALCE) Electronic Products and Systems Centre, a consortium linked to aerospace and military industries in the US. Currently the CALCE Tin Whiskers team (http://www.calce.umd.edu/lead-free/tin-whiskers/. Accessed 13 October 2009) is engaged in assisting consortium members in the implementation of Sn-based finish. Expression of this commitment is also the recent release of a software for assessing the tin whisker failure risk in electronic hardware [59].

be of special relevance in dictating the dominant stress source (e.g. tin coating for connectors).

- Conflicting results were reported in the early literature and are still being reported, most likely because of the multi-mechanism nature of whisker growth and the practical impossibility to control all the chemical and physical parameters influencing this phenomenon. The only way to master this inherent uncertainty is to provide for a sample size statistically meaningful.
- Whiskers growth goes through an incubation time which is unpredictable. A suitable test duration and/or an appropriate pre-conditioning must be allowed.
- There is no definitive consensus on mechanism, so that test conditions and possibly evaluation criteria may be still subject to revision.

6.7.1 JEDEC Test Method Standard

The JEDEC Standard JESD22A121 "Measuring Whisker Growth on Tin and Tin Alloy Surface Finishes" was released in May 2005 [112], after an intense work carried out under the guidance of the iNEMI Whisker Test Method Committee.

It recommends three test conditions, namely: two isothermal conditions with controlled humidity and a thermal cycling condition.

It is clearly stated that the information provided by the tests are inadequate to quantitatively predict whisker growth over long time period. This means that there is no correlation between the results of accelerated tests described in the Standard and in-service environmental conditions. The scope of the standard is in fact to provide a suite of test methods for measurement and comparison of whisker propensity for different finish processes, a consistent inspection protocol for tin whisker examination and a standard reporting format.

Test conditions have already experienced significant changes with respect to the initial proposal. When the "NEMI Test Method for Evaluating Tin Whisker Growth on Plated Surfaces", revision 6.1, of June 2004, was submitted as a new standard proposal to JEDEC, three tests were recommended to evaluate whisker growth propensity: storage at room ambient conditions: 30 ± 2°C/60 ± 3% RH; storage at 60 ± 5°C/93 ± 3% RH; air to air temperature cycling: −55 to 85°C or −40 to 85°C, 3 cycles/h.

Test conditions for high-temperature/humidity storage were later revised, due to concern about the risk of corrosion and its playing as a potential synergetic factor in whiskering, as pointed out by E3 members in the PROTIN project [51]. Damp-heat test conditions were then slightly mitigated and set to 60 ± 5°C, 87 +3/−2% RH, as given in the October 2005 revision of the Test Method. The influence of humidity and the potential synergy between corrosion and whisker formation was further investigated. This concern was stressed during the iNEMI Tin Whisker Workshop held on the 31st of May 2005 in conjunction with the 2005 ECTC conference. As stated in the press release: "a new factor in tin whisker generation

Table 6.1 Test conditions and inspection intervals prescribed by JEDEC Standard JESD22A121

Test	Conditions	Recommendations[a]	
		Inspection intervals	Minimum duration
Ambient *T/H* storage	30 ± 2°C/60 ± 3% RH	1,000 h	3,000 h
High *T/H* storage	55 ± 3°C/85 ± 3% RH	1,000 h	3,000 h
Temperature cycling	Min: −55 to −40 (+0/−10)°C Max: +85 (+10/−0)°C Air to air; 10 min soak; ∼3 cycles/h	500 cycles	1,000 cycles

[a] Total test duration is not specified; warning is given that if test duration is too short, whiskers will not be observed. Test durations are given in JEDEC Standard JESD201

that needs to be understood is heavy oxidation or «corrosion» of the tin in humid environments".[8] As new data became available [37], the iNEMI user group changed the high *T/H* test conditions, namely adjusting the temperature to 55°C and the relative humidity to 85%. The temperature cycling test duration was also changed to a total of 1,500 cycles, since this was shown to saturate whisker growth on matt Sn on copper lead-frame [113]. These changes were introduced in the Specification Standard JESD-201 [114].

The JEDEC Standard JESD22A121 provides methods and procedures to accelerate and characterize whisker growth on tin-based finishes; it does not define acceptance criteria for the purpose of technology/process qualification. Test conditions and inspection intervals, as defined in the Standard and later revised as in the Specification Standard JESD-201, are reported in Table 6.1. Optional test sample preconditioning treatments are also defined according to four options (paragraph 5.1.2 of the standard): none; ambient conditions storage (15–30°C, 30–80% RH) up to 4 weeks; simulated Sn–Pb assembly process; simulated Pb-free assembly process. Preconditioning should be agreed upon between user and supplier before testing.

It should be emphasized that the present test methods are not relevant for applications where mechanical stress and other potentially harmful effects play a role.

6.7.2 IEC Test Method Standard

The intention to propose a draft Tin Whisker Test Method to the IEC International Electrotechnical Commission was announced in the occasion of the Tin Whisker Joint Meeting held in Tokyo on May 15, 2003, by Soldertec, NEMI and JEITA.

8 See for meeting presentations: http://www.inemi.org/cms/newsroom/Presentations/tin_whisker_workshop_may_2005.html. Accessed 13 October 2009.

Table 6.2 Test conditions and duration dictated by the IEC Standard IEC 60068-2-82 Ed. 1.0: Environmental Testing—Part 2-82: Test–Test Tx: Whisker test methods for electronic and electric terminals

Test	Conditions	Duration
Ambient	(A) 30 ± 2°C/60 ± 3% RH	4,000 h
	(B) 25 ± 10°C/50 ± 25% RH	
Damp-heat test	55 ± 3°C/85 ± 5% RH[a]	2,000 h
Temperature cycle testing[b]	Low *T*: −55 ± 5°C/−40 ± 5°C	(Class P) 2,000 cycles
	High *T*: 85 ± 2°C/125 ± 2°C	(Class Q) 1,000 cycles

[a] Set test conditions so as to avoid dew condensation on test specimens

[b] Four classes of severity are specified, with changing low and high temperature, and two further severity classes according to the prescribed number of cycles. For all classes: minimum dwell time of 20 min; transfer time between extremes <30 s

The standard proposal was submitted by JEITA in March 2004 [115]; it was approved and introduced as new work item in the IEC/TC 91 work programme in January 2005.

The committee draft was circulated at the end of April 2005 and discussed at the TC91/WG3 meeting in Frankfurt, in September 2005. Based on the discussion results, a Committee Draft for Vote was prepared and circulated in December 2005, approved as of May 2006 and registered as FDIS, Final Draft International Standard, in November 2006.

The standard "IEC 60068-2-82 Ed. 1.0: Environmental Testing- Part 2-82: Test–Test Tx: Whisker test methods for electronic and electric terminals" was finally released in June 2007.

This document addresses the need for standardized whisker test methods: it does not define acceptance criteria for tin-based finishes on components, though it does provide guidelines for acceptance but only as informative material (Annex C and D).

Three test methods are proposed as the basis for a full protocol of whisker test conditions, similarly to the JEDEC Standard: an ambient test; a damp-heat test; a temperature cycling test. Test methods were defined on the basis of the tin whisker growth mechanism conceived by the JEITA Whisker subcommittee [115].

Preconditioning treatments are dictated for SMD and other components intended for soldering, namely a simulate reflow or wave-soldering process, respectively. Besides, the influence of all manufacturing processes should be considered, e.g. trim and form of lead-frames, and a suitable preconditioning is required to this purpose prior to testing. Test methods and conditions are summarized in Table 6.2.

The test method shall be selected depending on type of plating, underlayer and substrate material of components, according to a defined decision matrix. For example, see Table 6.2, the duration of the thermal cycling test shall be (P) for Fe–Ni substrate alloy plated with Sn or Sn alloy; (Q) for all other substrate/underlayer and plating combination.

The IEC standard does not specify tests for whisker growth under external mechanical stress[9] and more generally cannot be applied for the assessment of whiskering on other components of electronic and electrical equipment unless it is ascertained that the material system and the whisker growth mechanism are comparable to those addressed and assumed for the tests specified in the standard.

As already noted earlier, this standard is not a specification standard, and guidelines for acceptance are included only for informative purpose. Acceptance should be based on maximum whisker length observed on test samples, based on the assumption that the whisker risk is only related to short circuit, i.e. to the proximity of conductive elements and surfaces in electric or electronic circuitry. In fact, since there is no evidence of any specific risk related to whisker density, in addition, nor evidence of any correlation between whisker density and whisker length, there is no reason to introduce an acceptance criterion based on whisker density.

As default criterion, it is suggested that, unless otherwise specified in the relevant specification, the maximum whisker length at the end of the test period should not exceed 50 μm for any of the prescribed tests; however, according to the application, i.e., for higher packing density, higher value may also be specified.

On 21 October 2005, the JEDEC Standard JESD22A121 was submitted to TC47 Semiconductor Devices as PAS, Publicly Available Specification, and approved on January 27, 2006. The first edition of the IEC PAS 62483 was issued in September 2006. The agreement is apparently that as the test method standard was done by TC91, the acceptance standard will be done by TC47. In this frame, the approval of A121 as PAS was intended as a temporary solution waiting for the TC91 test method standard.

6.7.3 JEDEC Acceptance Standard

The main issue to face with tin whisker was the lack of a clear understanding of the mechanism. This primarily meant that it was not possible to rely on accelerated growth results obtained under test conditions (such as those of JESD22A121) for determining acceleration factors and, consequently, there was no room for the development of a qualification specification. However, under the pressure of the lead-free changeover, the need to develop a strategy to minimize the reliability risk brought about by the introduction of tin-based finishes was mandatory.

In this context, the E4 group established minimum requirements for product qualification to meet customer's specifications.[10]

---

[9] At the time of this writing a standard proposal has been submitted to IEC for a whisker test method designed for connectors of electronic equipment, based on the application of an external mechanical stress.

[10] E4 presentation: Lead-free packaging for semiconductor devices, 24 March 2005; available at http://www.st.com/stonline/leadfree/e4fsl.pdf. Accessed 13 October 2009.

The iNEMI Tin Whisker User Group was formed to work on the assessment of mitigation practices and definition of reliability criteria. During the development of the acceptance standard, this working group released two specification documents: "Recommendations on Lead-Free Finishes for Components Used in High-Reliability Products", first released in May 2005, and updated in December 2006 [116]; and "Tin Whisker Acceptance Test Requirements" (July 28, 2004).

The scope of these documents was, and still is in their more recent versions, to provide the industry with recommendations for components used in high-reliability products and to list alternatives and mitigation practices for eradicating or lowering the tin whisker risk on finished components, connectors and PCBs.

The above-mentioned document, in its original version, also served as the basis for the draft of the JEDEC/IPC joint publication JP-002, "Current Tin Whisker Theory and Mitigation Practices Guideline" (March 2006), which is a valuable source of background information on Tin Whiskers [78].

The draft document titled "Tin Whisker Acceptance Test Requirements" was submitted to JEDEC for formal standard creation and formed the basis of the JEDEC Standard JESD201 "Environmental Acceptance Requirements for Tin Whisker Susceptibility of Tin and Tin Alloy Surface Finish", released March 2006 [114]. An explanatory introduction to the JEDEC Acceptance Standard was later given in a short article authored by Joe Smetana and Rod Gedney [113].

The specification indicates the test methods for measuring tin whisker growth, the acceptance procedures for tin-based surface finishes, the acceptance criteria and an in-progress tin whisker evaluation programme. In particular, test duration and acceptance criteria are defined according to four different product classes, which should be agreed to between supplier and user. In addition, different test duration is prescribed for technology and process change acceptance.

Compared to the original user group recommendations, a few revisions were introduced during industry review of the standard proposal. In particular, as already mentioned previously, the high temperature/humidity storage test conditions were changed from 60 ± 5°C/87 +3/−2% RH to 55 ± 3°C/85 ± 3% RH in order to minimize the risk of condensation in the test chamber. This demonstrates that there was a general recognition of the necessity to preserve test samples from corrosion.

Recently, the iNEMI Tin Whisker Accelerated Test group completed its Phase 5 Evaluation, specifically addressing the influence of temperature and humidity in a wide range of conditions [117]. The final objective was the definition of phenomenological acceleration factors for whisker growth to provide ground for both the optimum choice of test conditions and the assessment of failure rate in field service. Additional points of debate were the storage tests duration and the sample size. Test duration was initially open since growth saturation was required for the test being completed. This condition was removed fixing at 4,000-h test duration for both storage conditions. As to sample size, a minimum of 18 samples per test type is prescribed, unless the number of terminations is less than 16, since a minimum number of inspected terminations is also prescribed. Moreover, additional samples shall be included for the high-temperature/humidity test to allow for terminations or components being skipped because of corrosion.

The outcome of Phase 5 evaluation was that both whisker growth and initiation of corrosion can be represented as a function of temperature and humidity. The development of such a phenomenological model is believed to allow the assessment of whisker growth in other temperature/humidity conditions based on the results of iNEMI tests. This extensive study also lead to the conclusion that a model for the incubation time of whisker growth, either in corroded or not corroded area, can be established, though with caution. A further significant result of the Phase 5 evaluation was also the remark that two distinct temperature/humidity test conditions are not necessary for the purpose of process qualification and acceptance requirements; in this respect, according to the iNEMI Tin Whisker Accelerated Test Project the optimal high-temperature/high-humidity test conditions for Sn-plated copper substrates are 60°C/87% RH.

6.7.4 GEIA Standard

The GEIA (now TechAmerica) standard GEIA-STD-0005-2, "Standard for Mitigating the Effects of Tin Whiskers in Aerospace and High Performance Electronic Systems", released June 2006, describes processes for mitigating the effects of tin finish in electronic systems used in military, aerospace and other high-reliability applications; it does not provide a formal risk assessment of whisker formation.[11] Several *Control Levels* are defined according to the type and extent of tin finish use restriction, as shortly outlined in Table 6.3, and guidance on selecting the appropriate level and product examples is provided in the Standard. Requirements are differentiated and detailed for each *Control Level* with reference to product and material documentation, monitoring plan and procedure, mitigation measures and risk assessment. Suppliers must meet the relevant requirements for the applicable *Control Level*.

The final normative section of the document is devoted to implementation techniques. It specifies control procedures over the introduction of tin finishes into production, methods for monitoring and testing for tin whiskers and mitigation strategies.

6.8 Closing Remarks

Tin-based solder finishes have been safely implemented in the consumer segment of the electronic industry. However, the tin whisker problem remains a potential

[11] For an outline of the scope and content of this document see: http://www.empf.org/empfasis/2006/mar06/geia-tin_whiskers.html. Accessed 13 October 2009.

Table 6.3 Control Levels defined in GEIA-STD-0005-2, "Standard for Mitigating the Effects of Tin Whiskers in Aerospace and High Performance Electronic Systems" for guidance in the use of Sn-based solder finishes in high-reliability products

Control level	Recommendations
Level 1	No restrictions on the use of tin finish
Level 2	Use of tin finish allowed under some circumstances
Level 2A	Use of tin finish without explicit control acceptable under most circumstances, but whisker risk assessment and mitigation strategies shall be documented. Tin finish may be prohibited in specific circumstances, according to contract agreements
Level 2B	Tin finish may be used but only under specified control measures and with customer's approval. Tin finish may be prohibited in some specific circumstances according to contract agreements. Application-specific tin whisker risk assessment will generally be applicable
Level 2C	Restricted use of tin finish. Tin finish is prohibited unless an exception with customer approval is made. Specific instructions and control measures shall be provided and reviewed on a case-by-case basis
Level 3	Use of tin finish is totally banned

risk that cannot be overlooked. Current and future research efforts in this field should address different and so far neglected aspects, both at the fundamental and applied level.

Fundamental understanding of whisker growth is essential to the development of reliable mitigation practices and prevention strategies, enabling larger use of tin-based finishes; it is also necessary to support and validate work on tin whiskers acceleration factors and risk assessment.

Chemical modification appears as the key factor in whisker formation, either if it is the case of intermetallic formation at interface, bulk phase transformation, surface or inter-granular oxide formation. According to the current interpretation, all these factors are ultimately responsible for inducing a compressive stress (gradient) into the tin or tin alloy layer; this compressive stress (gradient) is almost unanimously considered as the driving force for whisker growth. However, the current theory of whisker formation is not complete. In particular, for the sake of fundamental knowledge as well as for potential practical benefit, basic research should now concentrate on incubation and nucleation mechanisms.

From the industrial perspective, the need of long-term performance results and field studies for current accepted mitigation strategies is an obvious priority. Moreover, a unified set of test methods and acceptance criteria should be adopted globally. Eventually, special attention should be given to risk assessment. In this respect, material properties, such as internal stress in tin deposits, strength and resistance to mechanical shock and vibration of whisker crystals, should be further investigated.

Acknowledgments I am indebted to Pascal Oberndorff for his criticism and remarks to the draft manuscript. I am especially grateful to Jacob Klerk for his considerate encouragement during the short life of the Elfnet Tin Whisker Group.

References

1. Galyon GT (2005) Annotated tin whisker bibliography and anthology. IEEE Trans Electron Pack Manuf 28:94–122
2. Leidecker H and Brusse J (2006) Tin whiskers: a history of documented electrical system failures. Technical presentation to Space Shuttle Program Office. http://nepp.nasa.gov/whisker/reference/tech_papers/2006-Leidecker-Tin-Whisker-Failures.pdf. Accessed 30 Sept 2009
3. Arnold SM (1956) The growth and properties of metal whiskers. Technical proceedings of the 43rd annual convention of the American electroplaters society, pp 26–31
4. Dunn BD (1987) A laboratory study of tin whiskers growth. ESA Sci Tech Rev 223:1–50
5. Hilty RD, Corman N (2007) An electrical characterization of tin whisker shorting. Proceedings of the 2007 materials research society spring meeting 993:E02-02
6. Hada Y, Morikawa O, Togami H (1978) Study of tin whiskers on electromagnetic relay parts. 26th Annual National Relay Conference, 9.1–9.15
7. Courey KJ, Asfour SS, Bayliss JA, Ludwig LL, Zapata MC (2008) Tin whisker electrical short circuit characteristics—part I. IEEE Trans Electron Pack Manuf 31:32–40
8. Courey KJ, Asfour SS, Onar A, Bayliss JA, Ludwig LL, Wright MC (2008) Tin whisker electrical short circuit characteristics—Part II. IEEE Trans Electron Pack Manuf 32:41–48
9. Dunn BD (1987/1988) Mechanical and electrical characteristics of tin whiskers with special reference to spacecraft systems. ESA J 11–12:1–17
10. Okada S, Higuchi S, Ando T (2003) Field reliability estimation of tin whiskers generated by thermal cycling stress. Capacitor and resistor technology symposium (CARTS) Europe
11. Evans R (2006) Analysis of the effect of tin whiskers at high frequencies. IEEE Trans Electron Pack Manuf 29:274–279
12. Galyon GT, Gedney R (2004) Avoiding tin whisker reliability problems. Circuits Assembly 15(8):26–31
13. Fang T, Mathew S, Osterman M, Pecht M (2007) Assessment of risk resulting from unattached tin whisker bridging. Circuit World 33:5–8
14. Choi WJ, Galyon G, Tu KN, Lee TY (2004) The structure and kinetics of tin whisker formation and growth on high tin content finishes. In: Puttlitz KJ, Stalter KA (eds) Handbook of lead-free solder technology for microelectronic applications. Marcel Dekker, New York
15. Hilty RD (2009) Tin whiskers in electronic components. Tyco electronics. http://www.tyco-electronics.com/customersupport/rohssupportcenter/pdf/tyco_lf_customer_packet_whiskers.pdf. Accessed 10 October 2009
16. Nakadaira Y, Jeong S, Shim J, Seo J, Min S, Cho T, Kang S, Oh S (2008) Growth of tin whiskers for lead-free plated leadframe packages in high humid environments and during thermal cycling. Microelectron Reliab 48:83–104
17. Vicenzo A, Cavallotti PL, Crema P (2002) Electrochemical control of whisker growth on electrodeposited tin coatings. Trans IMF 80(3):79–84
18. Jiang B, Xian A-P (2007) Observations of ribbon-like whiskers on tin finish surface. J Mater Sci Mater Electron 18:513–518
19. Rinne CL, Hren JJ, Fedkiw PS (2002) Electrodeposition of tin needle-like structures. J Electrochem Soc 149:C150–C158
20. Nabarro FRN, Jackson PJ (1958) Growth of crystal whiskers. In: Doremus RH, Roberts BW, Turnbull D (eds) Growth and perfection of crystals. Wiley, New York
21. Hasiguti RR (1953) A tentative explanation of the accelerated growth of tin whiskers. Acta Metall 3:200–201
22. Franks J (1958) Growth of whiskers in the solid phase. Acta Metall 6:103–109
23. Sobiech M, Welzel U, Mittemeijer EJ, Hugel W, Seekamp A (2008) Driving force for Sn whisker growth in the system Cu–Sn. Appl Phys Lett 93:011906/1–011906/3

24. Sobiech M, Wohlschlögel M, Welzel U, Mittemeijer EJ, Hügel W, Seekamp A, Liu W, Ice GE (2009) Local, submicron strain gradients as the cause of Sn whisker growth. Appl Phys Lett 94:221901/1–221901/3
25. Choi WJ, Lee TY, Tu KN, Tamura N, Celestre RS, MacDowell AA, Bong YY, Nguyen L (2003) Tin whiskers studied by synchrotron radiation scanning X-ray micro-diffraction. Acta Mater 51:6253–6261
26. Lal S, Moyer TD (2005) Role of intrinsic stresses in the phenomena of tin whiskers in electrical connectors. IEEE Trans Electron Pack Manuf 28:63–74
27. Laurila T, Vuorinen V, Kivilahti JK (2005) Interfacial reactions between lead-free solders and common base materials. Mater Sci Eng R49:1–60
28. Tu KN, Thompson RD (1982) Kinetics of interfacial reaction in bimetallic Cu–Sn thin films. Acta Metall 30:947–952
29. Bhedwar H, Ray K, Kulkarni S, Balasubramanian V (1972) Kirkendall effect studies in copper–tin diffusion couples. Scr Metall 6:919–922
30. Onishi M, Fujibuchi H (1975) Reaction-diffusion in the Cu–Sn system. Trans Jpn Inst Met 16:539
31. Oh M (1994) Growth kinetics of intermetallic phases in the Cu–Sn binary and the Cu–Ni–Sn ternary systems at low temperatures. Doctoral dissertation, Lehigh University
32. Paul A (2004) The Kirkendall effect in solid state diffusion. PhD thesis, Eindhoven University of Technology
33. Ho CE, Yang SC, Kao CR (2007) Interfacial reaction issues for lead-free electronic solders. J Mater Sci Mater Electron 18:155–174
34. Vo N, Kwoka M, Bush P (2005) Tin whisker test standardization. IEEE Trans Electron Pack Manuf 28:3–9
35. Reynolds HL, Lee CJ, Smetana J (2007) Tin whiskers: mitigation strategies and testing. In: Bradley E, Handwerker CA, Bath J, Parker RD, Gedney RW (eds) Lead-free electronics. Wiley, Hoboken, NJ
36. The Facts About Tin Whiskers (2009) EIA Engineering Bulletin, August 2004. http://www.ec-central.org/pdf/Engineering/bulletins/EIA-ECA-CB19_13AUG04.pdf. Accessed 30 September 2009
37. Schroeder V, Bush P, Williams M, Vo NN, Reynolds HL (2006) Tin whisker test methods development. IEEE Trans Electron Pack Manuf 29:231–238
38. Dittes M, Oberndorff P, Crema P, Schroeder V (2003) The effect of temperature cycling on tin whisker formation. Proceedings of the electronics packaging technology conference, pp 183–188
39. Brusse J (2002) Tin whisker observations on pure tin-plated ceramic chip capacitors. Proceedings of the AESF SUR/FIN, pp 45–61
40. Compton KG, Mendizza A, Arnold SM (1951) Filamentary growths on metal surfaces whiskers. Corrosion 7:327–334
41. Kehrer H-P, Kadereit HG (1970) Tracer experiments on the growth of tin whiskers. J Appl Phys 16:411–412
42. Chang CY, Vook RW (1993) The effect of surface aluminium oxide films on thermally induced hillock formation. Thin Solid Films 228:205–209
43. Kawanaka R, Fujiwara K, Nango S, Hasegawa T (1983) Influence of impurities on the growth of tin whiskers. Jpn J Appl Phys 22:917–922
44. Keming C, Wilcox GD (2005) Observations of the spontaneous growth of tin whiskers on tin-manganese alloy electrodeposits. Phys Rev Lett 94:066104/1–066104/4
45. Chuang T-H (2006) Rapid whisker growth on the surface of Sn–3Ag–0.5Cu–1.0Ce solder joints. Scr Mater 55:983–986
46. Kim K-S, Matsuura T, Suganuma K (2006) Effects of Bi and Pb on oxidation in humidity for low-temperature lead-free solder systems. J Electron Mater 35:41–47
47. Hoffman EN, Barsoum MW, Wang W, Doherty RD, Zavaliangos A (2005) On the spontaneous growth of soft metallic whiskers. Proceedings of the 51st IEEE holm conference on electrical contacts, pp 121–126

48. Oberndorff P, Dittes M, Crema P, Chopin S (2005) Whisker formation on matte Sn influencing of high humidity. Proceedings of the electronic components and technology conference, pp 429–433
49. Oberndorff P, Dittes M, Crema P, Su P, Yu E (2006) Humidity effects on Sn whisker formation. IEEE Trans Electron Pack Manuf 29:239–245
50. Osenbach JW, De Lucca JM, Potteiger BD, Amin A, Shook RL, Baiocchi FA (2007) Sn corrosion and its influence on whisker growth. IEEE Trans Electron Pack Manuf 30:23–35
51. Dittes M, Oberndorff P, Crema P (2004) PROTIN final report
52. Osenbach JW, Shook RL, Vaccaro BT, Potteiger BD, Amin AN, Hooghan KN, Suratkar P, Ruengsinsub P (2005) Sn whiskers: material, design, processing, and post-plate reflow effects and development of an overall phenomenological theory. IEEE Trans Electron Pack Manuf 28:36–62
53. Handwerker C, Pedigo A, Blendell J (2007) Whisker to hillock transition: stress relaxation mechanisms in Sn, Sn–Cu, and Sn–Pb films. Tin Whisker workshop, Electronic components and technology conference, Reno
54. Hilty RD, Corman NE, Herrmann H (2005) Electrostatic fields and current-flow impact on whisker growth. IEEE Trans Electron Pack Manuf 28:75–84
55. Fukuda Y, Osterman M, Pecht M (2007) The impact of electrical current, mechanical bending, and thermal annealing on tin whisker growth. Microelectron Eng 47:88–92
56. Romm DW, Abbot DC, Grenney S, Khan M (2003) Whisker evaluation of tin-plated logic component leads. Application report SZZA037A. http://focus.ti.com/lit/an/szza037a/szza037a.pdf. Accessed 31 Aug 2009
57. Liu SH, Chen C, Liu PC, Chou T (2004) Tin whisker growth driven by electrical currents. J Appl Phys 95:7742–7747
58. Oberndorff P, Dittes M, Crema P (2004) Whisker testing: reality or fiction. Proceedings of the IPC/Soldertec 2nd international conference on lead free electronics, Amsterdam
59. Fukuda Y, Fang T, Ganesan S, Osterman M, Pecht M (2006) Tin whiskers in electronics. In: Ganesan S, Petch M (eds) Lead-free electronics. Wiley, Hoboken
60. Lindborg U (1976) A model for the spontaneous growth of zinc, cadmium, and tin whiskers. Acta Metall 24:181–186
61. Lee B-Z, Lee DN (1998) Spontaneous growth mechanism of tin whiskers. Acta Mater 46:3701–3712
62. Ellis WC, Gibbons DF, Treuting RJ (1958) Growth of metal whiskers from the solid. In: Doremus RH, Roberts BW, Turnbull D (eds) Growth and perfection of crystals. Wiley, New York
63. Smetana J (2007) Theory of tin whisker growth: the end game. IEEE Trans Electron Pack Manuf 30:11–22
64. Tu KN (1994) Irreversible process of spontaneous whisker growth in bimetallic Cu–Sn thin-film reactions. Phys Rev B 49:2030–2034
65. Glazunova VK, Kudryavtsev NT (1963) An investigation of the conditions of spontaneous growth of filiform crystals on electrolytic coatings. Zhur Priklad Khim 36:543–550 (translated from Russian)
66. Furuta N, Hamamura K (1969) Growth mechanism of proper tin-whisker. Jpn J Appl Phys 8:1404–1410
67. Vianco PT, Rejent JA (2009) Dynamic recrystallization (DXR) as the mechanism for Sn whisker development. Part I: a model. J Electron Mater 38:1815–1825
68. Vianco PT, Rejent JA (2009) Dynamic recrystallization (DXR) as the mechanism for Sn whisker development. Part II: experimental study. J Electron Mater 38:1826–1837
69. Buchovecky EJ, Du N, Bower AF (2009) A model of Sn whisker growth by coupled plastic flow and grain boundary diffusion. J Appl Phys 94:191904/1–191904/3
70. Franks J (1956) Metal whiskers. Nature 177:984
71. Tu KN, Li JCM (2005) Spontaneous whisker growth on lead-free solder finishes. Mater Sci Eng A 409:131–139

72. Galyon GT, Palmer L (2005) An integrated theory of whisker formation: the physical metallurgy of whisker formation and the role of internal stresses. IEEE Trans Electron Pack Manuf 28:17–30
73. Frye A, Galyon GT, Palmer L (2007) Crystallographic texture and whiskers in electrodeposited tin films. IEEE Trans Electron Pack Manuf 30:2–10
74. Moon K-W, Johnson CE, Williams ME, Kongstein O, Stafford GR, Handwerker CA, Boettinger WJ (2005) Observed correlation of Sn oxide film to Sn whisker growth in Sn–Cu electrodeposit for Pb-free solders. J Electron Mater 34:L31–L33
75. Poker DB, Schubert J, Stritzker B (1987) Ion-induced growth of whiskers on Sn films. MRS symposia proceedings, 74 beam-solid interactions and transient processes, Boston, pp 529–534
76. Osenbach JW, De Lucca JM, Potteiger BD, Amin A, Baiocchi FA (2007) Sn-whiskers: truths and myths. J Mater Sci Mater Electron 18:283–305
77. Kakeshita T, Shimizu K, Kawanaka R, Hasegawa T (1982) Grain size effect of electroplated tin coatings on whisker growth. J Mater Sci 17:2560–2566
78. JEDEC/IPC Joint Publication JP002 (2006) Current tin whiskers theory and mitigation practices guidelines. http://www.jedec.org/DOWNLOAD/search/JP002.pdf. Accessed 1 Aug 2009
79. Sakuyama S, Kutami M (2005) Substitute materials for complete elimination of hazardous substances. Study of whisker growth on lead-free plating. Fujitsu Sci Tech J 41(2):217–224
80. Boguslavsky I, Bush P (2003) Recrystallization principles applied to whisker growth in tin. IPC SMEMA Council APEX: S12-4, pp 1–10
81. Galyon GT, Xu C, Lal S, Notohardjono B, Palmer L (2005) The integrated theory of whisker formation—a stress analysis. Proceedings of the electronic components and technology conference, pp 421–428
82. Tu KN (1973) Interdiffusion and reaction in bimetallic Cu–Sn thin films. Acta Metall 21:347–354
83. Porter DA, Easterling KE (1992) Phase transformations in metals and alloys. CRC, New York
84. Dittes M, Oberndorff P, Petit L (2003) Tin whisker formation: results, test methods and countermeasures. Proceedings of the 53 IEEE electronic components and technology conference, pp 822–826
85. Tu K-N, Suh J-o, Wu A T-C, Tamura N, Tung C-H (2005) Mechanism and prevention of spontaneous tin whisker growth. Mater Trans 46:2300–2308
86. Moon K-W, Williams ME, Johnson CE, Stafford GR, Handwerker CA, Boettinger WJ (2001) The formation of whiskers on electroplated tin containing copper. In: Hanada S, Zhong Z, Nam SW, Wright RN (eds) Proceedings of the fourth pacific Rim international conference on advanced materials and processing. The Japanese Institute of Metals, pp 1115–1118
87. Williams ME, Moon K-W, Boettinger WJ, Josell D, Deal AD (2007) Hillock and whisker growth on Sn and SnCu electrodeposits on a substrate not forming interfacial intermetallic compounds. J Electron Mater 36:214–219
88. Xu C, Zhang Y, Fan C, Abys JA (2005) Driving force for the formation of Sn whiskers: compressive stress—pathways for its generation and remedies for its elimination and minimization. IEEE Trans Electron Pack Manuf 28:31–35
89. Sabbagh NAJ, McQueen HJ (1975) Tin whiskers: causes and remedies. Met Finish 73: 27–31
90. Oberndorff P, Chen CC, Yu E, van de Water J (2004) Whisker mitigation by postbake: test results. Proceedings of the IPC/Soldertec 2nd international conference on lead free electronics, Amsterdam
91. Pinsky DA (2008) The role of dissolved hydrogen and other trace impurities on propensity of tin deposits to grow whiskers. Microelectron Reliab 48:675–681
92. Sobiech M, Welzel U, Schuster R, Mittemeijer EJ, Hügel W, Seekamp A, Müller V (2007) The microstructure and state of stress of Sn thin films after post-plating annealing: an

explanation for the suppression of whisker formation? Proceedings of the 57th IEEE electronic components and technology conference, pp 192–197
93. Wei CC, Liu PC, Chen C, Lee JCB, Wang IP (2007) Relieving Sn whisker growth driven by oxidation on Cu leadframe by annealing and reflowing treatments. J Appl Phys 102:043521/1–043521/7
94. Zhang W, Egli A, Schwager F, Brown N (2005) Investigation of Sn–Cu intermetallic compounds by AFM: new aspects of the role of intermetallic compounds in whisker formation. IEEE Trans Electron Pack Manuf 28:85–93
95. Takeuchi M, Kamiyama K, Suganuma K (2006) Suppression of tin whisker formation on fine pitch connectors by surface roughening. J Electron Mater 35:1918–1925
96. Rodekohr CL, Flowers GT, Suhling JC, Bozack MJ (2008) Influence of substrate surface roughness on tin whisker growth. Proceedings of the 54th IEEE holm conference on electrical contacts, pp 245–248
97. Gorbunova KM, Glazunova VK (1984) Present state of the problem of spontaneous whisker crystals on electrolytic coatings. Prot Met 20:269–282
98. Britton SC (1974) Spontaneous growth of whiskers on tin coatings: 20 years of observation. Trans IMF 52:95–102
99. Kadesch JS, Brusse J (2001) The continuing dangers of tin whiskers and attempts to control them with conformal coat. NASA's EEE Links Newsletter, July 2001. http://nepp.nasa.gov/whisker/reference/tech_papers/kadesch2001-article-dangers-of-tin-whiskers-and-conformal-coat-study.pdf. Accessed 30 Sept 2009
100. Kadesch JS, Leidecker H (2000) Effects of conformal coat on tin whisker growth. Proceedings of the 37th IMAPS Nordic annual conference, pp 108–116
101. Woodrow TA, Ledbury EA (2005) Evaluation of conformal coatings as a tin whisker mitigation strategy. IPC/JEDEC 8th international conference on lead-free electronic components and assemblies, San Jose
102. GEIA Engineering Bulletin GEB-0002 (2003) Reducing the risk of tin whisker induced failures in electronic equipment
103. Novak J (1994) Physical analysis of dendrites within conformal coat. Proceedings of the IEEE aerospace applied conference, pp 381–390
104. Arnold SM (1959) The growth of metal whiskers on electrical components. Proceedings of the IEEE electronic components conference, pp 75–82
105. ASM Handbook (1992) Alloy phase diagrams, vol 3. ASM, Cleveland
106. Cavallotti PL, Nobili L, Vicenzo A (2005) Phase structure of electrodeposited alloys. Electrochim Acta 50:4557–4565
107. Raub B, Blum W (1955) Electrodeposition of Pb–Sn alloys. Metalloberflache 9A:54–57 (in German). English translation by Duffy G (1961) NIST report AD0843305
108. Raub E, Elser F (1978) Die inhibierte Elektrokristallisation von Zinn-bleilegierungen. J Less Common Met 62:431–445
109. Jordan M (1995) The electrodeposition of tin and its alloys. Eugen G. Leuze Publishers, Bad Saulgau, pp 219–226
110. Zhang W, Schwager F (2006) Effects of lead on tin whisker elimination. J Electrochem Soc 153:C337–C343
111. Boettinger WJ, Johnson CE, Bendersky LA, Moon K-W, Williams ME, Stafford GR (2005) Whisker and hillock formation on Sn, Sn–Cu and Sn–Pb electrodeposits. Acta Mater 53:5033–5050
112. JEDEC Standard JESD22-A121A (2008) Test method for measuring whisker growth on tin and tin alloy surface finishes. http://www.jedec.org/download/search/22a121A.pdf. Accessed 30 Sept 2009
113. Smetana J, Gedney R (2005) Tin whisker management guidelines. Printed Circuit Design Manuf 22(12):14–17
114. JEDEC Standard JESD201A (2008) Environmental acceptance requirements for tin whisker susceptibility of tin and tin alloy surface finish. http://www.jedec.org/download/search/JESD201A.pdf. Accessed 30 Sept 2009

115. Sakamoto I (2005) Whisker test methods of JEITA whisker growth mechanism for test methods. IEEE Trans Electron Pack Manuf 28:10–16
116. iNEMI user group recommendations on lead-free finishes for components used in high-reliability products, Version 4, 1 Dec, 2006. http://thor.inemi.org/webdownload/projects/ese/tin_whiskers/Pb-Free_Finishes_v4.pdf. Accessed 30 Sept 2009
117. Reynolds HL (2007) Accelerated tin whisker test committee update: Phase 5 evaluation. Tin whisker workshop, Electronic components and technology conference, Reno

Chapter 7
Electromigration in Solder Interconnects

R. Labie

7.1 Introduction

Due to microelectronics scaling, flip chip dimensions are reducing in size as well and electromigration effects become more important [2–12]. Additionally, a lower IC supply voltage while maintaining ever-increasing on-chip power consumption further increases the current flowing through the solder joint. While in the past electromigration was mainly a concern for failure of the chip metallization [13, 14], it can now be a risk for failure at the first packaging level as well. Particularly, since technology road maps [1] predict user currents of 0.2–0.4 A per connection. In the past years, due to the ban on Pb, extensive reliability studies have been carried out on Pb-free flip chip interconnections. However, major focus was put on the thermal–mechanical reliability of the solder joints [15, 16], the thermal interfacial reactions [17] or the solder—intermetallic (IMC) material properties [18]. Further research on the stability of the under bump metallization (UBM)–solder interface under electromigration is still needed.

Electromigration can be defined as the enhanced, net movement of atoms in the direction of the electron flow. Due to uncompensated metal movement, voiding will be induced at the cathodic side (entrance of the electron flow) and metal pile-up will occur at the anodic side (exit of the electron flow). This phenomenon was first discovered in the late sixties [13] when integrated circuits were introduced. While electromigration effects in Cu or Al chip metallization lines can be considered as relatively straightforward due to their homogeneous microstructure, the electromigration behaviour in multi-phase solder interconnections is much more complex. A solder interconnection involves a stack of different layers and phases where chemical gradients act as driving force for metal diffusion. Interdiffusion

R. Labie (✉)
IMEC, Leuven, Belgium
e-mail: Riet.Labie@imec.be

G. Grossmann and C. Zardini (eds.), *The ELFNET Book on Failure Mechanisms, Testing Methods, and Quality Issues of Lead-Free Solder Interconnects*,
DOI: 10.1007/978-0-85729-236-0_7,

results in intermetallic phase formation, which already starts at relatively low temperatures. On top of this, electrical current can act as an additional driving force for diffusion, which will effect the solder joint phase formations. Depending on the direction of the electron flow, intermetallic formation can be enhanced or slowed down. The impact of current on intermetallic growth has been investigated by different research groups [5, 6]. Apart from this, enhanced diffusion of metal atoms can lead to an increased UBM (under bump metallization) consumption. When the UBM layer is completely consumed (by intermetallic formation or dissolution), the adhesion with the underlying chip layers can no longer be guaranteed and joint failure will be induced. Furthermore, accumulation of voids can lead to an open joint or can facilitate crack growth since voiding deteriorates the mechanical integrity of the joint.

For degradation mechanisms ascribed to atom movement, it can be easily understood that the MTTF (mean time to failure) is inversely proportional to the atomic velocity [2]. This velocity can be described by a Maxwell–Boltzmann equation which can be explained by the fact that atom movement is proportional to the amount of atoms that have sufficient energy to overcome the energy barrier (Q) to jump to a neighbouring site. In case a driving force F is active, this will reduce the energy barrier Q for movement in one direction (along the driving force) and increase the energy barrier in the other direction (opposite to the driving force). The atomic velocity can then be written as Eq. 7.1 and after further mathematical deduction, Eq. 7.3

$$v = v^+ - v^- = v_0 \cdot \left[\exp\left(-\frac{Q - aF}{kT}\right) - \exp\left(-\frac{Q + aF}{kT}\right)\right] \quad (7.1)$$

$$v = v_0 \cdot \exp\left(-\frac{Q}{kT}\right)\left(2 \cdot \sinh\frac{aF}{kT}\right) \quad (7.2)$$

$$v = A \cdot F^n \cdot \exp\left(-\frac{Q}{kT}\right) \quad (7.3)$$

With electromigration, the driving force F is proportional to the current density j. The MTTF for electromigration can then be written as Eq. 7.4, which is known as Black's law or Black's equation:

$$\mathrm{MTTF}^{-1} = A \cdot j^n \cdot \exp\left(-\frac{Q_{\mathrm{EM}}}{kT}\right) \quad (7.4)$$

As mentioned before, the electromigration process occurring in solder joints is not an isolated mechanism and will be combined with chemical and thermal processes, which simultaneously alter the microstructure and can either amplify or compensate the electromigration damage process. The activation energy reflects a mixture of all these mechanisms and is therefore sensitive to the testing conditions range and sample geometries. The sample geometry may have a large effect on a phenomenon known as back stress. The net diffusion flux that occurs due to a certain force can be counteracted by a reverse flux to compensate for the compressive stress build-up at the anodic side. Due to a gradient of vacancy concentration, an atom flux can be generated in the opposite direction, which could cancel out the former net diffusion. A clear overview of all physical, chemical and mechanical driving forces active in a solder joint and their individual contributions to the net atom migration is given by Ohring [2].

7.2 Experimental Guidelines

7.2.1 Statistical Evaluation

For mechanisms of which the MTTF depends exponentially on activation energy (Black's law) and assuming that this activation energy is Gaussian (normally) distributed, the failure mechanism can be considered as lognormal distributed. It is therefore valid to characterize the failure distribution of experimental electromigration data by a lognormal plot. This distribution is in widespread use to predict lifetimes for electromigration testing of on-chip conductor tracks and is more recently also used for describing the electromigration mechanism in flip chip joints.

7.2.2 Testing Conditions

Extensive experiments with variable testing conditions (temperature T and current density j) are needed in order to fit Black's equation and accurately define the electromigration activation energy Q_{EM} and the current density exponent n with high statistical relevance. In different literature references, temperature values range between room temperature and 170°C and current densities vary from 0.4 mA/μm^2 in [3] down to 0.06 μA/μm^2 in [10]. In order to reduce the total test time, testing conditions are often chosen at the higher limit, which means high temperatures and high current densities. However, as with all testing conditions, they should give rise to identical failure mechanisms as observed during service life. An unchanging failure mechanism is needed when one wants to extrapolate test results to operational conditions and perform lifetime predictions. It is found that different failure mechanisms can occur for different testing conditions: Ding et al. [3] have reported varying failure mechanisms involving different locations for void formation below and above 120°C for Cu UBM solder joints stressed with a current density of 0.4 mA/μm^2. For similar testing conditions, contradicting failure mechanisms have been reported for Ni UBM solder joints being UBM dissolution [3] and solder depletion at the solder–IMC interface [12]. A careful analysis of the microstructural evolution of joints subjected to electromigration is therefore needed and one should be cautious when extrapolating failure times for conditions outside the testing range.

The testing procedure for electromigration testing is described by the JEDEC standards (EIA/JESD61 or JESD22-A108-B). The resistance of the electromigration test structure should be monitored by a 4-point measurement, also known as a Kelvin test structure. Such a test structure has separate contact pads for the voltage and current path and allows a very accurate measurement of the resistance value. Ding et al. [3] propose the use of a Wheatstone bridge circuit in order to increase the voltage measurement sensitivity.

7.2.3 Flip Chip Electromigration Test Structures

Electromigration testing can be performed on single bump structures or on daisy chains (which are serpentine connections from top to bottom device/substrate and vice versa). These latter structures are of interest for the statistical evaluation of a specific bump configuration since they take into account a large amount of bumps in only one measurement. By testing more bumps, the experiment becomes more sensitive to early failures as well. The disadvantage of daisy chains is the large total resistance value (combination of bumps and conductor tracks), which makes small resistance variations occurring in a single bump difficult to distinguish. Furthermore, higher resistance values increase the risk for Joule heating. And finally, while the direct link between resistance changes and bump failure location is missing for daisy chains, for single bump measurements, the monitored resistance variations can be ascribed to a local bump microstructural change that facilitates the failure analysis.

A schematic cross section of a single bump Kelvin test structure is shown in Fig. 7.1. Although the electromigration current flows through several bumps (from bottom to top substrate and vice versa), the resistance of only one single bump is monitored through a 4-point measurement. In order to induce failure in this bump, it is advised to divide the reverse current (from top to bottom substrate or vice versa) over several joints to reduce their local current density. The monitored resistance remains a combination of several layers and interfaces, which is schematically shown in Fig. 7.1 (bottom).

A novel single bump test structure (also called modified test structure or cross-linked test structure) [19] is described in more detail as case study in Sect. 7.5. This structure allows separating the electrical measurements of the anodic and cathodic interface in the solder joint. Since electromigration is a typical asymmetrical mechanism creating different effects at the different electrodes such a test structure can easily identify the most critical interface and allows a better understanding of the electrical behaviour.

In order to allow in situ monitoring of the electromigration experiment, the chosen test structure should be packaged and connected to a current source and measurement box. Packaging the sample can however add additional critical

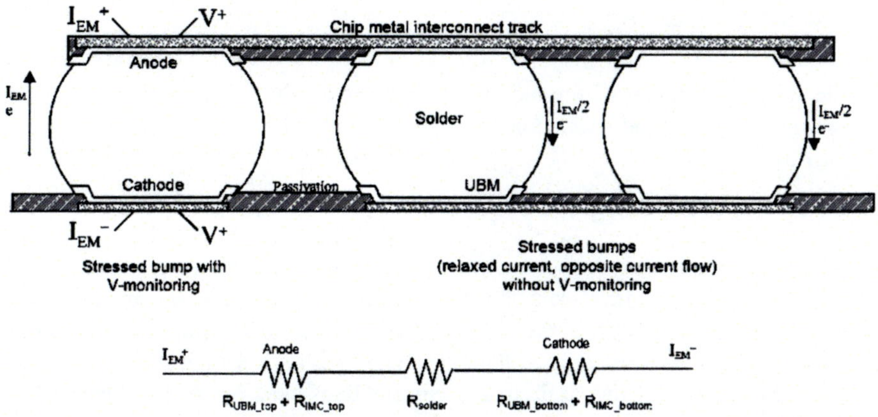

Fig. 7.1 Cross section (*top*) and resistance scheme (*bottom*) of a single bump Kelvin structure (4-point measurement)

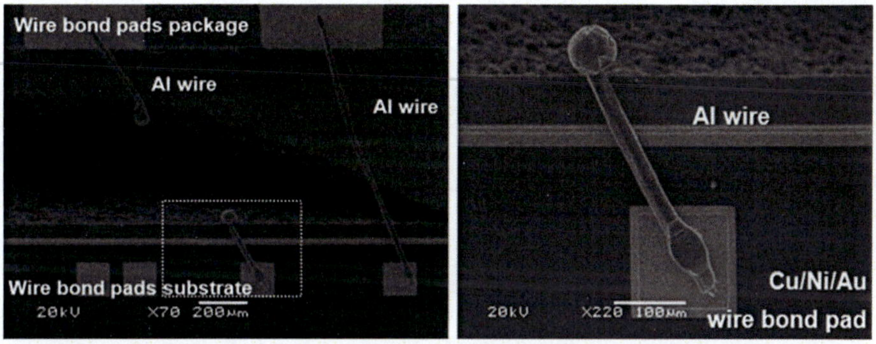

Fig. 7.2 Wire bond failure (25 µm diameter Al) after a few hours of electromigration testing with 0.5 A at 150°C

points since each component (wire bond connections, metal routing lines,...) on itself should withstand the electromigration current. Early failures of other components than the flip chip joint mask the real electromigration failure time of the solder interconnections. Therefore, every conductive component of the current track should be carefully designed and the applied current density should stay well below the threshold value for electromigration. In other words, the MTTF for the inspected flip chip bumps should be much below the MTTF of the other components.

An example of a packaging failure is shown in Fig. 7.2. An open wire bond connection is created by melting and fracture of the current-carrying Al wire suggesting an extremely high local temperature, which can be induced by Joule heating and poor heat transfer.

7.2.4 Joule Heating

Joule heating is the heat production Q that occurs due to an electrical current flowing through a conductor. It depends on the applied current I, resistance R and time t by:

$$Q = R \cdot I^2 \cdot t \tag{7.5}$$

For electromigration experiments, Joule heating is induced by the resistance of the total current path starting from the package current entrance, the metal routing lines, the solder joint itself to the current exit. Since the solder joint dimensions are still relatively large compared to the chip metallization lines, large current values are needed to obtain the desired test current densities. Apart from the risk of inducing electromigration at locations other than the flip chip joint, one should try to minimize the Joule heating as well. The temperature rise due to Joule heating is depending on the environmental surroundings. Depending on the thermal resistance of the package, the resulting temperature rise may be different. This temperature increase should be taken into account when defining the actual testing temperature.

An example of such Joule heating is shown in Fig. 7.3 where sudden resistance jumps can be observed as soon as higher testing currents are applied through a packaged flip chip bump. The according temperature rise can be deduced from these resistance jumps when the TCR (coefficient of thermal resistance) value of the conductor is known. This coefficient indicates the resistance of a material to a temperature increase. By heating up a conductor, the atoms in the lattice start to vibrate which increases the scattering with the conducting electrons and increases the resistance. When the resistance is measured as a function of the temperature, the TCR value can be derived from the slope $\Delta R/\Delta T$ and the initial resistance value according to following definition:

$$\text{TCR} = \frac{\Delta R}{\Delta T} \frac{1}{R_0} \tag{7.6}$$

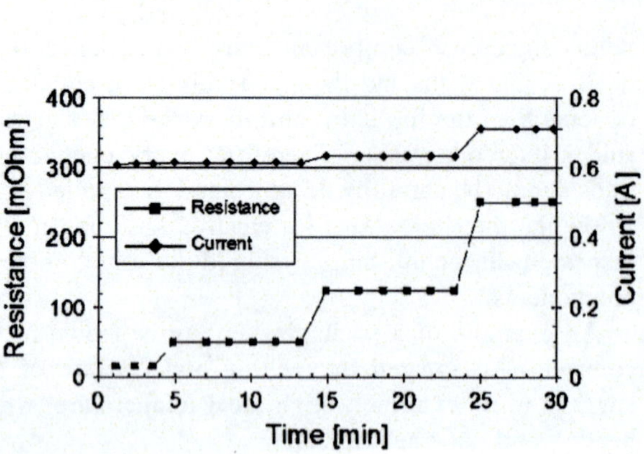

Fig. 7.3 Resistance rise of a conductor (solder flip chip bump) due to Joule heating induced by applying an electrical current

The same formula can be used for calculating the temperature rise for a given resistance jump and a known TCR value.

7.2.5 Failure Criterion

The failure criterion could be arbitrarily chosen: it can be a relative value (a specific percentage of resistance increase is allowed) or it can be an absolute value (resistances above a specific value are considered as failed). It is observed that resistances of single bumps can easily rise a few hundreds per cent before an actual open failure occurs (Fig. 7.4, left). Therefore, the chosen failure criterion can have a large impact on the defined MTTF as can be seen in Fig. 7.4, right.

When structures are tested till open failures, one should take into account the compliance value of the current source of the electromigration testing system. With a testing current value and system compliance value of for instance 0.5 A and 20 V, respectively, the test will continue until a current track resistance of 40 Ω is measured. Allowing such high compliance values may lead to dramatic power generation in the package, which further damages the test structure and complicates post-mortem cross-sectional analysis.

An example of such excessive heat generation prior to or during bump failure can be observed by the cross section of a failed flip chip bump (Fig. 7.5, left). The original bump shape can be derived from the underfill surface. A large amount of Sn has disappeared and indicates solder melting and even retraction of the solder from the bond pad. The combination of an open failure criterion and large system compliance has led to a bump melting side effect, which masks the actual electromigration failure mechanism. This mechanism is revealed when inspecting at intermediate testing times (before open joint), which shows solder depletion and void growth at the intermetallic interface of the cathodic side of the joint (Fig. 7.5, right).

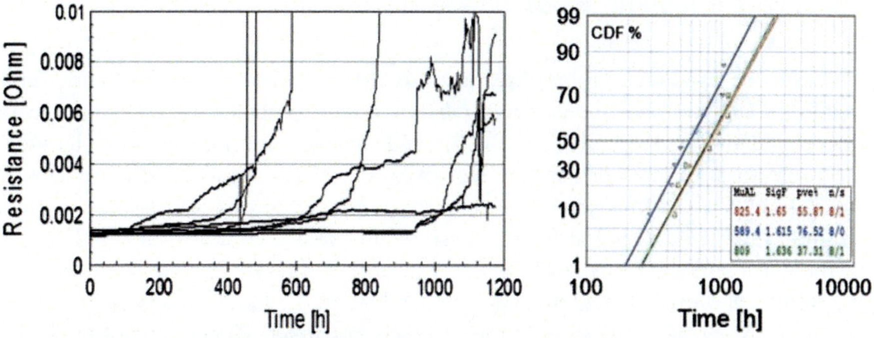

Fig. 7.4 Resistance change during electromigration testing of single bumps (*left*)—lognormal failure distribution of this experiment with variable failure criteria (*right*)

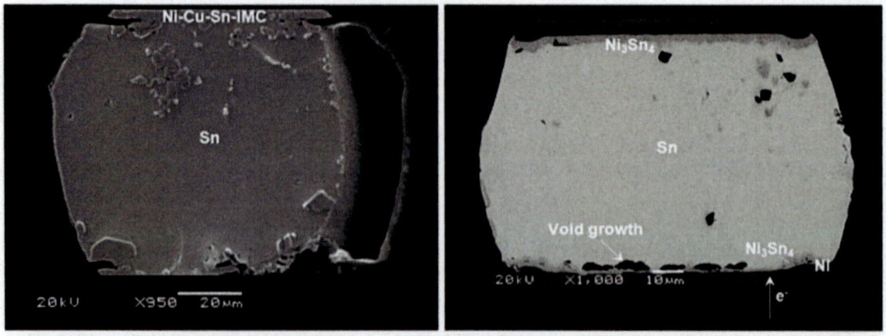

Fig. 7.5 Cross-sectional inspection of 0.5 A current-stressed flip chip bump showing re-melting of bump, excessive Sn migration and total UBM consumption (*left*), cross-sectional inspection before open failure showing void formation at cathodic intermetallic interface (*right*)

7.2.6 Some Experimental Results

The amount of published investigations on electromigration failures in solder flip chip joints is recently increasing. The scaling down of package interconnects has made these structures more vulnerable to this failure mechanism. Unlike scaling, the switch to Pb-free solder alloys seems to have a beneficial effect on the electromigration resistance. Several authors have observed a significant increase (up to 40 times) of the MTTF value when using Pb-free solder rather than eutectic Sn–Pb [7, 8]. Most investigations compare different UBM solder material combinations in terms of MTTF and failure mechanism but not always show conclusive results. Depending on the applied testing conditions, one UBM layer may be beneficial over the other and vice versa [3]. It is therefore recommended to keep in mind all test and sample parameters that may affect the global reliability behaviour.

7.3 Finite Element Modelling Aid

FEM can be used to examine the current density distributions in the solder connection. Only the path for carrying current should be taken into account, while non-conductive parts, such as the Si chip and substrate, do not have to be included for predicting the current density. However, these parts are nevertheless needed when FEM is used for trying to predict the temperature distribution due to Joule heating. A detailed view for the fine mesh on the cross section is shown in Fig. 7.6, right. The mesh density should be fixed in each direction (*X*, *Y* and *Z*) since the magnitude of current crowding is also dependent on the mesh density. In principle, its value always increases when a finer mesh is employed. The total current density I_{total} is a vector, whose value equals to the square root of the sum of the squares of the current components in *X*, *Y* and *Z* direction as I_x, I_y and I_z.

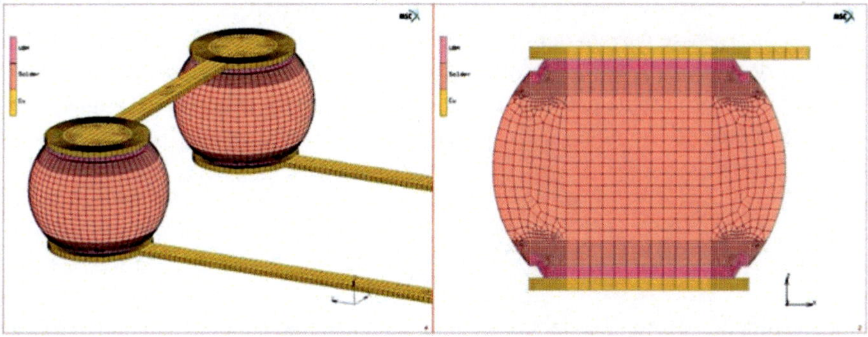

Fig. 7.6 FEM model for current stressing (*left*) including all conductive parts, detailed view of fine mesh in the joints cross section (*right*)

7.3.1 Current Crowding Simulations

A current distribution in one quarter of the model is shown in Fig. 7.7. Current crowding is observed at places where the current goes in and out. For the solder joint with a maximum bump diameter of 120 μm, the average current density should be 0.04 mA/μm^2 (for 0.5 A). This average value is however not visible in (Fig. 7.7, left) when the current crowding in the Cu interconnect track is dominant. A more detailed current density distribution in the solder joints is better visualized in Fig. 7.7 (right) after removing the Cu interconnect track and UBM parts. It reveals that the current has a concentration at the corner of the joint. The current density value varies from 0.25 mA/μm^2 to less than one tenth of this value. The current density gradient also suggests the desirable path for the electron flow: that is from the bottom corner where the current comes in to the upper corner where the current goes out. In fact, a direction change of the current results in current crowding at the turning point. The magnitude of current crowding is influenced by

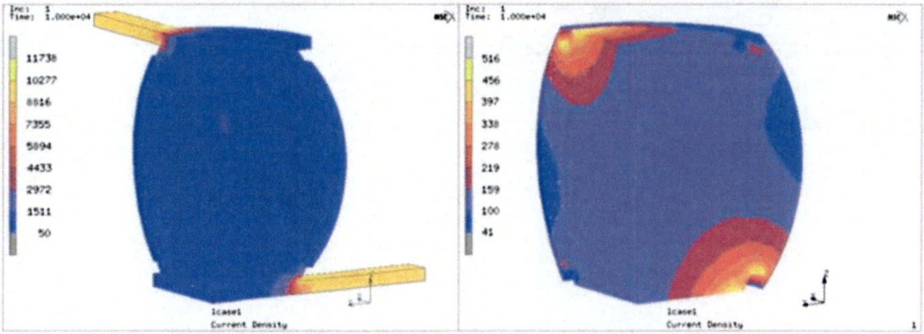

Fig. 7.7 Current crowding occurs at the places where the current goes in and out (*left*), detailed view of the current distribution in one quarter of a joint (UBM and Cu track removed for better visualization)

each component involved, such as their material properties and geometry. While usually an average current density value is chosen to fit Black's law, the simulated current density distributions clearly show large deviations from the mean value. Taking into account the presence and location of current density peaks may help to better understand the damage evolution.

7.4 Case Study: Development of a Novel Test Structure

As mentioned before, Kelvin single bump electromigration test structures monitor the resistance of the complete bump stack. The measured resistance value is the sum of the resistances of the different layers and interfaces in the bump. These are the chip (top substrate) metallization, the chip UBM, the intermetallic compounds at the chip side, the solder, the intermetallic compounds at the package side (bottom substrate), the package UBM and finally the package metallization (Fig. 7.1, bottom). Electromigration causes a typical asymmetrical behaviour depending on the current direction. It is expected to cause voiding at the cathodic interface (current input) and metal pile-up at the anodic interface (current output). By using a standard bump 4-point measurement, it is not possible to discriminate the electrical resistance values of the different interfaces. A novel test structure is therefore developed that allows separate resistance measurements of both the current input and output interface (Fig. 7.8) [10, 19].

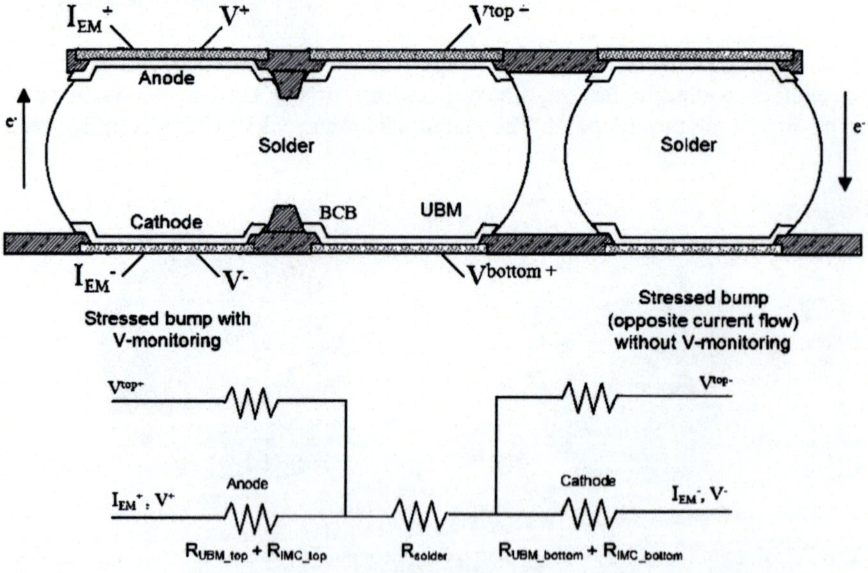

Fig. 7.8 Cross section and resistance scheme of modified bump 4-point measurement

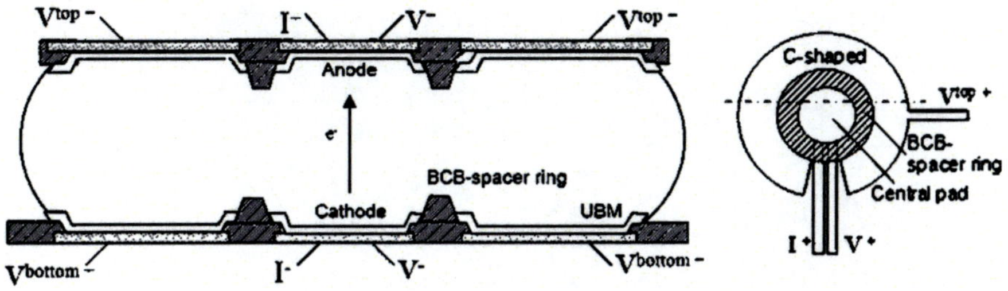

Fig. 7.9 Cross section and *top view* of modified bump 4-point measurement with concentric side-bump design

The novel test structure consists of two 'cross-linked' bumps of which the processing is as follows: two neighbouring bumps are placed in such a way that after solder deposition and reflow, bridging occurs and an electrical connection is formed along the bulk solder (Fig. 7.8, top). This cross-linking of bumps is obviously a not-wanted situation in regular bump processing since this leads to a short circuit. However, in this case, it is deliberately done to be able to execute multiple 4-point measurements over each interface. In order to have a homogeneous current distribution, the second bump or 'side' bump is placed in a C-shape or concentric way around the first one (Fig. 7.9).The electromigration current I_{EM} and the measurement current I are applied through the central bump. The principle of a 4-point measurement is that the resistance value of only the common part of both the (I^+, I^-) and (V^+, V^-) tracks is measured. With the novel test structure, two (V^+, V^-) tracks can be combined with one (I^+, I^-) track which makes multiple 4-point measurements possible. As mentioned before, the electromigration current is applied through the central bump. By measuring the voltage over this central bump, a standard flip chip resistance measurement, which gives the resistance value of the whole stack, is still possible. When measuring however the voltage from a side bump to one interface of the central bump (for instance: V^{top+}, V^-), the resistance of that specific interface can be measured separately. Since the current only flows through the central bump, the measured resistance will be defined by the common part only. This is a sum of the solder and the bottom interface resistance, which consists of the under bump metallurgy and the formed intermetallic compound. Obviously, the same method can be applied to the top interface. With this novel test structure, it is now possible to monitor the resistance changes of the cathodic and anodic interfaces of an electromigration bump separately. This way the most critical interface in a bump connection can be defined.

7.4.1 Experimental Set-Up

This novel test structure is validated by performing electromigration experiments on pure Sn solder flip chip bumps. Two different UBM choices are investigated: either 5 μm Cu or 3/2/0.15 μm Cu/Ni/Au. The metallization is always symmetrical

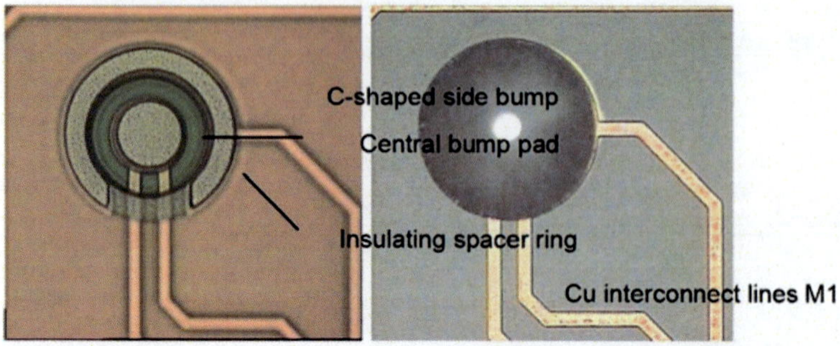

Fig. 7.10 Top view LOM picture of cross-linked bump after UBM deposition (still in resist) (*left*) and after solder plating and reflow (*right*)

which means that it is identical for the top and bottom substrate. The experiments are executed with a constant ambient oven temperature of 150°C, the actual joint temperature is unknown. The electromigration current is set at 120 mA which corresponds to a current density of 0.06 mA/μm^2 with a central bond pad diameter of 50 μm^2. The used test conditions are comparable with general flip chip requirements defined by bump supplier and consumer companies [20] who indicate that bumps should withstand at least 1,000 h with 350 mA (0.05 mA/μm^2) at 125°C ambient temperature and 150°C joint temperature.

The proposed structure is implemented on a (non-active) oxidized Si wafer. Cu interconnect lines of 5-μm thickness are deposited by electroplating in a resist pattern on a Ti/Cu seed layer. After removal of resist and etching the seed layer, a 5-μm BCB (benzo-cyclo-butene) passivation layer is applied on top of the Cu structures. The bump contact pad UBM is deposited, and a BCB ring is applied as insulating ring in order to prevent wetting amongst the pads and avoid current leakage. A top view picture of this stage of the processing is shown in Fig. 7.10 (left). Next, the solder material is deposited and reflowed. Visual inspection and X-ray investigation is used in order to confirm good wetting on the central and the C-shaped side pad. The opposite side of the cross-linked bump is processed in an identical way.

The remaining of the die surface is filled with dummy bumps with a standard circular footprint of 90 μm diameter. Most of the bumps are placed for mechanical reasons but some of them are connected with contact pads and can be used for 4-point measurements. These bumps are considered as thermal reference bumps for measuring the resistance change by thermal stress only.

7.4.2 Electrical Measurements for Cu–Sn–Cu Interconnections

Three test chips are tested until one flip chip connection failed after 560 h at 150°C. After stopping the test, the samples are cross-sectioned for microstructural

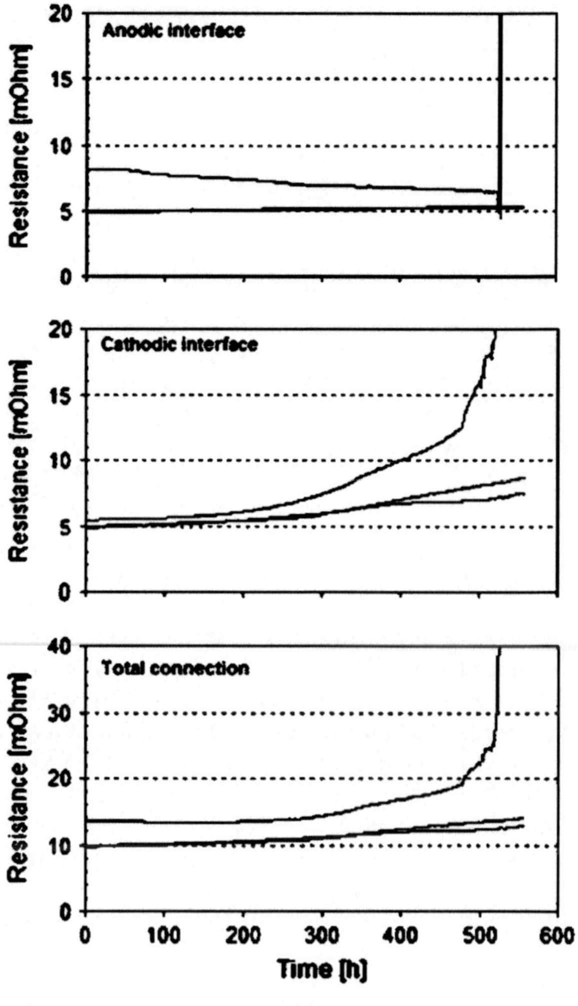

Fig. 7.11 Electrical 4-point measurements of different interfaces of modified Cu–Sn–Cu joints stressed with an electromigration current of 120 mA at 150°C

analysis. The resistance measurements of both the top and bottom interface and the total bump interconnection are shown in Fig. 7.11.

The initial resistance value is around 10 mOhm for the total bump connection and 5 mOhm for each interface separately (cathodic and anodic interface). While the cathodic interface shows a gradual resistance rise until failure, the anodic interface shows hardly any change. When the average time to a 5% resistance increase is used as measure for comparison, a large time difference is obtained when comparing the cathodic interface (5% increase is reached after 100 h) with the anodic interface (reached after 300 h). This difference can be attributed to the microstructural changes detected at anode and cathode and are described in the next paragraph.

7.4.3 MicroStructural Inspection for Cu–Sn–Cu Interconnections

Figure 7.12 (right) shows a cross section of a thermal reference bump. The black spots on the image are defects due to sample preparation. Pieces of Si break-off during grinding and get stuck in the soft solder material. These defects are observed on most of the following cross sections. A symmetrical Cu consumption is seen when comparing the top and bottom Cu UBM metallization. After 560 h at 150°C, the 5 μm Cu UBM is almost completely consumed and transformed into Cu_3Sn and Cu_6Sn_5. The Cu interconnect pads are not affected yet. A little amount of spalled intermetallics is seen in the bulk solder. This amount of intermetallic formation is more than what should be formed by thermal ageing only. Based on the formed intermetallic thickness of the thermal reference bumps and making use of the diffusion parameters for Cu–Sn interdiffusion [21], the total average temperature of the package would have been 155°C. Joule heating in the electromigration test structures caused a temperature rise of the whole package.

A cross section of an electromigration bump is shown in Fig. 7.12 (left). The black spots are again caused by sample preparation. In this case, a clear asymmetrical Cu consumption is detected. A larger magnification of the central bump interfaces is shown in Fig. 7.13 for the cathodic interface (left) and for the anodic interface (right). For the cathodic interface, besides the 5 μm Cu UBM, also the 5 μm Cu interconnect pad is consumed. The anodic interface shows almost 5 μm Cu UBM consumption and no consumption of the Cu interconnect pad. Opposite to these observations, the amount of Cu–Sn intermetallic formation is smaller at the cathode than at the anode. Despite the more extensive Cu consumption, a limited amount of intermetallics, more particularly Cu_6Sn_5, is formed at the cathode. This contradiction can only be explained by Cu migration. The Cu_3Sn intermetallic is not detected at the cathodic interface what points out a restricted amount of available Cu. At the anode side, the behaviour is similar to the thermal

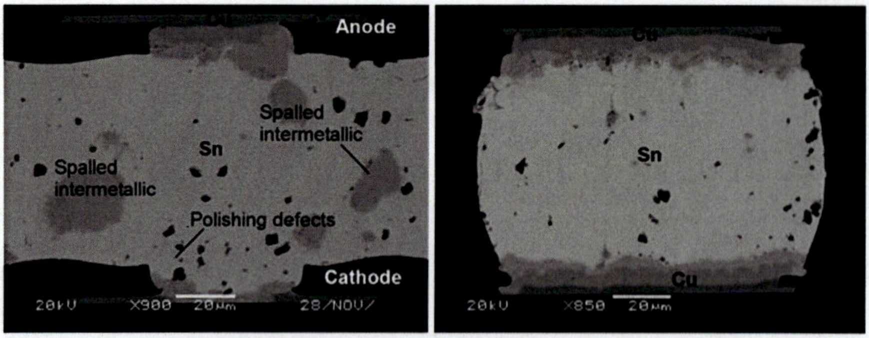

Fig. 7.12 Cross section of central part of cross-linked bump with cathode at the *bottom* and anode at the *top* interface (*left*), thermal reference bump (*right*)

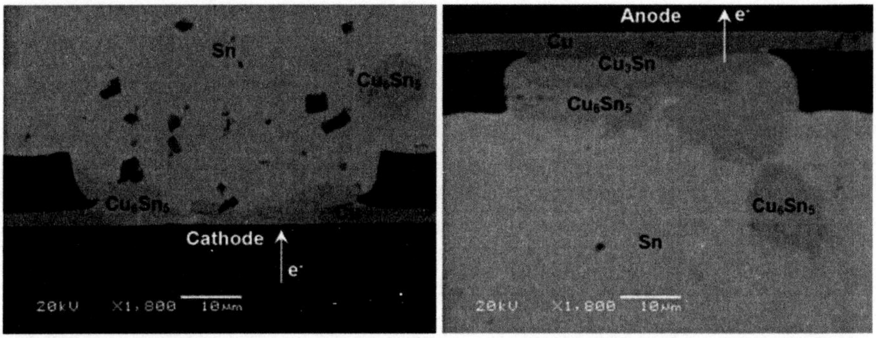

Fig. 7.13 Cathodic interface (*left*) and anodic interface (*right*) of cross-linked bump

reference bump: 5 μm Cu is transformed into 4.5 μm Cu_3Sn and 5.5 μm Cu_6Sn_5. In the bulk solder, large areas of spalled intermetallics can be observed suggesting excessive Cu dissolution leading to precipitation once the solubility limit is reached.

When the electrical measurements and cross section inspections are put side by side, one can see that the growth of an intermetallic layer does not affect the resistance. This is concluded from the resistance monitoring of the anodic interface: although significant growth of the intermetallic layer exists, no change of resistance is seen for at least 2 out of 3 test samples. The third sample shows a gradual resistance decrease which is believed to be caused by annealing effects. For the cathodic interface, a gradual increase in resistance is noticed for all samples. When compared with the microstructural changes, this increase can be related to the consumption of the metal interconnect pad and some void formation in the remaining intermetallic layer.

7.4.4 Electrical Measurements for Cu/Ni/Au–Sn–Cu/Ni/Au Interconnections

A similar behaviour as with the Cu–Sn–Cu sample is seen for a Cu/Ni/Au UBM joint. Electrical measurements show a gradual increase in the resistance value of the cathodic interface, while the anodic interface is characterized by a constant resistance value until sudden failure. In fact, the resistance increase in the anodic interface has no physical meaning since it only occurs after failure of the cathodic interface which ends the measurement. It can therefore not be considered as a failed interface. In case of an identical bond pad metallization at the cathode and at the anode, degradation and failure of the flip chip bump occurs at the cathodic interface.

7.4.5 MicroStructural Inspection for Cu/Ni/Au–Sn–Cu/Ni/Au Interconnections

The differences in electrical behaviour between the cathodic and anodic interface can again be ascribed to microstructural changes. A faster UBM consumption is noticed at the cathodic interface: after 2,032 h, the 3/2/0.15 μm Cu/Ni/Au UBM as well as the 5 μm Cu interconnect track underneath is consumed (Fig. 7.14). At the anodic interface, only part of the UBM is consumed and the formed intermetallic is based on the $(Cu,Ni,Au)_6Sn_5$ stoichiometry. Despite the larger amount of cathodic metal consumption, the formed amount of intermetallics is comparable with the anodic side. This suggests metal movement from cathode to anode. Additional evidence of asymmetrical metal migration is the presence of voids between the intermetallic layer and solder material at the cathodic interface. Also large areas of spalled intermetallics are found in the solder material (Fig. 7.15, left), which is not seen with the thermal reference bumps (Fig. 7.15, right). Due to extensive metal migration between cathode and anode, the solubility limit is reached and

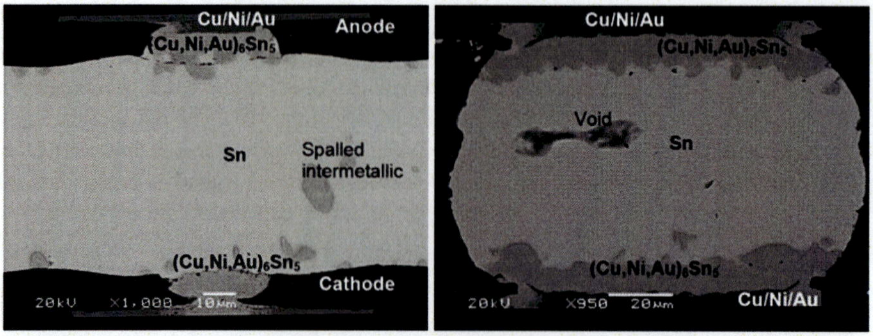

Fig. 7.14 Cross section of central part of Cu/Ni/Au–Sn–Cu/Ni/Au cross-linked bump with cathode at the bottom and anode at the top interface (*left*), thermal reference bump (*right*)

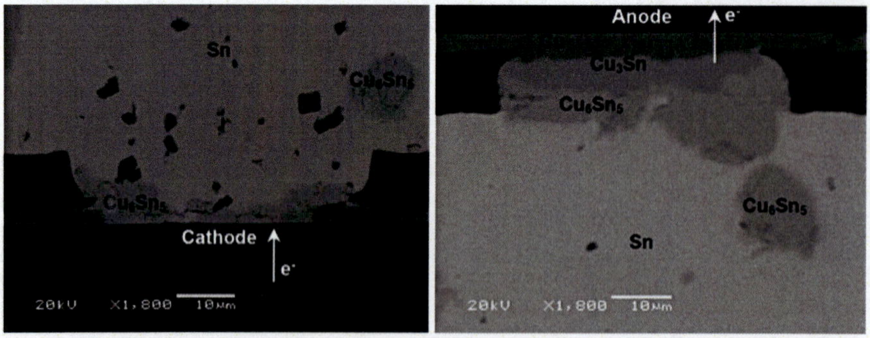

Fig. 7.15 Cu/Ni/Au cathodic interface (*left*) and anodic interface (*right*) of cross-linked bump

intermetallic precipitation occurs. The crack seen at the anodic interface is caused by a polishing defect. A symmetrical UBM consumption is seen when comparing the top and bottom of a thermal reference bump.

Again, growth of the intermetallic phase does not influence the interface resistance which can be concluded from both the resistance monitoring of the anodic interface as from the monitoring of the thermal reference sample. Consuming the metal interconnection pad and the formation of occasional voids in between the solder and the intermetallic interface, results in a gradual resistance increase.

7.5 Conclusions

As general guideline, one can use Black's equation for the lifetime prediction of solder flip chip bumps stressed under electromigration. It has shown to be useful for comparing different material combinations or geometries. Due to the complexity and dynamic nature of the solder microstructure, different mechanisms can be involved during the electromigration process. It is therefore likely to find different activation energy values for largely deviating testing conditions. One should be cautious when extrapolating lifetimes to user conditions and confirm having similar failure mechanisms.

Both single bump structures and daisy chains can be used for performing electromigration experiments on flip chip bumps. In either case, the resistance monitoring during testing should be accurately measured by a Kelvin structure. A novel test structure is proposed and tested [19] which allows the separate measurement of the cathodic and anodic interface. Since electromigration results in an asymmetrical response, the resistance behaviour is expected to be different at the different interfaces. The cathodic interface of the solder joint (entrance of electrons) is shown to be the most critical part of the solder joint, and the measured resistance rise corresponds to cathodic UBM consumption, sometimes combined with void formation.

Performing electromigration experiments on flip chip joints requires careful design of the test structure, the packaging and the test equipment. It is shown that Joule heating can lead to a dramatic temperature increase which not only can lead to a significantly higher testing temperature than originally aimed for, it can also destroy other test sample parts before electromigration has taken place in the solder joint itself (for instance packaging failure) or it can complicate the post-mortem failure analysis by masking the actual failure mechanism (for instance by melting of the solder).

References

1. ITRS Roadmap
2. Ohring M Reliability and failure of electronic circuits

3. Ding M, Wang G, Chao B, Ho PS (2005) A study of electro-migration failure in Pb-free solder joints. In: Proceedings IEEE 43rd annual international reliability physics symposium, pp 518–523
4. Choi WJ, Yeh ECC, Tu KN (2001) Electro-migration in flip chip solder bump on Cu/Ni(V)/Al thin film UBM. In: Proceedings of ECTC, pp 1201–1205
5. Gan H, Tu KN (2002) Effect of electro-migration on intermetallic compound formation in Pb-free solder–Cu interfaces. In: Proceedings ECTC, pp 1206–1212
6. Chae SH, Chao B, Zhang X, Im J, Ho PS (2007) Investigation of intermetallic compound growth enhanced by electromigration in Pb-free solder joints. In: Proceedings ECTC, pp 1442–1449
7. Lee TY, Tu KN, Frear DM (2001) Electromigration of eutectic SnPb and SnAg3.8Cu0.7 flip chip solder bumps and under-bump metallization. J Appl Phys 90:3189–3193
8. Ebersberger B, Bauer R, Alexa L (2004) Qualification of SnAg solder bumps for Pb-free flip chip applications. In: Proceedings ECTC, pp 683–691
9. Lin JK, Wang JW, White J (2003) Characterization of solder joint electro-migration for flip chip technology. In: Proceedings ECTC, pp 816–821
10. Labie R, Webers T, Swinnen B, Beyne E (2005) Electro-migration behavior of Pb-free flip chip bumps. In: Proceedings EMPC
11. Chae SY, Zhang X, Lu KH, Cha HL, Ho PS, Ding M, Su P, Uehling T, Ramanathan LN (2007) Electromigration statistics and damage evolution for Pb-free solder joints with Cu and Ni UBM in plastic flip chip packages. J Mater Electron 18:247–258
12. Kwon YM, Paik KW (2007) Electromigration of Pb-free solder flip chip using electroless Ni-P/Au. In: Proceedings of ECTC, pp 1472–1476
13. Black JR (1967) Current limitations of thin film conductors. In: Presented at 20th annual international reliability physics symposium
14. Loyd JR, Clemens J, Snede R (1999) Copper metallization reliability. Microelectron Reliab 39:1595–1602
15. Dunford S, Primavera A, Meilunas M (2002) Microstructural evolution and damage mechanisms in Pb-free solder joints during extended −40 to 125°C thermal cycles. In: Proceedings IPC
16. Pang JHL, Xiong BS, Low TH (2004) Low cycle fatigue study of Pb-free 99.3Sn–0.7Cu alloy. Int J Fatigue 26
17. Laurila T, Vuorinen V, Kivilahti JK (2005) Interfacial reactions between lead-free solders and common base materials. Mater Sci Eng R 49:1–6
18. Chromik R (2003) Nano-indentation measurements on Cu–Sn and Ag–Sn intermetallics formed in Pb-free solder joints. J Mater Res 9:2251–2263
19. Labie R, Webers T, Beyne E, Mertens R, Van Humbeeck J (2006) A modified electro-migration test structure for flip chip interconnections. IEEE Trans on CPMT 29:508–511
20. Horner V, Balkan H (2001) Reliability manual, report from FlipChip technologies. http://www.flipchip.com/
21. Labie R, Ruythooren W, Van Humbeeck J (2007) Solid state diffusion of Cu–Sn and Ni–Sn diffusion couples with flip chip scale dimensions. Intermetallics 15:396–403

Chapter 8
Impact of Black Pad and Intermetallic Layers on the Risk for Fractures in Solder Joints to Electroless Nickel/Immersion Gold

P.-E. Tegehall

8.1 Introduction

When frequent fractures were observed in solder joints to ball grid array (BGA) components in the late nineties, these were first associated with a plating defect in electroless nickel/immersion gold (ENIG) finishes called "black pad". The inclination for fractures is then caused by a poorly developed intermetallic layer on solder lands affected by the black pad defect. Later, it has been realised that fractures to ENIG finishes may occur even when a proper intermetallic layer is formed. The fractures then occur as fractures in the intermetallic layer or between two intermetallic phases in the intermetallic layer. However, it may often be difficult to determine whether a fracture to an ENIG surface is caused by the black pad defect or by a fracture in the intermetallic layer. Since the two failure mechanisms cannot be completely separated from each other, both will be described. The characteristics of fractures due to the black pad defect and the mechanisms causing black pad defects will first be discussed. The mechanism for fractures in the intermetallic layer will then be discussed.

8.2 Failure Mechanisms

8.2.1 Fractures Due to "Black Pad" Defects

The defect "black pad" on an ENIG finish was first reported in 1990 for flip chips soldered to electroless nickel/immersion gold-coated substrate pads [1]. When the flip

P.-E. Tegehall (✉)
Swerea IVF AB, Box 104, 431 22 Mölndal, Sweden
e-mail: per-erik.tegehall@swerea.se

G. Grossmann and C. Zardini (eds.), *The ELFNET Book on Failure Mechanisms, Testing Methods, and Quality Issues of Lead-Free Solder Interconnects*,
DOI: 10.1007/978-0-85729-236-0_8,

chips were soldered to the substrate pads, dewetting of the nickel surfaces was observed. Since the exposed nickel surfaces had a dark grey to black appearance, the defect was referred to as the "black pad" condition. It was determined that the dark colour was due to Ni_3P present on the nickel surface and it was concluded that it was this phase that caused the dewetting. Therefore, it was initially believed that the black pad defect was caused by the phosphorus content in the electroless nickel and that a remedy would be to decrease the phosphorus content. However, most investigations that later have been performed indicate that a decrease in the phosphorus content rather increases the risk for the black pad defect [2–4]. In fact, Cordes and Huemoeller obtained very poor shear strength for solder joints to surfaces with immersion gold on electrolytic nickel, i.e. a phosphorus-free nickel, but also excellent shear strength of solder joints to surfaces with electrolytic gold on electroless nickel [3]. These results indicate that the enrichment of phosphorus on the nickel surfaces is not the cause of the increased risk for solder joint failures. Instead, the enrichment is a consequence of the mechanism causing the increased risk for failures.

Very large efforts have been made to understand what causes the black pad defect and how to prevent it. Although the frequency of the black pad defect has been reduced due to various actions performed, black pad defects still occurs and there are differing opinions of what causes it and how it should be prevented.

The known characteristics of the black pad defect give important clues to the mechanisms causing the defect but also for identifying when a fracture is due to the black pad defect. The most typical characteristics are as follows:

- poorly developed and/or locally missing intermetallic layer in the solder joints
- corrosion damages of the electroless nickel plating beneath the gold coating
- a layer with enriched phosphorus beneath the gold coating
- a gold coating that is slightly thicker than normal
- a thin layer of nickel oxide on top of the gold coating
- an increased risk for the black pad defect when the phosphorus content is decreased in the electroless nickel
- plating of immersion gold on electrolytic nickel may cause a similar effect (although the fractured surfaces are then not black) but not plating of electrolytic or autocatalytic gold on electroless nickel
- black pad defects are often located to certain solder pads on a specific board design and
- black pad defects occur mainly on boards coated with solder masks

By understanding the chemistry of the plating reactions, all these characteristics and the mechanism causing the black pad defect can be explained.

8.2.1.1 Chemistry of Metal Plating

The plating process of a metal can be divided into two reactions, one cathodic and one anodic reaction. The metal is deposited by the cathodic reaction in which metal ions dissolved in the plating bath are reduced to metal on the surface to be plated. The reaction can be written:

$$Me(1)^{n+} + ne^{-} \rightarrow Me(1) \quad (8.1)$$

The anodic reaction may vary. It may be a reversed cathodic reaction:

$$Me(2) \rightarrow Me(2)^{n+} + ne^{-} \quad (8.2)$$

where Me(2) may be the same metal as Me(1) or it may be another metal. The anodic reaction may, for example, also be oxidation of water:

$$2H_2O \rightarrow O_2 + 4H^{+} + 4e^{-} \quad (8.3)$$

At *electrolytic plating*, the plating is achieved by applying a voltage between the surface to be plated (cathode) and a counter electrode (anode). Since the applied voltage is the driving force for the plating process, the plating rate is mainly determined by its magnitude. The quality of the plating is to a part determined by the applied voltage but primarily by various compounds added to the plating bath. Some added compounds directly affect the plating reactions, whereas others more indirectly affect them. Examples of the latter type are salts that are added to increase the electric conductivity and buffer systems that are added to stabilise the pH. Additives that directly affect the plating reactions are compounds that form strong complexes with the metal ions and/or are co-deposited in the plating. Compounds that form strong complexes with metal ions are often negatively charged inorganic or organic ions, for example halides, cyanide and organic acids, but may also be non-charged organic compounds, for examples amines. Other metal ions may also be added to be co-deposited in the plating. The additives affect the crystal structure of the deposit and are divided into groups depending on the dominating effect. Brighteners give a more mirror-like surface, grain refiners give a more fine-grained structure and stress reducers reduce the stress in the plating.

Plating may also be achieved by non-electrolytic plating, i.e. without an applied voltage. The most common process is *autocatalytic plating* or electroless plating as it is also called. The plating is then achieved by adding a reducing agent to the plating bath. The reaction can be written:

$$Me^{n+} + Red \rightarrow Me + Ox \quad (8.4)$$

where Red is the reducing agent and Ox the oxidised form. In an autocatalytic process, the deposit metal catalyses the reaction and it will continue as long as the metal surface is exposed to the solution. Common reducing agents are hypophosphite, formaldehyde, borohydride, hydrazine and thiourea. Plating processes for electroless (autocatalytic) nickel use almost exclusively hypophosphite as reducing agent. The reaction when hypophosphite is used is very complex but it can be simplified and written as: [5]

$$Ni^{2+} + H_2PO_2^- + H_2O \rightarrow Ni + H_2PO_3^- + 2H^+ \quad (8.5)$$

In addition, the following reaction occurs simultaneously:

$$3H_2PO_2^- \rightarrow H_2PO_3^- + H_2O + 2OH^- + 2P \quad (8.6)$$

As a result of the latter reaction, phosphorus is co-deposited in the nickel coating. The phosphorus content may vary from 1 to 13 wt%.

The plating rate of an autocatalytic process, which is determined by the temperature and composition of the plating bath, is essentially constant as long as the plating conditions do not change.

Another non-electrolytic plating process is *immersion plating*, which is caused by a galvanic displacement reaction. When the substrate metal is less "noble" than the plating metal, the uppermost metal layer of the substrate metal may be substituted with the plating metal. A well-known example is immersion plating of copper on iron in an acidified copper sulphate solution:

$$Fe + Cu^{2+} \rightarrow Fe^{2+} + Cu \quad (8.7)$$

The total reaction is the same as occurs when copper is electroplated with a counter electrode made of iron. The main differences are that during immersion plating, it is the difference in electrochemical potential (galvanic cell) for the two metals that is the driving force for the reaction and that anodic and cathodic reactions occur on the same surface.

Although immersion plating can be achieved with many combinations of substrate and plating metals, the depositions are usually porous and have poor adhesion as the example with immersion plating of copper on iron. There are two reasons for this. First, since the plating is achieved through a displacement reaction, the dissolution of the substrate metal under the plated metal may lead to that the latter lose adhesion. Second, and perhaps most important, the deposition rate is basically determined by the difference in electrochemical potential for the substrate and the plated metal (difference in nobility) and thus is difficult to control. If the difference in electrochemical potential is large, the initial plating rate will be too high to get a well-structured and adherent metal coating. In fact, it is often necessary to perform a strike plating during electrolytic plating in order to prevent that an initial immersion plating ruins the adhesion of the electroplated coating. The strike plating is performed in a bath with low activity of the plated metal (low concentration of the metal and high concentration of complexing agents) to diminish immersion plating. Therefore, good adhesion of immersion platings can usually only be achieved if the difference in electrochemical potential is small.

The electrochemical potential of the substrate and plating metals can to some extent be affected by the addition of complexing agents to the bath and, by this means, the plating rate can be affected. Compounds that form strong complexes with the plated metal will reduce the activity of the plating metal which will decrease the plating rate, whereas compounds that form strong complexes with the substrate metal will enhance the dissolution of it and thereby increase the plating rate.

Nevertheless, it is a difficult balance act to optimise the bath composition. When the displacement reaction starts, the plated metal will be deposited as small islands on the substrate metal. These islands will be cathodic surfaces, whereas the rest of the substrate metal will be anodic surfaces. That is, initially, the anodic surfaces will be much larger than the cathodic surfaces. As more and more of the surface is coated with the plating metal, the cathodic areas increase in size and at the end of the process, they are much larger than the anodic areas. As a consequence, the plating rate per unit area at the cathodic surfaces will decrease with time since the amount of metal reduced at the cathodic areas must balance the amount of metal oxidised at the anodic areas. This also means that the oxidation rate per unit area at anodic surfaces will increase as more of the area is coated with the plated metal and may finally become very high.

Ideally, an immersion plating process will lead to a completely coated substrate metal and, in contrast to autocatalytic plating, the plating process will cease by itself when the coating is completely covering. In reality, it is difficult to achieve a completely coated surface. If the electrochemical potentials are balanced to give a low initial plating rate, the coating may become very thin and it will be difficult to achieve complete coverage. On the other hand, if the initial plating rate is too fast, the coating may become porous. Since the anodic reaction is an etching process and the plated metal will function as an etch resist, a porous coating and an aggressive plating solution may lead to over-etching. The effect will be that the surface of non-coated substrate metal will reach a minimum and then increase again when over-etching exposes new surfaces beneath the plated metal layer. The best solution to avoid this type of over-etching would probably be to perform a sort of strike plating. That is, start the plating in a bath balanced to give slow initial plating in order to get a low-porous layer and then continue in a bath balanced to give faster plating reactions to get a high final coverage.

8.2.1.2 Immersion Gold Plating on Electroless Nickel

Usually, there is a large difference in the electrochemical potentials for gold and nickel. Thus, it is difficult to get good adhesion of immersion gold on nickel. However, electroless nickel offers a possibility to decrease the plating rate of immersion gold since the reactivity of electroless nickel is very much affected by the phosphorus content; the higher the phosphorus content, the lower the reactivity of the nickel coating.

Cordes and Huemoeller have examined how the deposition rate of gold and the adhesion of the gold deposit were affected by the phosphorus content in the electroless nickel [3]. They found a clear correlation between the phosphorus content and the deposition rate; the higher the phosphorus content, the lower the deposition rate (see Table 8.1). It was not described how the deposition rate was determined but obviously it must have been measured as the average value of deposition rate for a defined time period. The initial deposition rates were probably much higher. The adhesion of the gold layers was tested by a tape test. Again, they found a clear

Table 8.1 Effect of nickel composition on the deposition rate and adhesion of the gold deposit [3]

Phosphorus content (%)	Deposition rate (μm/min)	Adhesion—tape test (% peeling)
0 (electrolytic Ni)	0.025	75
3	0.019	5–10
5	0.015	1–2
7	0.010	0
9	0.005	0

correlation; the higher the deposition rate, the poorer the adhesion. Poor adhesion indicates that over-etching beneath the gold layer has occurred. The effect of the phosphorus content on the shear strength of reflowed solder balls was also investigated. Concurrently with the decreased adhesion of the gold layer when the phosphorus content was decreased, the shear strength of reflowed solder balls degraded.

It can be noted that the deposition rate was highest on electrolytic nickel, i.e. a nickel coating containing no phosphorus. The adhesion of the gold layer was also worse to the electrolytic nickel and the shear strength of reflowed solder balls to this coating was very poor. On the other hand, excellent ball shear results were obtained for samples with electrolytic gold on electroless nickel [5]. These results clearly show that it is not the phosphorus in the nickel coating that is the cause of the black pad defect.

On electrolytic nickel, over-etching beneath the gold coating may degrade the solderability in several ways. Exposed nickel surfaces will be oxidised. Residues from the plating bath may be trapped beneath the gold coating. These residues will make it difficult to wet the surface by a solder. They may also contribute to more extensive oxidation of the exposed nickel surfaces. Corrosion products may also migrate through the porous gold coating and deposit on top of the gold surface. McFaddin has reported that, by using low-vacuum scanning electron microscope (LVSEM) and energy-dispersive X-ray spectroscopy (EDS) with a low electron beam voltage, a very thin film of nickel oxide often can be detected on top of the gold layer on surfaces affected by the black pad defect [6]. If this film is thick enough, it may prevent wetting the gold layer by solder. However, it is usually no problem to wet the gold layer on surfaces affected by the black pad defect. Thus, this film is normally too thin for preventing wetting. It is residues and corrosion products beneath the gold layer that hinder good wetting on electrolytic nickel. However, the thin nickel oxide film offers an opportunity to non-destructively detect surfaces affected by the black pad defect.

On electroless nickel, the dissolution of nickel during the immersion plating causes an enrichment of phosphorus at the surface. Chan et al. have reported that phosphorus is enriched in the uppermost 150 Å of the nickel layer when 0.7–0.8 μm immersion gold is deposited on electroless nickel with 8–10 wt% phosphorus [4]. The phosphorus content in the enriched layer was maximum 20 wt%. This is in agreement with the results from other investigations that have shown that the phosphorus content in the enriched layer is usually 2–3.5 times higher than in the electroless nickel [4, 7–10].

The enrichment of phosphorus has a negligible effect on the wettability of the nickel surface in a soldering process. However, when the black pad defect occurs, extensive etching of the nickel may take place leading to a much thicker layer with enriched phosphorus. This increase in thickness of the layer with enriched phosphorus may be the reason for the difficulties to properly wet surfaces with black pad defects. However, entrapped bath residues and the oxidation of exposed nickel surfaces are probably the main reasons. This is supported by the finding of higher oxygen content associated with black pad regions [11].

An autocatalytic nickel coating consists of nickel nodules that have been growing upward from the underlying copper surface. The nodules can be described as closely packed columns. Microsections of solder pads with the black pad defect often show gold spikes that penetrate downward from the surface along the nickel nodule boundaries [7, 9, 10, 12]. The gold spikes are associated with corrosion damages that can be seen as dark spikes that penetrate even deeper along the nickel nodule boundaries than the gold spikes. In worst case, they may penetrate through the nickel coating down to the underlying copper. This type of corrosion damages is probably caused by crevices present between some nickel nodules prior to the coating with immersion gold [13]. The bath solution that fills a crevice will not contain enough gold ions to completely coat the sides of the nickel nodules. When the solution has been depleted with gold ions, the non-coated surfaces will be anodic surfaces through the remainder of the plating time. This type of corrosion damages can probably only be prevented by assuring that there are no crevices between the nickel nodules. That can be achieved by a slow rate of nickel deposition which reduces the sizes of the crevices [13]. It can also be achieved by increasing the phosphorus content in the nickel. According to Lamprecht et al., a smoother surface without crevices between the nickel nodules is achieved if the phosphorus content is increased to 11.2 wt% [14]. The extent of the corrosion damages during immersion gold coating will also be affected by the plating time and by the chemistry of the plating bath.

In other cases, corrosion of the nickel is observed as a more or less continuous black band beneath the gold coating [7, 9]. The black band may be confined to a few nodules or cover a large area of the pad and have a thickness of 1 μm. When the gold layer is stripped away from areas with a thick dark band, the nickel surface has an appearance of cracked mud. If a cross section is made using FIB, many very narrow voids perpendicular to the surface can be observed in the black band [7]. This type of corrosion damages is probably caused by a porous gold deposit, i.e. a too fast initial gold deposition rate. To prevent this type of corrosion damages, the chemistry of the immersion gold bath must be balanced to give the optimum plating rate with respect to the reactivity of the electroless nickel coating. As discussed previously, it is probably easier to balance an immersion gold bath for electroless nickel if the phosphorus content is high. Eslambolchi et al. noted a correlation between phosphorus content and immersion gold thickness where higher phosphorus content led to a thinner gold layer [2]. A thinner gold layer indicates that it is less porous.

Obviously, it is possible to balance an immersion gold process for electroless nickel coatings with different phosphorus content. Thus, it is only possible to specify optimal phosphorus content in the electroless nickel for a defined gold-plating bath. If the chemistry of the immersion gold bath is changed, the optimal phosphorus content in the electroless nickel may also be changed. Furthermore, a changed reactivity of the plating bath or of the electroless nickel with time may put the balance out of order. Whereas it was previously recommended that the phosphorus content was in the range 6–10 wt% [3, 4, 15], it has during the last years been recommended to increase the phosphorus content to 9.5–13 wt% [8, 14]. According to Johal et al., the maximum thickness of the gold layer that can be achieved on high-phosphorus electroless nickel is 0.06 μm [14].

In contrast to the findings by most others, Roepsch et al. have reported that they found an increased risk for the black pad defect with increasing phosphorus content in the electroless nickel [11]. In their investigation, they compared samples from different batches from various printed circuit board manufactures. That is, the phosphorus content was not purposely varied in the samples. The phosphorus content in electroless nickel normally increases with the number of metal turn over (MTO) in the plating bath but also the reactivity of the nickel increases with the number of MTO [5, 16]. Thus, increased phosphorus content in the samples may have been due to that they have been plated in baths with a high MTO, which has caused a higher reactivity of the nickel. This may explain the correlation between increased risk for the black pad defect and a high phosphorus content.

Various compounds are added to the electroless nickel bath. They have a number of functions, for example to stabilise the bath against spontaneous nickel reduction and to control the quality of the plating at edges [5, 13, 15]. The main stabilisers used are sulphur-containing compounds and lead [12]. Although they are added to concentrations of only 0.5–1.0 ppm, they have a significant impact on the plating rate and the properties of the plated nickel coating including the phosphorus content. Even an increase in the concentration of the sulphur compound from 0.3 to 1.0 ppm will cause a decrease in the phosphorus content and, more important, a considerable increase in the reactivity of the nickel. Also an increase in the lead concentration will increase the reactivity of the nickel but the effect is not as large as it is for the sulphur compound. Thus, it is very important to assure a constant concentration of the stabilisers in order to have control of the reactivity of the nickel coating.

According to Crouse and Cullen, solder masks may leach sulphur-containing compounds (photo-initiators) that have the same effect as the stabilisers added to nickel baths [12]. By soaking a solder mask in deionised water and then using this water for making up an electroless nickel bath, they could prove that the solder mask leached substances that caused increased plating rate and increased reactivity of the nickel coating and even caused black pad defects when it was followed by immersion gold plating. Solder masks are often intentionally under-cured when applied to PCBs in order to increase the pliability and reduce the risk that they crack or detach during ENIG processing. Under-curing makes it easier for compounds to leach out. Crouse and Cullen showed that the impact of contaminants leaching from solder masks could be reduced by additional curing and by more effective cleaning of them. Since the black pad defect has been reported to occur mainly on solder mask-coated PCBs [12], leaching of sulphur compounds from the solder mask is probably one of the major causes of black pad defects. The fact that black pad defects tend to appear more frequently and more severely at the corners and edges of pads versus the centre of the pad [11] supports this conclusion.

The black pad defect is often stated to be due to a hyperactive corrosive immersion gold process [4, 7, 8]. A too high temperature of the gold bath or a too high gold concentration can cause a too aggressive attack [13]. It can be noted that when the black pad defect was first observed, the problem was dissolved by increasing the thickness of the gold layer [1]. Thereafter, the ceramic substrate was heat treated at 600–650°C to convert the gold coating into an alloy region with nickel which eliminated the black pad defect. The thicker gold coating was achieved by using a loosely complexed gold bath, i.e. a more aggressive bath.

However, in most cases, it is probably not an increased activity of the gold-plating bath but an increased reactivity of the nickel surface that causes the excessive corrosion. It is difficult to maintain constant phosphorus content in electroless nickel. The phosphorus content is affected not only by additives to the bath but also by temperature, pH, MTO and loading factor [2]. Even if the phosphorus content is kept constant, the reactivity of the nickel may vary significantly as discussed. Hence, it may be necessary to regularly check the reactivity of the nickel to assure that the immersion gold process will not cause a too fast plating rate. This may be done by testing the corrosivity of the nickel coating. One method used is to measure the weight loss after immersion in concentrated nitric acid for 30–60 s [12]. Another method is to measure the time it takes before the nickel surface turns black when it is dipped in nitric acid [16].

To achieve an even gold coating, it is important that the nickel surface is clean and has a minimum of oxides. Since nickel oxidises very fast, it is not possible to have an oxide-free nickel surface but the time between nickel and gold plating should be kept at a minimum. Chan et al. has shown that the number of corrosion defects increases with increasing hold time between electroless nickel and post-rinse steps [4]. Contamination and oxide layers may cause poor adhesion of the gold layer and that it becomes porous. However, it may cause poor quality of the coating also in a more indirect way. Contamination and oxide layers with uneven thickness on the nickel surface may affect the chemical potential of the surface leading to that some areas will be predominantly cathodic areas and other areas will be predominantly anodic areas during the gold-plating process. Areas that become predominantly anodic areas will be more extensively corroded during the plating process and more likely to show black pad defects.

A similar effect may be seen if areas located at different parts of a PCB are electrically connected to each other and especially if they have different sizes. Due to variation in bath agitation and consumption of bath chemicals, or temperature gradients on PCBs having a large thermal mass, a potential difference may arise between the areas that causes one area to become predominantly cathodic and the other predominantly anodic. This is probably the reason why black pad defects do not occur randomly but is often localised to specific pads on a PCB [7]. Surfaces affected by the black pad defect are often connected to larger surfaces not affected by the black pad defect [17] and usually they are found on boards with complex circuitry [12]. The fact that black pad defects appears to occur more frequently on finer pitched parts with smaller pads than on larger pads [18] may also be caused by leakage from the solder mask which could have a larger influence on a small pad.

Since the plating process never ceases by itself when black pad defects occur, the gold coating on areas with corroded nickel is often thicker than normal [2, 7, 9, 19], typically about twice as thick as normal [11]. This increase in gold thickness is not large enough to be used as an indication of black pad defects. By interrupting the gold-plating process before the surface is completely covered with gold, the extent of eventual black pad defects can probably be reduced. This is a method that is used but the gold coating will then likely have less protective properties [5].

8.2.1.3 Effect of Black Pad Defects on the Reliability of Solder Joints

In cases of severe black pad defects, wetting will not occur to the nickel surface during soldering. Initially, the solder wets the gold surface but when the gold has been dissolved in the solder, dewetting will occur on parts of the solder pad or, in worst case, on the whole pad. Rework of the solder joint fails since it is not possible to wet the surface unless the surface is mechanically abraded. However, in most cases with black pad defects, solder wets the nickel surface and the solder joints look completely normal when visually inspected [2]. The defect will then not be discovered until a failure occurs which may happen when the assembly is exposed to stress. A fracture will then take place between the intermetallic layer and the nickel surface. Little to no remaining Ni–Sn intermetallic phases are observed on the nickel surface [2, 7, 8].

When a solder joint with the black pad defect is micro-sectioned, it is often observed that the intermetallic layer is poorly developed and is missing at areas with corrosion damages in the nickel, but this is not always observed. The presence of corrosion damages in the form of a rather thick "black band" at the nickel surface is a strong indication of a black pad defect. Corrosion damages of the nickel in the form of spikes between nickel nodules are also a clear indication of the black pad defect but according to Chan et al., there is no evidence that such spikes have any effect on solder joint strength [4]. Corrosion damages in the form of a band at the surface are much more detrimental than corrosion spikes since it will result in poor wetting of a larger area. Though, if corrosion spikes extend down to the copper, copper may be dissolved in the solder and intermetallic phases with copper will be formed. According to Champaign et al., these copper-containing intermetallics create a brittle zone and the corrosion spikes act as stress risers assisting in the formation of cracks [10].

8.2.2 Brittle Failures in Intermetallic Layers

It has been recognised that also a second failure mode may cause fractures to ENIG [20, 21]. According to IPC's standard IPC-7095A, Design and Assembly of Process Implementation for BGAs, interfacial fracture may happen between the nickel surface and the nickel–tin intermetallic layer under a high level of both applied strain and strain rate even if hyperactive corrosion does not take place [20]. However, for properly formed solder joints, it may be difficult to clearly distinguish this failure mechanism from that caused by black pad defects and it is likely that the two failure mechanisms interact.

Since these fractures have appearances of brittle fractures, they are often referred to as brittle fractures. In some cases, they are true brittle fractures but in other cases, they are rather interfacial fractures between two layers with different compositions.

According to Sohn et al. [22] and Mattila and Kivilahti [23], the interfacial cracking in solder joints on electroless nickel not having a black pad defect occurs in a very thin Ni–Sn–P layer formed on top of the layer with enriched phosphorus. That is, for both failure mechanisms, the fracture occurs between the Ni–Sn intermetallic layer and the layer with enriched phosphorus on the nickel surface. Thus, very little Ni–Sn intermetallic phases will remain on the nickel surface in both cases.

Neither is the presence of a nickel layer with enriched phosphorus a clear indication of a black pad defect. The formation of an intermetallic layer on electroless nickel during soldering causes a selective solution of nickel atoms leading to an enrichment of phosphorus in the remaining nickel, just as black pad defects do. Furthermore, immersion gold plating inevitably leads to the formation of a layer with enriched phosphorus although it may be less than 20-nm thick under optimal conditions [4]. Thus, wetting during soldering must always occur to a nickel surface with enriched phosphorus. Hence, the formation of a nickel surface with enriched phosphorus as such does not seem to cause a non-wettable surface. Most probably, it is the over-etching beneath the gold layer leading to entrapment of bath residues and exposure of the nickel surface to oxidation that causes the surface to become non-wettable. The finding by Cordes and Huemoeller that immersion gold plating of electrolytic coating resulted in solder joints with very poor shear strength is hard to explain in any other way [3]. Since the function of the gold layer is to preserve the solderability by preventing oxidation of the nickel surface, over-etching will reduce that function.

The composition of the phosphorus-enriched layer formed during soldering is usually specified to be Ni_3P which corresponds to a phosphorus concentration of 15 wt%. In comparison, the phosphorus content in the enriched layer formed during immersion gold plating is normally specified to be between 10 and 20 wt%, i.e. in the same order. Hence, the properties of the phosphorus-enriched layer are probably very similar regardless of how it was formed.

Whereas there seems to be consensus that electroless nickel with high phosphorus content is to be preferred in order to minimise the risk for black pad defects, there are different opinions how a high phosphorus content affects the reliability of the solder joints. According to Bulwitch et al. [8] and Lee [16], high-phosphorus nickel increases the risk for brittle fractures caused by phosphorus enrichment during soldering and subsequent thermal excursions. On the other hand, Johal et al. have reported that high-phosphorus nickel not only decreases the risk for the black pad defect but also improves the reliability of the solder joints [14]. They tested how the phosphorus content affected the shear strength of Sn–Ag–Cu balls before and after 1,000 temperature cycles between –55 and +125°C. Before temperature cycling, shear tests of solder balls to nickel coatings with 8.0 and 11.2 wt% P gave consistent results with high values and little spread in shear force, whereas solder balls to a nickel coating with 4.2 wt% P showed a wide spread in shear force with considerably lower shear force for some solder balls. After temperature cycling, also the solder balls to the nickel coating with 8.0 wt% P gave a large spread in shear force, while the solder balls to the nickel coating with 11.2 wt% P still showed high values and little spread in shear force. When examined using cross-sectioning, the fractures to the nickel coating with 8.0 wt% P showed partial brittle fractures, whereas the fractures to the nickel coating with 11.2 wt% P showed ductile fractures. The structures of the intermetallic layers were examined by stripping off the solder. The intermetallic layer on the nickel coating with 8.0 wt% P had a structure that was courser and less dense compared to the intermetallic layer on the nickel coating with 11.2 wt% P. They proposed that the difference in structure was due to differing dissolution rate of nickel during soldering and ageing. Less nickel should be dissolved from the more corrosion-resistant high-phosphorus nickel.

As discussed previously, recently performed investigations indicate that fractures on electroless nickel not affected by the black pad defect occur in a thin Ni–Sn–P layer between the Ni_3P and Ni–Sn intermetallic layers [22, 23]. In the investigation done by Mattila and Kivilahti, a Sn–Ag–Cu solder was used to solder chip scale packages (CSP) to ENIG and a porous microcrystalline or amorphous thin Ni–Sn–P layer with the composition $Ni_{55}Sn_{45}P_{10}$ was formed between the Ni_3P layer and a layer of $(Cu,Ni)_6Sn_5$. Cracking occurred in the Ni–Sn–P layer when the assembled CSPs were exposed to a drop test. Sohn et al. performed shear testing of test vehicles with both Sn3.5Ag and Sn3.0Ag0.5Cu solder joints to ENIG. In their study, brittle fractures only happened in the Ni–Sn–P layer for the test vehicles soldered with the Sn3.5Ag solder. The Ni–Sn–P layer in this case was formed between the Ni_3P layer and a layer of Ni_3Sn_4, it was nanocrystalline with Kirkendall voids and had a composition of Ni_3SnP. The reason why brittle cracking only occurred for the Sn3.5Ag solder was claimed to be due to spalling of the Ni_3Sn_4 layer and thereby larger consumption of nickel during soldering. This resulted in an increased growth of the Ni–Sn–P layer. The shear rate used for testing was only 0.2 mm/min. A higher shear rate may have caused brittle fractures also for the Sn3.0Ag0.5Cu solder joints.

Obviously, solder joints to electroless nickel are prone to brittle fractures regardless if the intermetallic layer consists of Ni_3Sn_4 or $(Cu,Ni)_6Sn_5$. However, the growth rate of the intermetallic layer and its thickness may be important for the brittleness of the solder joint. It will affect the thickness of the Ni–Sn–P layer and probably the extent of voids in this layer. Thus, measures that would decrease the dissolution rate of nickel during soldering ought to decrease the brittleness of the solder joints. That may explain the finding by Johal et al. that high-phosphorus nickel improves the resistance against brittle fractures. Consequently, if the conclusion is correct, it will mean that some enrichment of phosphorus in the nickel surface during immersion gold plating may be beneficial for the reliability of the solder joint since it will decrease the dissolution rate of the nickel. That is under condition that proper wetting occurs to the nickel surface.

8.3 Test Methods for Detecting Black Pad Defects and Assessing the Risk for Brittle Fractures

8.3.1 Detection of Black Pad Defects

The method using LVSEM and EDS with a low electron beam voltage to detect a thin film of nickel oxide on top of the gold layer reported by McFaddin is a non-destructive method that can be used to detect black pad defects on PCBs [6]. This is a good indication of black pad defects. However, the size of the LVSEM chamber will set a limit of the size of the PCBs that can be non-destructively analysed.

A destructive but probably more reliable method to detect black pad defects on PCBs is to dissolve the immersion gold layer using a cyanide strip and then examine the nickel surface using SEM [7, 10, 15]. Areas with black pad defects will then have a mud-cracked appearance with black regions around the nickel nodules.

Another destructive method that can also be used for assembled boards is cross-sectioning of the solder lands [6, 7, 9–11, 19]. Corrosion damages of the nickel layer in the form of spikes and black bands can then be detected and for assembled boards, poorly developed intermetallic layers, all indicators of black pad defects.

Another possibility is to expose assembled boards to environmental stress screening (ESS). By performing ESS tests utilising fast temperature changes and/or exposure to vibration, black pad defects can be stimulated to precipitate as failures. Thereby, assemblies having black pad defects can be detected.

8.3.2 Tests for Assessing the Risk for Brittle Fractures in Intermetallic Layers

If a test for assessing the risk for brittle fractures shall be adequate, there must be a clear correlation between the results from such a test and failures occurring in field conditions. Too often, tests are used because they are easy to use, inexpensive and/or give fast results, or just because they have been standardised, without considering whether they are relevant or not.

According to IPC-7095A, interfacial fracture happens on nickel surfaces under a high level of both applied strain and strain rate [21]. Nevertheless, the brittleness of solder joints is usually evaluated using shear and pull tests with low strain rate [14, 22, 24–28]. A few investigations have been performed where the results from tests with high strain and strain rate have been compared with results from traditional shear and/or pull tests [29–35]. They have all shown that especially traditional shear tests but also pull tests are not adequate for testing the risk for brittle fractures.

According to Sykes, a shear rate needs typically to be greater than 1 m/s and a pull rate in excess of 0.5 m/s to test the resistance against brittle fractures [29]. Newman has presented results using equipment for high-speed shear and pull testing [30]. Using the shear test, only at shear rates of 0.1 m/s or higher did interfacial fractures occur. For Sn–Pb solder joints, interfacial fractures mainly occurred to ENIG surfaces. A few interfacial fractures were observed to electrolytic Ni/Au at shear rates of 1 and 4 m/s but none to copper. A considerably higher risk for brittle fractures to ENIG was observed when testing was performed within 2 days after reflow compared to testing after 1–6 months. Sn–Ag–Cu solder joints had a significantly higher risk for brittle fractures at high-shear rate than Sn–Pb solder joints. For the lead-free solder, interfacial fractures dominated to both electrolytic Ni/Au and copper even at a shear rate of 0.1 m/s. No data were presented for Sn–Ag–Cu solder joints to ENIG. Similar results were obtained when performing pull tests except that interfacial fractures occurred at lower pull rates. For Sn–Ag–Cu solder joints, brittle fractures occurred at a pull rate of 0.005 m/s when testing after 1–6 months.

8.3.3 Standards Developed for Assessing the Risk for Brittle Fractures

JEDEC has developed a standard for drop testing of components for handheld electronic products, JESD22-B111 [36]. For portable products, dropping to the ground usually results in board bending. If a product with the assembly fixed in the corners is dropped with the assembly in a horizontal position, the interconnection stress due to board bending is 2 orders of magnitude higher than that due to acceleration according to Wong et al. [33]. For that reason, the JEDEC standard requires that the test board be mounted on a base plate with the components facing down. The board shall be fixed to the base plate in the four corners using 10-mm-long standoffs to allow bending of the board. The base plate with the assembly is then dropped with the assembly in a horizontal position.

In another JEDEC standard, JESD22-B110 [37], guidance is given for how to simulate shipping- and handling-related shock of electronic subassemblies.

IPC and JEDEC have developed a joint standard for characterising the fracture strength of a component's board-level interconnects when exposed to monotonic bending [38]. The monotonic bend characterisation results provide a measure of fracture resistance to flexural loading that may occur during conventional non-cyclic board assembly and test operation. A four-point bend test is used to characterise the fracture strength. Since storage conditions and storage duration affects the fracture resistance, it is required that testing is performed minimum 8 h and maximum 168 h after soldering.

In another standard developed by IPC and JEDEC, guidelines are given for how to perform strain gage testing, and in order to assess the strain and strain rates, a surface mount technology (SMT) package is subjected to during PCB assembly, test and operation [39, 40]. Manufacturing steps that can cause strain-induced failures and therefore may need to be characterised include:

- SMT assembly process: board depanelisation, all manual handling processes, all rework and retouch processes, and connector and component installation.
- Board test processes: in-circuit test and board functional test.
- Mechanical assembly: heat sink assembly, board support/stiffener assembly, system board integration, peripheral component interconnect or daughter card installation and dual in-line memory module installation.
- Shipping environment.

References

1. Puttlitz KJ (1990) Preparation, structure, and fracture modes of Pb–Sn and Pb–In Terminated flip chips attached to gold-capped microsockets. In: Proceedings of 40th electronic components and technology conference, pp 360–366
2. Eslambolchi A, Johnson P, Kaufmann M, Mei Z (1998) Electroless Ni/immersion Au evaluation—final program report. Electronic Assembly Development Center, Hewlett-Packard, Palo Alto
3. Cordes F, Huemoeller R (1999) Electroless nickel–gold: is there a future? Electroless Ni/Au plating capability study of BGA packages. In: proceeding of IPC Printed Circuits Expo, Paper S13-1
4. Kwok RWM, Chan KCM, Bayes MW (2004) Development of an electroless nickel immersion gold process for PCB final finishes. Circuit World 30(3):37–42
5. Johal K (2001) Are you in control of your electroless nickel/immersion gold process. In: Proceedings of SMTA International
6. McFaddin WE (2003) Non-destructive analysis method for detection of the "black pad defect" on PCB surfaces. In: Proceedings of SMTA International, pp 395–403
7. Biunno N (1999) A root cause failure mechanism for solder joint integrity of electroless nickel/immersion gold surface finishes. In: Proceedings of IPC Printed Circuits Expo, Paper S18-5
8. Bulwitch RA, Trosky M, Pichione LM, Hug D (2002) The "black pad" failure mechanism—from beginning to end. Global SMT Packag 2(6):9–13
9. Jay R, Kwong A (2001) Dealing with the "black pad defect"—a failure analyst's perspective. In: Proceedings of SMTA International

10. Champaign RF, Roepsch JA, Downey MR (2003) Afraid of the dark? Circuits Assemb 14(1):22–25
11. Roepsch JA, Champaign RF, Waller BM (2003) Black pad defect: influence of geometry and other factors. In: Proceedings of SMTA International
12. Crouse K, Cullen D (2002) A key failure mode resulting in interfacial fracture of soldered ENIG surfaces. PC FAB Mag (Feb):22–32
13. Milad G, Martin J (2000) Electroless nickel/immersion gold, solderability and solder joint reliability as functions of process control. CircuiTree 13(10):56–62
14. Johal K, Lamprecht S, Schreier H-J, Roberts H (2004) Impacts of bulk phosphorous content of electroless nickel layers to solder joint integrity and their use as gold- and aluminum-wire bond surfaces. In: Proceedings of SMTA Pan Pacific Microelectronics Symposium
15. Goosey M (2002) Factors influencing the formation of "black pad" in electroless nickel-immersion gold solderable finishes—a processing perspective. Circuit World 28(3):36–39
16. Lee B (2003) Implementing a simple corrosion test method to detect "black pad" phenomenon in electroless nickel/immersion gold plating. CircuiTree 16(11):40–46
17. Houghton B (2000) Solving the ENIG black pad problem: an ITRI report on round 2. Future Circuits Int 6:121–128
18. Houghton B (2000) ITRI project on electroless nickel/immersion gold joint cracking. Circuit World 26(2):10–16
19. Tomia Y, Wu Q, Maeda A, Baba S, Ueda N (2000) Advanced surface plating on the organic FC-BGA package. In: Proceedings of 50th electronic components and technology conference, pp 861–867
20. Bansal A, Yoon S, Mahadev V (2005) Flexural strength of BGA solder joints with ENIG substrate finish using 4-point bend test. In: Proceedings of SMTA Pan Pacific Microelectronic Symposium
21. Design and assembly of process implementation for BGAs (2005) IPC Standard IPC-7095A. http://www.ipc.org
22. Sohn Y-C, Yu Y, Kang SK, Shih D-Y, Lee T-Y (2005) Effect of intermetallics spalling on the mechanical behavior of electroless Ni(P)/Pb-free solder interconnection. In: Proceedings of 55th electronic components and technology conference, pp 83–88
23. Mattila TT, Kivilahti JK (2005) Failure mechanisms of lead-free chip scale package interconnections under fast mechanical loading. J Electron Mater 34(7):969–976
24. Paik K-W, Jeon Y-D, Cho M-G (2004) Interfacial reactions and bump reliability of various Pb-free solder bumps on electroless Ni-P UBMs. In: Proceedings of 54th electronic components and technology conference, pp 675–682
25. Zheng Y, Hillman C, and McCluskey P (2002) Intermetallic growth on PWBs soldered with Sn3.8Ag0.7Cu. In: Proceedings of 52nd electronic components and technology conference, pp 1226–1231
26. Liu CM, Ho CE, Chen WT, Kao CR (2001) Reflow soldering and isothermal solid-state aging of Sn-Ag eutectic solder on Au/Ni surface finish. J Electron Mater 30(9):1152–1156
27. Coyle RJ, Hodges Popps DE, Mawer A, Cullen DP, Wenger GM, Solan PP (2003) The effect of modifications to the nickel/gold surface finish on assembly quality and attachment reliability of a plastic ball grid array (Peer Review Version). IEEE Trans Compon Packag Technol 26(4):724–732
28. Chouta P, Santos D, Srihari H, Sammakia B, Andros F, DiPietro M (2001) A shear strength study of lead-free solder spheres for BGA applications on different pad finishes. In: Proceedings of SMTA Pan Pacific Symposium
29. Sykes B (2005) Lead-free BGA reliability: high-speed bond testing and brittle fracture detection. Global SMT Packag 5(9):20–23
30. Newman K (2005) BGA brittle fracture—alternative solder joint integrity test methods. In: Proceedings of 55th electronic components and technology conference, pp 1194–1201
31. Chiu T-C, Zeng K, Stierman R, Edwards D, Ano K (2004) Effect of thermal aging on board level drop reliability for Pb-free BGA packages. In: Proceedings of 54th electronic components and technology conference, pp 1256–1270

32. Gregorich T, Holmes P, Lee JCB, Lee CC (2004) SnNi and SnNiCu intermetallic compounds found when using SnAgCu solders. In: Proceedings of IPC/Soldertec global 2nd international conference lead free electron
33. Wong EH, Rajoo R, Mai YW, Seah SKW, Tsai KT, Yap LM (2005) Drop impact: fundamentals and impact characterisation of solder joints. In: Proceedings of 55th Electronic Components and Technology Conference, pp 1202–1209
34. Hasegawa K, Takahashi A, Noudou T, Nakaso A (2004) Electroless Ni-P/Pd/Au plating for semiconductor package substrates. Plat Surf Finish 91(11):20–25
35. Date M, Shoji T, Fujiyoshi M, Sato K, Tu KN (2004) Impact reliability of solder joints. In: Proceedings of 54th Electronic Components and Technology Conference, pp 668–674
36. JESD22-B111 (2003) Board level drop test method of components for handheld electronic products. JEDEC http://www.jedec.org
37. JESD22-B110A (2004) Subassembly mechanical shock. JEDEC http://www.jedec.org
38. IPC/JEDEC-9702 (2004) Monotonic bend characterization of board-level interconnects. IPC http://www.ipc.org
39. IPC/JEDEC-9704 (2005) Printed wiring board strain gage test guidelines. IPC http://www.ipc.org
40. Ahmad M, Duggan R, Hu T, Ong B, Ralph C, Sethuraman S, Shangguan D (2005) Strain gage testing: standardization. Surf Mt Technol 19(7):36–42

Chapter 9
Reliability of Electronic Assemblies Under Mechanical Shock Loading

T. T. Mattila, T. Laurila, V. Vuorinen and J. K. Kivilahti

9.1 Introduction

Reliability of portable electronic devices is becoming an important factor of success in the highly competitive global markets due to the fact that increasing heat dissipation, current densities and mechanical loadings of power components make electrical interconnections more vulnerable to failures. Portable electronic products encounter diverse environments in ordinary daily use and therefore their reliability should be studied with tests that simulate real-use strains and stresses as realistically as possible. Portable electronic products are exposed to temperature fluctuations due to either internally generated heat dissipation or the external operational environment, but they are especially prone to failure due to mechanical shock loads caused by accidental drops. These loadings are simulated with standardized reliability tests that also allow comparative studies of different material combinations.

The usage of lead-free solders, components under bump metallizations (UBM) or lead metallizations, and printed wiring board (PWB) protective coatings add to the complexity of the interconnection metallurgies. The number of different material combinations increased markedly as the traditionally used SnPb-based solders and protective coatings were replaced with different lead-free alternatives. Because microstructures significantly affect the reliability of soldered interconnections, reliability of each material combination is likely to be different. For these reasons, more extensive and comprehensive reliability testing of new products has become ever more important.

T. T. Mattila (✉), T. Laurila, V. Vuorinen and J. K. Kivilahti
Department of Electronics, School of Electrical Engineering, Aalto University,
P.O. Box 13340, FI-00076 Aalto, Finland
e-mail: Toni.Mattila@hut.fi

G. Grossmann and C. Zardini (eds.), *The ELFNET Book on Failure Mechanisms, Testing Methods, and Quality Issues of Lead-Free Solder Interconnects*,
DOI: 10.1007/978-0-85729-236-0_9,

Because reliability of electronic assemblies is chiefly determined by the ability of electrical interconnections to withstand various loadings during products' operational lifetime, it is of primary importance to investigate systematically metallurgical reactions in the effective joining region and the resulting microstructures within the solder interconnections. Lead-free solder interconnections can contain more complex intermaterial layers such as phosphides, which can markedly weaken solder interconnections. Moreover, the microstructures formed during soldering are not stable and will evolve during the operation of products. Hence, in order to ensure the best possible reliability of electronic assemblies against various loading conditions, much better understanding of interconnections microstructures, including interfacial reaction products, and their evolution is needed.

9.2 Effect Loading Rate on the Reliability of Solder Interconnections

Portable electronic equipment experiences during their use many different kinds of loading conditions, in which mechanical shocks and thermomechanical loadings are perhaps the most critical ones. These two fundamentally different loading conditions will evoke different failure mechanisms leading to dissimilar failure modes. Under thermomechanical loading, nucleation and propagation of cracks are controlled by the microstructures formed during soldering and their recrystallization behavior during use [1–5]. Mechanical shock impacts, on the other hand, are known to produce entirely different kinds of failure modes [6–13]. Cracks in the newly soldered interconnections do not propagate through the bulk solder of the interconnections, but mainly in the brittle intermetallic compound layers formed between solder and contact metallizations [4, 5, 7].

Several studies carried out with commercial portable electronic products have shown that impact forces generated when products are dropped onto the ground are transmitted through the product casing to the component boards and make the boards bend and vibrate excessively. The results from product-level tests have been used to develop board-level drop tests [14]. For example, JEDEC has published the JESD22-B111 board-level drop test standard for handheld electronic products.

The drop tester is composed of a mechanism to drop the board repeatedly in a specified orientation and a high-speed data acquisition system to record deceleration, strains on the component boards and the number of drops-to-failure. The condition B of the JESD22-B111 standard defines the deceleration pulse that has the shape of a half-sine with 0.5 ms width and the maximum at 1500 G. The shape of the deceleration pulse is not only a function of the drop height but depends on the characteristics of the strike surface: drop height determines the maximum deceleration and strike surface the pulse width.

The standardized layout of the test board along with its dimensions is presented in Fig. 9.1. The test board is attached horizontally on a fixture, and the components

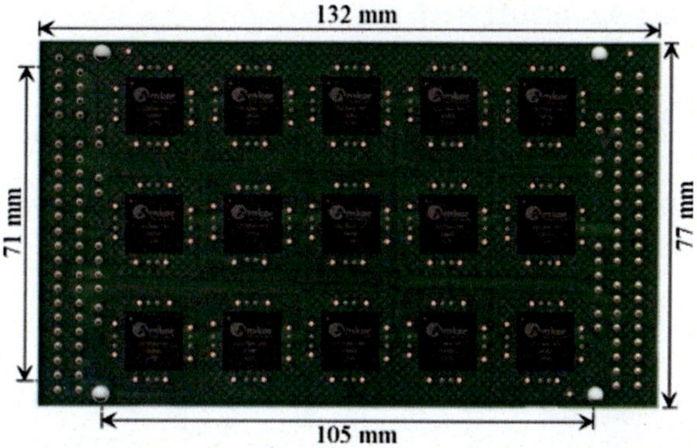

Fig. 9.1 Layout and dimensions of the drop test board

are facing downward during the test (Fig. 9.2b). The fixture is mounted on a sledge that is dropped down on a rigid surface from a specified height in a controlled manner with the help of guiding rails (see Fig. 9.2a). This is because the most detrimental factor for the assembly caused by dropping is not the mechanical shock itself, but the subsequent bending and vibration of the board [13, 15]. Placing the component boards horizontally achieves maximum flexure of the board and onsets the natural vibrating motion. Bending causes displacement between the board and the components resulting in component, interconnection, or board failures. The drop test can be considered an accelerated test because the test structure lacks the support provided by product casings and other adjacent structures and, therefore, during testing the test board experiences larger than real-use bending amplitudes.

Since the component board is allowed to bend freely at the shock impact, there are numerous different modes in which it can bend. The natural mode of the component board describes the shape in which the board bends, and the natural frequency describes how fast the bending takes place. Figure 9.3 shows three of the most significant natural modes of the JESD22-B111 compliant test board and their associated natural frequencies. Each of these natural modes vibrates at a characteristic frequency, and the total bending and frequency of the component board is their sum since the different natural modes act simultaneously. The shapes of the natural modes depend on the support structure of the component board, whereas the natural frequencies depend on the stiffness and mass of the component board. The natural modes with highest frequencies are usually not of great importance because their amplitude is relatively small and vibrations are attenuated quickly. Only the lowest frequencies, in the case of the JESD22-B111 board the lowest three, can be considered significant. Owing to the simultaneous action of many different natural modes and frequencies, and the fast attenuation of the vibration amplitude, the strain distribution on the test board changes very rapidly. Thus, the location of the highest stress changes quickly also. Figure 9.4 shows the longitudinal strain measured in the middle of the board layout on the opposite side of the board from the components. Figure 9.4a shows how the strain develops after the drop impact. The macroscopic oscillation is due to the natural mode with the

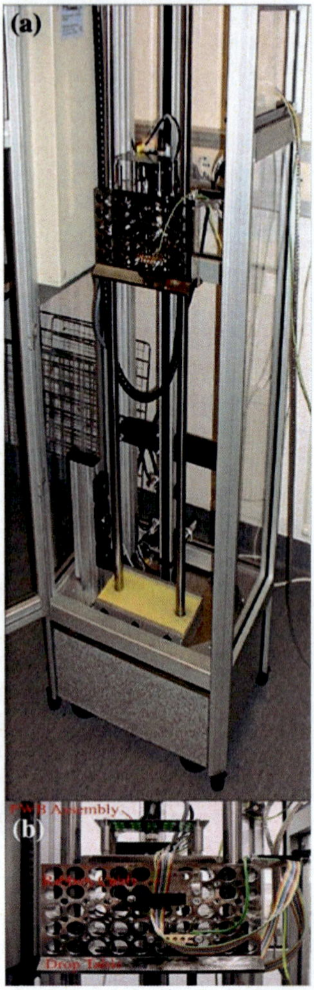

Fig. 9.2 JESD22-B111 compliant drop tester: **a** Structure of the tester, **b** Test board is attached from the four corners on a support pins and components are facing downward

lowest frequency. Oscillations at higher frequencies are embedded in the larger strains. The strains due to natural modes with higher frequencies are shown in Fig. 9.4b, which presents only the 0–5 ms time interval from Fig. 9.4a.

The failure mechanism under mechanical shock loading differs greatly from that under thermomechanical loading, where the localized recrystallization of the interconnections enables the nucleation and propagation of cracks in the bulk solder [1–5]. Besides temperature the most important difference between drop tests and thermal cycling tests is the deformation rate. The drop tests are typically carried out at room temperature ($\sim$295 K), which is relatively high (0.6 T_m) when compared to the melting point of the solder ($\sim$500 K). At such high (homologous) temperatures, the plastic behavior of the solder is strongly strain-rate dependent. As shown in Fig. 9.5, the solder becomes remarkably stronger as the strain rate

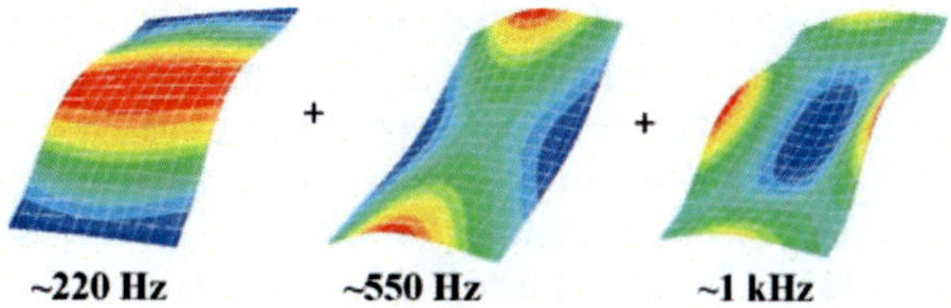

Fig. 9.3 Total flexure of the test board is the sum of the different natural modes. Three of the most significant natural modes of the JESD22-B111 compliant test board. *Different shades* of *gray* represent vertical displacement [15]

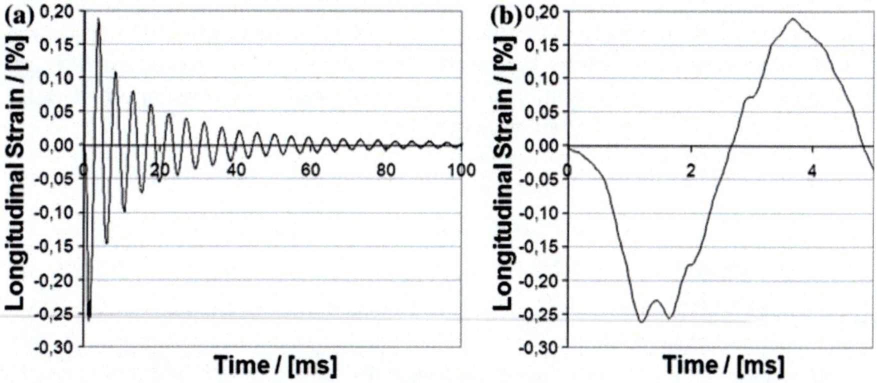

Fig. 9.4 Measured longitudinal strain at the center of the board

Fig. 9.5 Effect of strain rate on strength of Sn and two Sn-based solders [16–18]

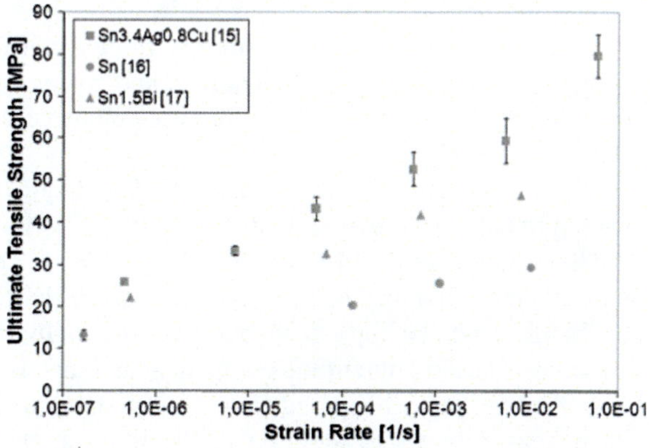

increases from that used in thermal cyclic tests ($\sim 10^{-6}$–10^{-5} 1/s) to that used in drop tests ($\sim$1–10 1/s). As can be seen, the flow stress of solder is about two to three times as high when the strain rate is increased from that occurring in thermal cycling to that occurring in drop tests. Both the ultimate tensile strength and the yield strength increase with strain rate, but the yield stress is typically even more strain-rate sensitive. Subsequently, the magnitudes and distributions of the stresses in the solder interconnections are different under thermal cycling and drop test conditions. Our finite element calculations have shown that as the strain rate increases not only stresses in solder interconnections increase but also they become more concentrated on the component side edges of the interconnections [6, 15]. Due to the much higher flow stress of the solder interconnections in the drop tests, the intermetallic compound layers will experience significantly higher stresses than those in thermal cycling. The same calculations showed that stresses at the solder|pad interface on the PWB side are less than half of that on the component side. The tensile strength of the solder increases above the fracture strength of the IMC, and this ultimately makes the fractures propagate inside the IMC layers, instead of the bulk solder. In thermal cycling, where the strain rates are relatively low, the cyclic thermomechanical loading of the interconnections generates plastic deformation, which ultimately leads to propagation of fatigue fracture through the solder interconnections.

9.3 Reliability of Lead-Free CSPs under Mechanical Shock Loading

In this study, the drop reliability of high-density electronic assemblies was studied by applying standardized JESD22-B111 test procedure for test vehicles that represent the technologies and lead-free materials typically used in portable electronic products. A large number of component boards were assembled in a full-scale production line to enable proper statistical and fractographic analyses. The test boards were assembled with different printed wiring board protective coatings, component under bump metallizations and solder pad structures. The component boards were drop tested and the times-to-failure of the various combinations were statistically analyzed.

The component used was a lead-free Sn0.2Ag0.4Cu (wt%)-bumped chip scale packaged (CSP) component with 144 bumps (500 μm in diameter) and bump pitch of 800 μm (denoted CSP144 in the following). The height of the bumps was 480 μm and the under bump metallurgy consisted of a ~0.6–0.8 μm-thick-electrochemical Ni, on top of which there originally was a thin gold layer before bumping. The structure of the component and materials used are shown in Fig. 9.6. The PWB was a double-sided (1 + 6 + 1) stack-up multilayer FR4 board with six inner copper layers in addition to the topmost resin-coated copper layers. The component boards for the drop tests were designed according to the JESD22-B111 drop test standard. The pad and conductor patterning on the board was the same on the two sides except that one side of the board had microvias in all of the soldering pads and the other did not. The PWBs were manufactured with either electroless Ni plating containing about 9 wt% (16 at%) P with flash Au finish on top (denoted Ni(P)|Au) or organic solderability preservative (denoted Cu|OSP) protective

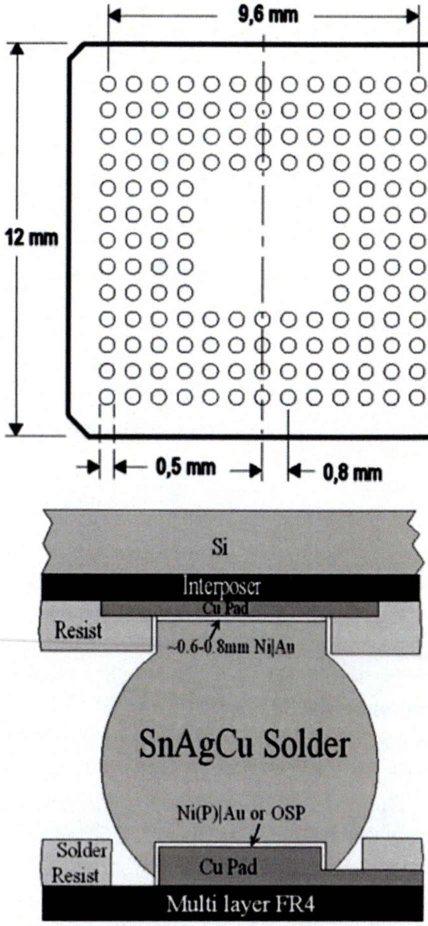

Fig. 9.6 Structure and dimensions of the CSP144 component

coating on the Cu soldering pads. The test boards were assembled by using the Sn3.8Ag0.7Cu solder paste. Although the board was double-sided, components were mounted on the non-microvia side only. The drop height was set to 82 cm in order to achieve the required peak deceleration of 1500 G for the duration of 0.5 ms (half-sine pulse). A failure was recorded when the resistance through the daisy chain network exceeded the 1.0 kΩ threshold resistance for more than 200 ns three times in a sequence of five drops.

Figure 9.7 presents the average drops-to-failure of the different material combinations of the experimental design. The significance of the differences between the material combinations was studied with the nonparametric Wilcoxon Rank-Sum Test (the numerical data failed the Shapiro–Wilk test for normality). The effect of both the factors, the *PWB coating* and the *component side metallization*, was statistically significant: the copper metallization on the component side was

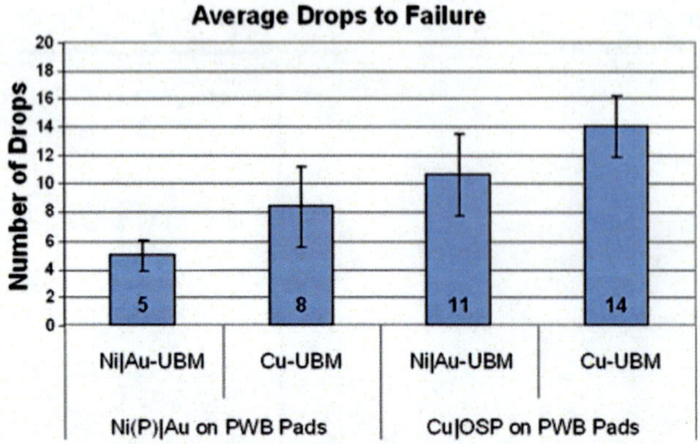

Fig. 9.7 Average drops-to-failure of the different material combinations with respective standard deviations

more reliable than the electrochemical nickel ($\alpha = 4.7\%$) and, on the PWB side the Cu|OSP was more reliable than the Ni(P)|Au ($\alpha = 0.2\%$).

The Weibull characteristic life times (η) of the Ni|Au-metallized and Cu-metallized components on the Ni(P)|Au-coated PWBs were 6 and 10 drops, respectively, while those on the Cu|OSP-coated PWBs were 12 and 16, respectively. Similarly the shape parameters (β) of the Nu|Au-metallized and Cu-metallized components on the Ni(P)|Au-coated PWBs were 1.7 and 1.5, respectively, while those on the Cu|OSP-coated PWBs were 1.9 and 3.7, respectively.

The primary failure modes identified at the failure analysis stage were different from those typically observed after the thermal cycling tests with similar components and the same lead-free materials [1, 3]. In addition, the modes were also different between the interconnections on the Ni(P)|Au and the Cu|OSP. The solder interconnections on the Cu|OSP-coated PWBs failed at the component side, where cracks propagated through the $(Cu,Ni)_6Sn_5$ or the Cu_6Sn_5 reaction layer depending on the component under bump metallization: Cu_6Sn_5 on the copper under bump metallization (see Fig. 9.8) and $(Cu,Ni)_6Sn_5$ on the Ni|Au metallization (see Fig. 9.9). (In the case of Cu-UBM, there should thermodynamically be a layer of Cu_3Sn between the Cu and the Cu_6Sn_5 but the thickness of the Cu_3Sn layer is too small to be observed in the images). Interconnections on the Ni(P)|Au failed from the PWB side interface, where cracks propagated between the $(Cu,Ni)_6Sn_5$ and the Ni(P) metallization (see Fig. 9.10). Cracks in the Cu|OSP interconnections typically nucleated at the corner of the interconnections, a varying distance away from the intermetallic compound (IMC) layers in the bulk solder (see e.g. Figs. 9.8, 9.9), but jumped very quickly into the IMC layers, which obviously provided a more favorable path for the crack to propagate due to the brittle nature of the compound. The fractures in the Ni(P)|Au interconnections propagated very close to the nickel metallizations underneath the $(Cu,Ni)_6Sn_5$ intermetallic layer, as shown in Fig. 9.10.

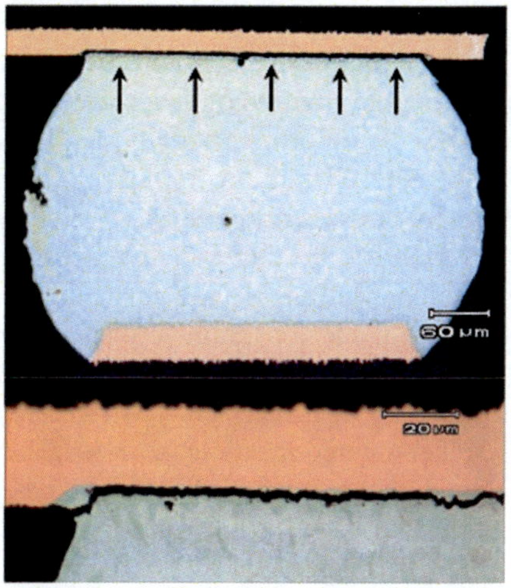

Fig. 9.8 The primary failure mode of the Cu-UBM interconnections on the Cu|OSP-coated PWBs

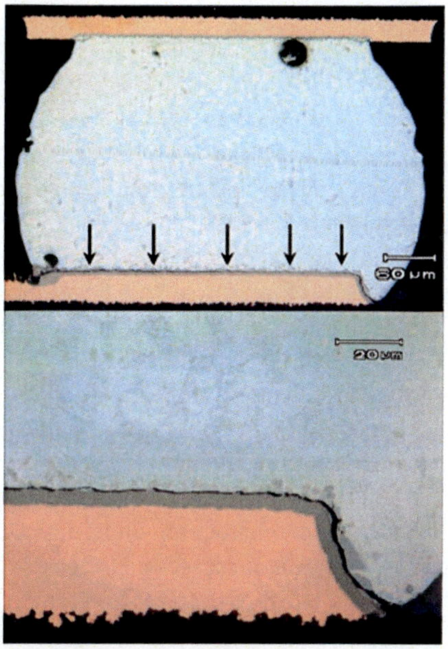

Fig. 9.9 The primary failure mode of the Ni-UBM interconnections on the Cu|OSP-coated PWBs

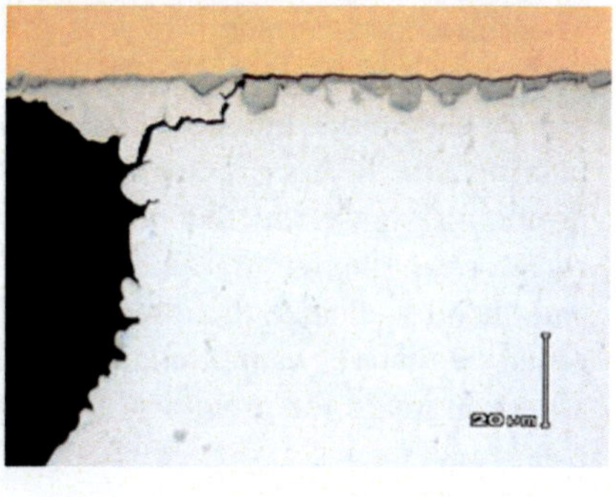

Fig. 9.10 The primary failure mode of the Cu-UBM as well as the Ni|Au-UBM interconnections on the Ni(P)|Au-coated PWBs

No recrystallization was observed in the drop-tested samples, even after several months of storage at room temperature. This is most likely because during drop testing the strength of the solder interconnections increases and the solder does not markedly deform plastically. As the strain rate is increased twinning mechanism is activated. Twins are typically observed in the regions of the interconnections where stresses are highest.

Despite the fact that the stresses at the PWB side interfacial regions of the interconnections are much smaller than at the component side [6, 15], the primary failure mode for the interconnections soldered on Ni(P)|Au was a fracture on the PWB underneath the $(Cu,Ni)_6Sn_5$ intermetallic layer. This emphasizes the weak nature of this interface. The fracture typically propagated completely through the solder interconnections at a single or very few impacts. The fracture path was always very smooth and straight, when compared with the Cu|OSP interconnection fractures discussed earlier.

9.4 Effects on Small Additions of Ni in the SnAgCu solder and in the Drop Reliability of BGA Component Boards

At the dawn of the adoption of the lead-free electronics manufacturing legislation (RoHS) at the EU region, compositions of the currently used lead-free solder alloys were decided mainly on the basis of their ability to replace the eutectic Pb-containing solder alloys in the existing assembly processes. However, recently the emphasis of research and development work has shifted toward the reliability issues related to the newly adopted compositions. Reliability improvements have been sought by modifications of the microstructures and mechanical properties of the near-eutectic SnAgCu solders by adding minor elements and optimizing their compositions. In addition to Ni, the effects of elements such as Co, In, Sb and Zn have been investigated [19].

As discussed earlier, the increased strength of solder interconnections caused by the strain-rate hardening under fast deformation rates is perhaps the most important reason for the increase in stresses at the interfacial regions of the solder interconnections above the fracture strength of the intermetallic layers (cracking of the interfacial layers) [4, 7]. Silver is know to be a very effective strengthener of tin, and the yield strength of the alloy increases with increasing silver content [20, 21]. Therefore, lower strength of low-silver SnAgCu alloys (lower than the about 3.5 wt%) may be advantageous for drop reliability because it allows plastic deformation of the solder interconnections and, thereby, the maximum stresses near the intermetallic layers are smaller. Furthermore, modifications of the intermetallic layers, for example by the introduction on minute amounts of nickel in the SnAgCu solders, can change the morphology and properties of interfacial reaction layers and, thereby, change their performance under reliability tests. It has recently been observed that very small addition of nickel, even below 0.1 wt%, in the SnAgCu alloy can retard the growth of the initially very thin layer of Cu_3Sn [20–23].

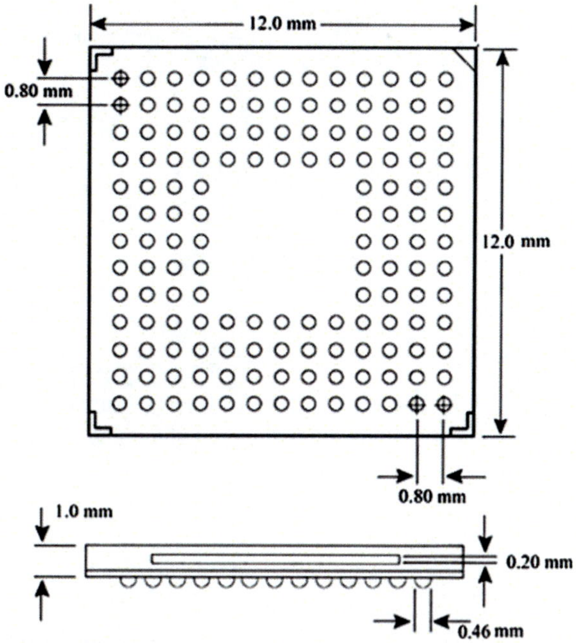

Fig. 9.11 Layout and dimensions of the component

Therefore, drop reliability of the CSP/BGA component boards soldered with different interconnections compositions was investigated. The commonly used near-eutectic SnAgCu solder alloy (Sn3.1Ag0.5Cu) was compared to the reliability of the component boards that were soldered with a low-silver nickel containing SnAgCu alloy (Sn1.1Ag0.5Cu0.1Ni). The reliability was studied by applying the JESD22-B111 drop test standard (condition B: 1500 G with 0.5 ms pulse width and the single component on board configuration). The component used in this study was a 12 mm × 12 mm chip scale sized BGA with 144 bumps, the pitch of 0.8 mm and a bump diameter of 0.5 mm (see Fig. 9.11). The component was bumped with either Sn3.0Ag0.5Cu (SAC305) or Sn1.2Ag0.5Cu 0.05Ni (LF35). There was no under bump metallization between the solder bumps and copper pads of the component. The printed wiring board (PWB) layout was designed according to the JESD22-B111. The copper soldering pads of the PWBs (1 + 6 + 1 build-up FR4) were coated with organic solderability preservative (Cu|OSP).

The two different interconnection compositions were formed by using different commercial solder pastes: (a) the SAC305-bumped components were reflow solder with the Sn3.8Ag0.7Cu to form the Sn3.1Ag0.5Cu and (b) the LF35-bumped components with the Sn0.7Cu0.1Ni to form the Sn1.1Ag0.52Cu0.01-0.1Ni after reflow compositions of the solder interconnections. The after reflow nominal composition of the interconnections was calculated by assuming complete filling of the stencil openings, perfect paste release from the stencil, and mixing of the bump and paste material volumes during the soldering. The component boards

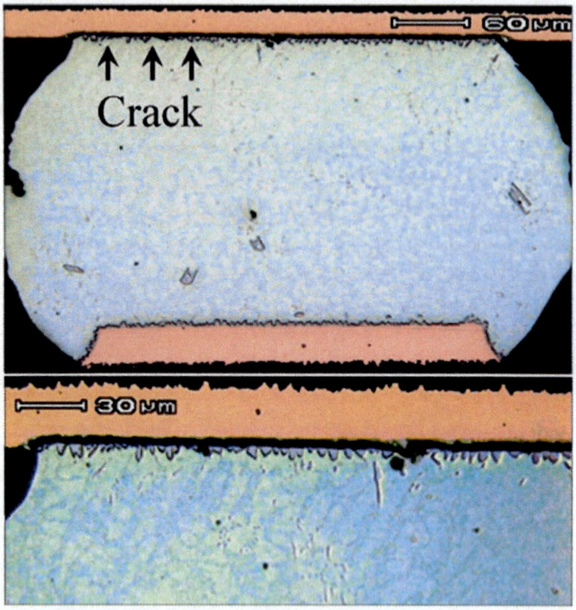

Fig. 9.12 Primary failure mode of the SnAgCu solder interconnections

were given a second reflow in order to give all test boards the same reflow treatment before testing. All together 46 component boards (23 replications of each interconnection compositions) were drop tested.

The reliability of the component boards was dependent on the solder composition. The average number of drops-to-failure of the component boards with the SnAgCu solder interconnections was 53 drops while that of the component boards with the SnAgCuNi interconnections was 71 drops. The SnAgCuNi solder interconnections are more reliable than the SnAgCu interconnections at 9.3% risk level (Wilcoxon Rank-Sum Test). The failure analysis of the drop-tested component boards revealed that the SnAgCu interconnections failed by cracking of the component side intermetallic layers, where cracks propagated inside the Cu_6Sn_5 intermetallic compound layer, as shown in Fig. 9.12. The primary cracks, i.e. the cracks that had failed the components electrically, had nucleated and propagated on the component side, while only some secondary cracks (cracks that do not cause electrical failures) were observed at the PWB side of the solder interconnection.

The SnAgCuNi interconnections, on the other hand, failed by cracking of the copper traces at the edge of the soldering pads on the PWB side of the interconnections. The copper traces had failed at the vicinity of the corner interconnections as shown in Fig. 9.13. Figure 9.14 shows a secondary failure mode, cracking of the bulk solder that was characteristic only to the low-silver, nickel containing interconnections. They were not observed to have caused electrical failures, and cracking of interfacial intermetallic layers was not observed in any of the SnAgCuNi solder interconnections.

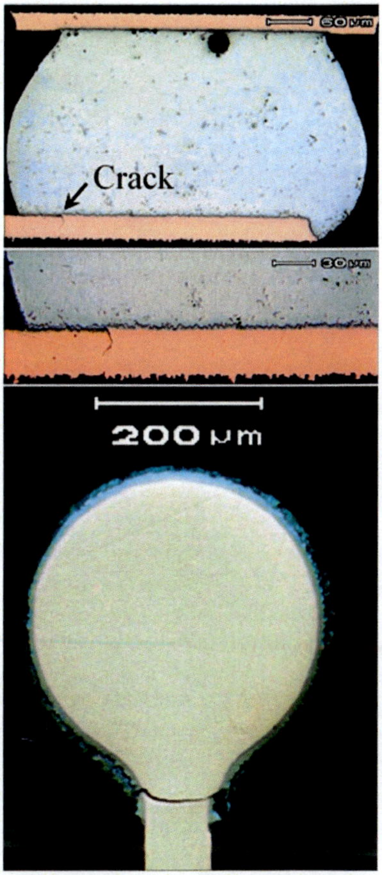

Fig. 9.13 Primary failure mode of the SnAgCuNi solder interconnections

The interfacial intermetallic layers were inspected with the help of the polarized light microscopy. Polarized light images make individual grains visible by showing grains in different grain orientations in different colors (asymmetric crystal structures have different refraction indices in different crystal directions and, therefore, cross-sections of grains in different orientation are seen in different colors when the reflected plane polarized light is cross-polarized). The micrographs in Fig. 9.15 show distinct differences in the morphology of the reaction layers. The interfacial reaction between the SnAgCu solder and copper pads forms a two-layered structure: Cu|Cu_3Sn|Cu_6Sn_5|solder. The Cu_3Sn layer in the SnAgCu interconnections (see Fig. 9.15) is clearly visible but not distinguishable in the SnAgCuNi interconnections. It seems that the nickel addition has not to have affected the total thickness of the reaction layers considerably but the grain size is notably reduced and the reaction layers are, consequently, much more uniform. The scallop-shaped morphology has changed to more granular-type morphology. It should be noted that in the case of the SnAgCu interconnections cracking took

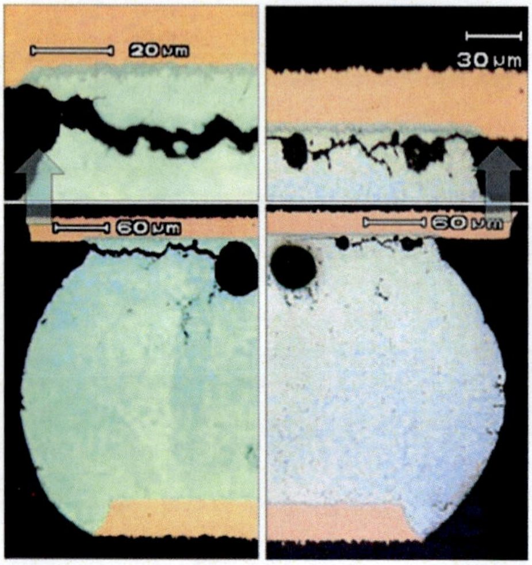

Fig. 9.14 Secondary failure modes of the SnAgCuNi interconnections

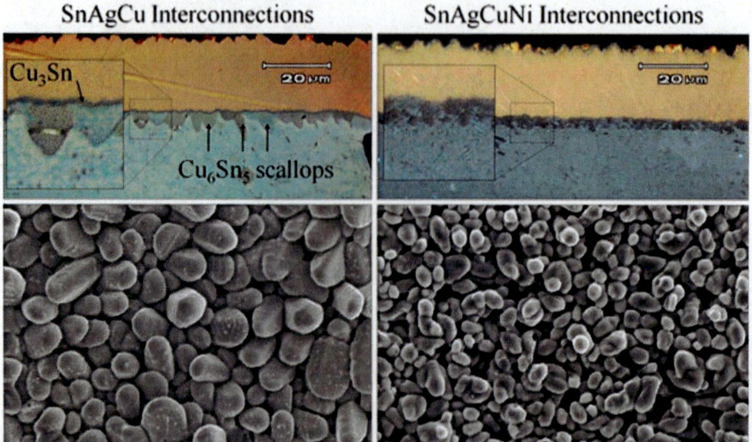

Fig. 9.15 Morphology of the interfacial reaction layers of the SnAgCu (*left*) and SnAgCuNi layers (*right*); *Top row*: cross-section of the intermetallic layer taken with cross-polarized light (different grain orientations appear in *different colors*), *bottom row*: top view of the intermetallic reaction layer taken with the SEM

place in the intermetallic layers, whereas in the case of SnAgCuNi interconnections cracks propagated in the copper trace of the PWB or in the bulk solder but never in the intermetallic layers.

9.5 Intermetallic Compounds Layers: Their Structure and Formation

Because the reliability of electronic assemblies under mechanical shock loading is very often dependent on the ability of intermetallic layers to withstand the stresses produced in their operation conditions, the formation and properties of intermetallic compound layers are discussed next. All the common base materials (coatings or metallizations) such as Cu, Ni, Ag, Ag–Pd and Au, form intermetallic compounds (IMCs) with Sn, which is the generally accepted major element in most lead-free solders. Chemical reactions occur between solders and conductor metals during soldering (i.e. component metallizations, boards surface finishes and underlying conductors), and IMCs will nucleate and grow at the solder/conductor interfaces (provided that the solder alloys do not contain such elements that can prevent the formation of IMC layer). It is well known that the presence of the IMCs between solders and conductor metals is an indication of good metallurgical bonding. A thin, continuous and uniform IMC layer is an essential requirement for good bonding. However, due to their inherent brittle nature and the tendency to generate structural defects, too thick IMC layer at the solder/conductor metal interface may degrade the reliability of the solder joints.

Even though there are some characteristic differences between the systems commonly used in electronics devices (Cu–Sn, Ni–Sn, Ag–Sn and Au–Sn) the intermetallic reaction layers are formed, in principle, in three consecutive stages: (a) dissolution of contact metallization to molten solder, (b) chemical reaction i.e. the formation of IMCs, and (c) solidification of the interconnections. Although the relative importance of each stage may vary between different systems depending on the solubility of conductor metal in tin. The general sequence of events during a soldering operation can be described as follows. Immediately after the flux has removed the oxides and permits metallurgical contact of solder with the conductor metal, the contacted metal starts to dissolve to the molten solder. Initially, the rate of dissolution is very high, particularly if the solder is not alloyed with the metal in question and, therefore, very high concentrations of solute elements can be realized locally. After a short period of time, the layer of molten solder that is adjacent to the contacted metal becomes supersaturated with the dissolved metal throughout the interface. Thermodynamically, at the local (metastable) equilibrium solubility, the solid IMC starts to form in this part of the solder. The formation of the IMC takes metal solutes out of the saturated liquid solder and causes some further dissolution of the contacted metal, especially if the intermetallic layer is not uniform on top of the substrate. During storage or in use of electronic devices, the IMCs generated during soldering grow further in thickness or increase in number, especially if the operational temperatures are well above the ambient. Therefore, both solid/liquid and solid/solid systems must be studied to have better understanding of the reliability of soldered assemblies. It should be noted that the microstructures formed during soldering are under continuous microstructural evolution during the use of the devices. The local equilibrium conditions in the solder interconnections will change locally owing to the consumption of one or more of the components. This may then change the phases that can exist in local equilibrium accordingly and result in new interconnection microstructures. The microstructural changes thereby can affect the modes and mechanisms by which interconnections fail under loading. Thus, knowledge of the solder|conductor interaction and phase evolution in the solder interconnections is important not only to understand the reliability issues of the solder interconnections but also to the optimization of the soldering process from the metallurgical standpoint.

The lead-free materials used in the reliability tests were chosen to represent those typically used in portable electronic products. Ni(P)|Au or organic soldering preservative (OSP) protective coatings were used on the printed wiring boards (PWB), and Ni|Au or bare Cu were used as the under bump metallization (UBM) on the component side. SnAgCu-bumped components were reflow solder with different commercial near-eutectic SnAgCu solder pastes. Thus, there are three metallurgically different interfacial regions and each of them will be discussed in the following. A detailed review about the formation of intermetallic compound layers in different systems is available [24] and therefore only binary Cu–Sn, Ni–Sn and ternary Ni–P–Sn cases are shortly presented here.

9.5.1 Cu–Sn

Out of the material combinations studied in this work, the interconnections with copper in contact with the solder formed the most reliable interconnection structure under the mechanical shock loading. In the temperature range of interest (e.g. below 260°C), the interfacial reaction with molten Sn-based solder results in the formation of Cu_3Sn (ε) and Cu_6Sn_5 (η) layers.

In general, at temperatures typical for the soldering processes, i.e. below 260°C, Cu_6Sn_5 is the first phase to form at the liquid Sn/Cu conductor interface. The first stage of the reaction is the dissolution of Cu to liquid solder, until the solder becomes supersaturated with Cu more or less uniformly at the Cu/liquid interface (i.e. reaches the metastable solubility limit). The metastable solubility is important, because it essentially determines the dissolution rate of metal to liquid solder. The metastable solubility is always higher and typically it is 2–3 times the stable solubility. Here, it indicates the largest possible amount of Cu that can dissolve in the liquid without precipitating back as pure Cu. It is to be noted that when calculating metastable solubilities, one has to use the data being assessed both stable and metastable phases. If the data have not been optimized for metastable phase, the liquids are generally far too stable in their metastable regions and so the evaluation gives too high values for the metastable solubility.

When Cu comes into contact with molten Sn, it starts to dissolve rapidly [25, 26]. Initially, the dissolution is a non-equilibrium process and locally very high concentrations of Cu can be realized in the very vicinity of the Cu/liquid interface. Since there is large driving force for the chemical reaction between Cu and Sn

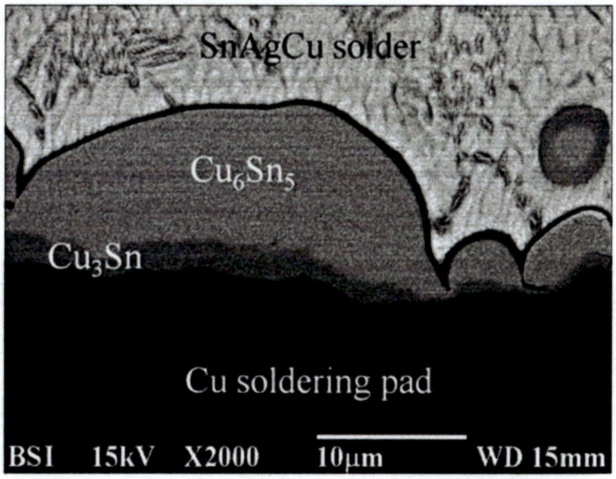

Fig. 9.16 Cu|SnAgCu interface annealed at 250°C for 1 h in order to detect the Cu_3Sn layer in between the Cu pad and the Cu_6Sn_5

atoms at the metastable composition, Cu_6Sn_5 crystallites can form very fast by the heterogeneous nucleation and growth at the Cu/liquid interface. In addition to the more or less uniform scallop-type Cu_6Sn_5 layer (uniphase), the (Cu_6Sn_5+Sn) two-phase layer can form next to the uniphase layer, most likely enhanced by the local constitutional supercooling of liquid. The final thickness and the morphology of the reaction layer is determined primarily by the rate of the dissolution of Cu in liquid and the chemical reaction between Sn and Cu and, secondarily, by the diffusion of Cu in the liquid. The metastable solubility of Cu in liquid solder is therefore the most important factor. On the other hand, the rate of diffusion of Cu atoms in the liquid determines the extent and morphology of the two-phase zone in front of the uniphase layer. Thermodynamically, there should also be a layer of Cu_3Sn layer between Cu and Cu_6Sn_5 (see Fig. 9.16). However, the thickness of the Cu_3Sn layer is much smaller than that of the Cu_6Sn_5 layer after reflow or wave soldering, and its formation usually requires extended contact times.

When one considers solid state reactions between Sn and Cu, it can be concluded that from room temperature up to 50–60°C only the η'-phase grows with an observable rate and the reaction is controlled by the release of Cu atoms from the Cu lattice. The main diffusing species at room temperature is Cu. Above 60°C, the ε-phase starts to grow with measurable rate and its fraction out of the total intermetallic layer increases as annealing continues. Between 60 and 200°C, diffusion of Sn starts to control the growth of the η-phase. This is rational, as the temperature is increased, volume diffusion starts to dominate and grain boundary and interstitial diffusion of Cu do not play such a big role anymore. During this temperature interval, the ε-phase continues to grow at the expense of the η-phase and both Cu and Sn are mobile during its growth. Consistent with the Cu_3Sn rule, Cu is the more mobile species in the ε-phase. As interdiffusion in the η-phase at temperatures 160–220°C requires higher activation energy, the fraction of the ε-phase out of the total intermetallic layer thickness is larger at temperatures 160–200°C than at 220°C. It should also be noted that if solid state aging follows

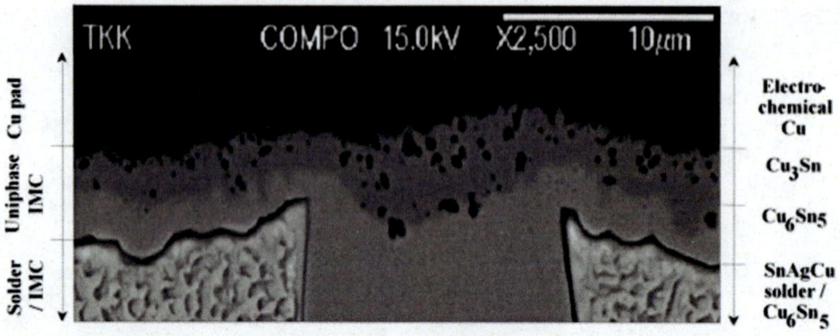

Fig. 9.17 Electrolytic Cu|Sn diffusion couple annealed at 125°C for 1000 h

soldering, the resulting morphology after aging is dependent on the initial morphology formed during the solid–liquid contact. Finally, excessive void formation and redistribution of impurities in Cu_3Sn phase already at low temperatures, as sometimes detected, induces some reliability concerns for Cu conductors and under bump metallization (see Fig. 9.17).

9.5.2 *Ni–Sn*

Ni is often used as a diffusion barrier layer between Cu and Sn, since the reaction rate of Ni with liquid Sn is typically smaller than that of Cu with Sn. This, in turn, results in thinner IMC layers in the Sn–Ni system than in the Sn–Cu system. The Sn–Ni system is well characterized. There are stable three IMCs (Ni_3Sn, Ni_3Sn_2 and Ni_3Sn_4) at the temperatures of our interest i.e. below 260°C. It has been reported that the Ni_3Sn_4 phase (or metastable phases) forms during soldering at the interface between liquid non-Cu-containing solders, such as SnPb, SnAg, or SnSb, and Ni metallization. The Ni_3Sn_4 phase forms even though it is thermodynamically the least stable of all the Ni–Sn compounds and it possesses the most complicated crystal structure. Similar arguments regarding the initial periods of dissolution and Ni_3Sn_4 nucleation are valid here as already discussed in the case of Cu. However, the solubility of Ni is much smaller and, therefore, its dissolution rate is much lower.

On the other hand, when using Pb-free solders that include even small amounts of Cu, the first phase to form at the interface changes so that it is the $(Cu,Ni)_6Sn_5$ instead of the Ni_3Sn_4. This was the case with the drop test results presented in Sect. 9.3 above where the component had Ni|Au under bump metallization and the bump was SnAgCu alloy. The incorporation of Ni in the Cu_6Sn_5 was observed to weaken the mechanical properties of intermetallic layer by producing severe cracking of the intermetallic phase (see Fig. 9.18). The cracking of the $(Cu,Ni)_6Sn_5$ is sometimes visible even after reflow.

Hence, to determine the critical Cu concentration (shown in Fig. 9.19), which causes the formation of $(Cu,Ni)_6Sn_5$ instead of $(Ni,Cu)_3Sn_4$, is of practical significance. The critical amount of Cu required to change the first forming phase is

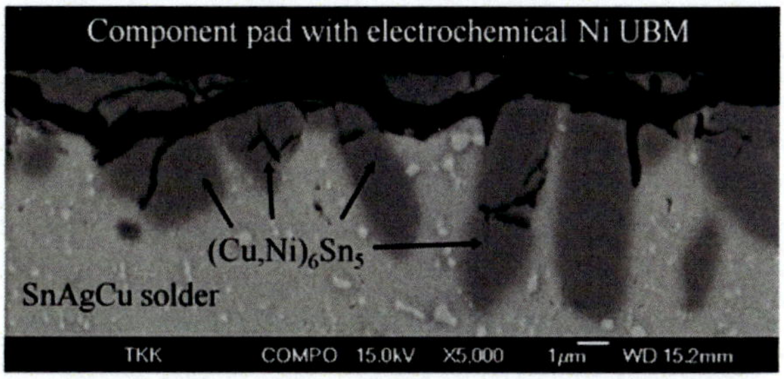

Fig. 9.18 Cracking of the $(Cu,Ni)_6Sn_5$ reaction layer is characterized by relatively large number of secondary cracks along the crack path

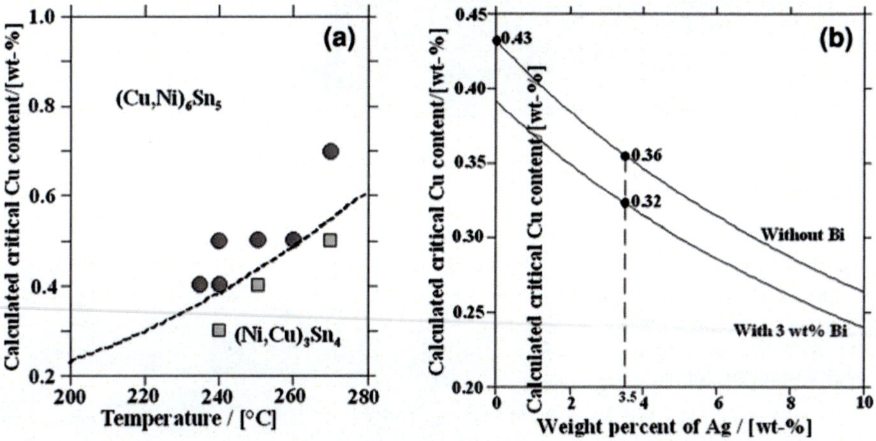

Fig. 9.19 **a** Calculated critical Cu content in liquid Sn to change interfacial reaction product from $(Ni,Cu)_3Sn_4$ to $(Cu,Ni)_6Sn_5$, as a function of temperature together with the experimental points. **b** Effect of alloying Ag and Bi to Sn on the critical Cu content at $T = 250°C$ [43]

dependent on temperature and other elements in the alloy. As shown in Fig. 9.19a, the critical Cu concentration increases with increased peak reflow temperature. On the other hand, the alloying elements, for example Ag and Bi shown in Fig. 9.19b, decrease the critical Cu content. The detailed thermodynamic explanation for this transition has been explained elsewhere [24, 43] and is therefore not repeated here.

9.5.3 Ni–Sn–P

During recent years, the electroless Ni|immersion Au coatings (denoted Ni(P)|Au) have been extensively used in high-density component assemblies as a surface finish on printed wiring boards (PWB) because they have many advantages over other surface finishes such as hot air solder levelling (HASL) SnPb finishes: Ni(P)|Au coatings provide desired flat and uniform pad surface and they maintain good wettability, even after multiple reflows. Furthermore, Ni(P)|Au coatings provide higher mechanical strength and resistance against thermal fatigue of

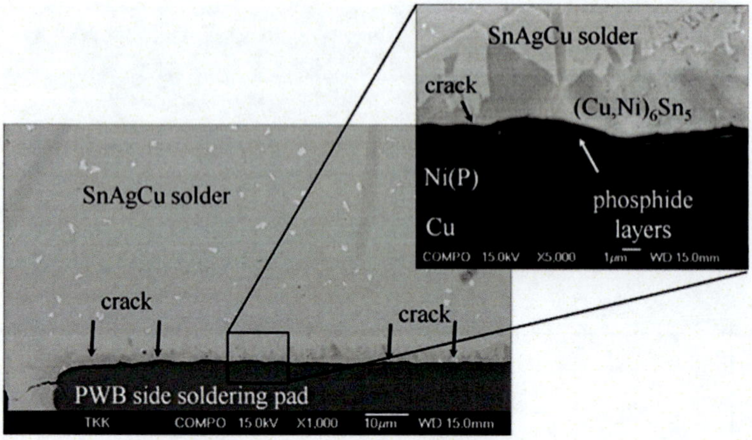

Fig. 9.20 Cracking of the PWB side interfacial reaction layers between the Ni(P)|Au and SnAgCu solder

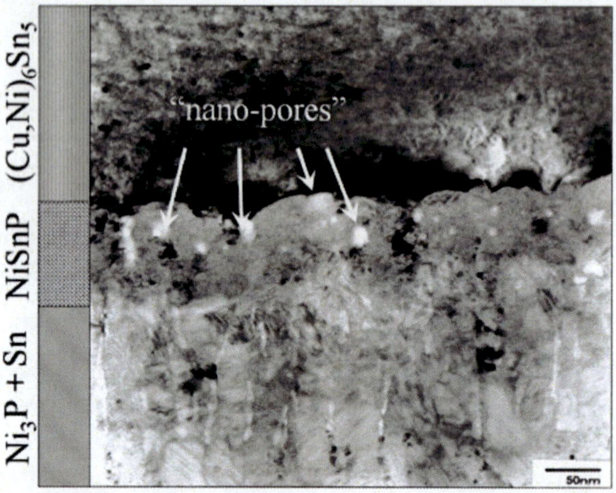

Fig. 9.21 Transmission electron micrograph from the interfacial reaction layer in the interconnections on the Ni(P)|Au metallization

lead-free solder interconnections than can be achieved when using organic solder preservatives on Cu pads. Moreover, as an insulating material OSP has a disadvantage when connectors are assembled on the same board. Finally, nickel acts as a diffusion barrier layer between Sn-based solders and copper conductors and prevents the underlying copper from reacting with Sn.

On the other hand, reliability problems have been reported when using Ni(P)|Au coatings with Sn-based solders. Under mechanical shock, loading cracks propagate in a very narrow reaction zone (NiSnP in Fig. 9.20) between the two-phase layer and the $(Cu,Ni)_6Sn_5$ intermetallic layer, as shown in Fig. 9.21. During the electroless coating process, Ni is deposited on Cu together with phosphorous, because hypophosphite is used as a reducing agent in plating baths. It is the presence of phosphorus in the surface finish layers that has been observed to be associated with the above-mentioned reliability problems. Although the wetting occurs properly and the chemical reaction between Sn and Ni is evident, the interfacial strength is not adequate. The weakest interfacial reaction product readily fractures under mechanical stress even during cooling from soldering temperatures and leaves behind an open circuit. The formation mechanism of the interfacial reaction products that causes the reliability problem has not yet been identified with certainty.

When one considers reaction between liquid Sn (or SnAgCu solders) and Ni(P)|Au surface finish, the following reaction sequence takes place: Immediately after the solder has melted Au, Ni and P start dissolving rapidly into the liquid solder, which—due to the existence metastable liquid miscibility gap in the Sn–P–Ni system—is divided into Sn-rich liquid (L_1) and (Ni,P)-rich liquid (L_2). Because of high Ni content, the L_2 is unstable and it transits into the nanocrystalline NiSnP, which provides the substrate for the formation of the $(Cu,Ni)_6Sn_5$. Subsequently, the metastable NiSnP layer starts to transform into the columnar Ni_3P. In this transformation, Sn atoms and impurities, which do not dissolve into the Ni_3P, diffuse toward the remaining NiSnP layer. The impurities in this ternary layer are revealed as numerous small "pores" detected with the high-resolution transmission electron microscopy (see Fig. 9.21). More detailed description of the above presented reaction sequence can be found from the literature [27].

Under mechanical shock loading, the reliability of interconnections on the Ni(P)|Au is inferior to that of interconnections on the Cu|OSP due to the formation of the complex P-rich reaction layers between the Ni(P) coating and the solder. Cracks nucleate and propagate in the porous and highly brittle NiSnP layer between the columnar two-phase (Ni_3P+Sn) layer and the $(Cu,Ni)_6Sn_5$ intermetallic layer.

9.5.4 Effect of Alloying and Impurity Elements on Cu–Sn IMC Formation and Properties

Owing to the reasons discussed earlier, there is a continuous interest to better understand and influence the properties of IMC layers in order to increase the reliability of solder interconnections. One way to influence the interfacial reactions and the resulting product layers in a given system is to alloy either metallization (conductor) or solders with small amounts of additional elements. It is to be noted that also the presence of impurities in the interconnection system may have marked effects on the properties and growth of IMC layers. Additional elements can basically have three major effects on the reaction between the solder and the conductor metal: (a) they can increase or decrease the reaction/growth rate, (b) additives can change the physical properties of the phases formed (in the case of Cu and Sn, Cu_3Sn and Cu_6Sn_5), and (c) they can form additional reaction layers at the interface or they can displace the binary phases that would normally appear and form other reaction products instead. Further, these elements can be divided roughly into two major categories: (a) elements that only change the activities of

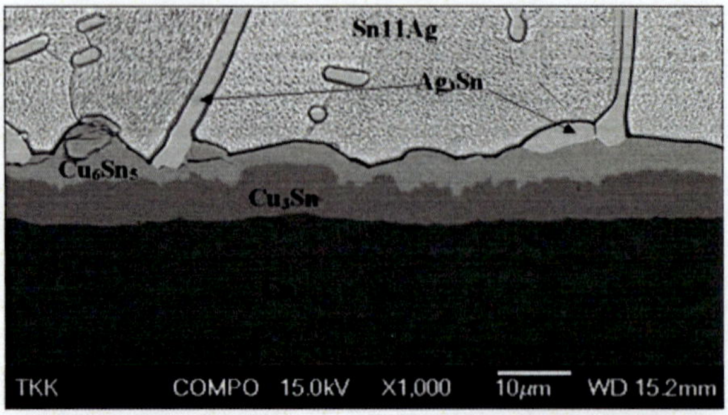

Fig. 9.22 Reaction layers formed in the reaction between Sn11Ag and Cu after annealing at 150°C for 2560 h

species taking part in the interfacial reaction and do not participate themselves (not extensively soluble in IMC layer) and (b) elements that take part in the interfacial reaction in question (generally show marked solubility in IMC layer). The elements belonging to the latter category usually have more pronounced effect on the IMC formation. Because the Sn–Cu system is the most important binary reaction couple associated with soldering, we will take it as the base system and consider the effect of one element from each category (a) Ag and (b) Ni on the binary reaction.

9.5.4.1 Effect of Ag

Figure 9.22 shows the interface between Sn11Ag (at%) and Cu after annealing at 150°C for 2560 h. The large Ag_3Sn IMCs are also clearly seen in the micrograph. The plot of IMC thickness as a function of square root of time for this couple is shown in Fig. 9.23.

As can be seen (see Figs. 9.23, 9.25) the growth behavior is very similar to that of 100Sn–Cu system (see Fig. 9.24) as shown also in previous publications [28–30] and indicated by the Sn–Ag–Cu phase diagram and thermodynamic data [31, 32].

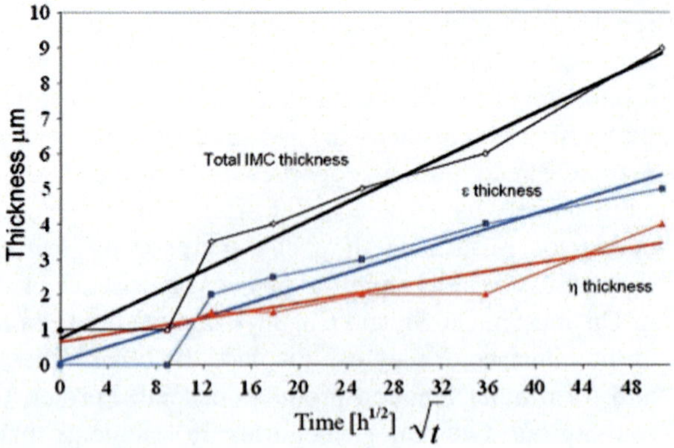

Fig. 9.23 The plot of IMC thickness as a function of square root of time (hours) for the Sn11Ag–Cu couple

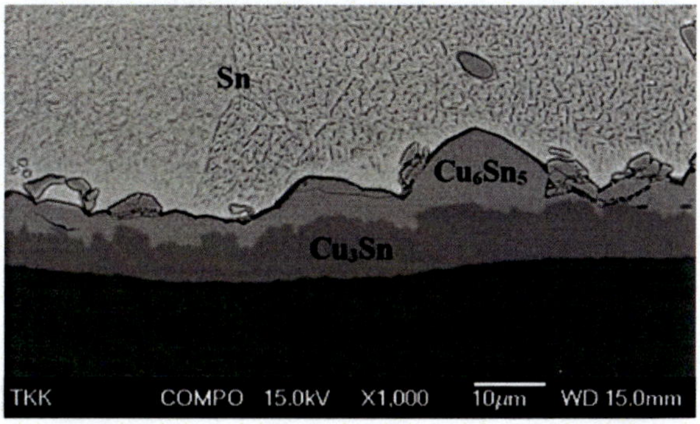

Fig. 9.24 Reaction layers formed in the reaction between pure 100Sn and Cu after annealing at 150°C for 2560 h

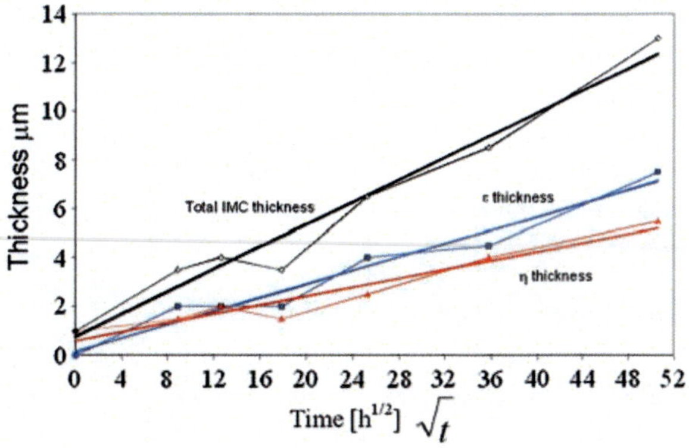

Fig. 9.25 The plot of IMC thickness as a function of square root of time (hours) for the 100Sn–Cu couple

The reasons for the observed behavior can be interpreted with the help of thermodynamic-kinetic information. As Ag does not dissolve in Cu_6Sn_5 or Cu_3Sn, the only way for it to influence the interfacial reactions is by influencing the activities of the reacting species, in this case that of Sn [32]. The presence of Ag slightly decreases the activity of Sn in the solder and, thus, reduces the driving force of the diffusion of Sn through the Cu_6Sn_5 layer. As Sn is the main diffusing species in Cu_6Sn_5 at these temperatures, the resulting decrease in the Sn flux through the layer favors the growth of Cu_3Sn, where Cu is the main diffusing species [24, 33–40]. It is to be noted, however, that the IMC growth behavior in the Cu–Sn system is not only strongly temperature dependent as already discussed in Sect. 9.5.1. The relative thickness of Cu_6Sn_5 and Cu_3Sn is changed as a function of temperature due to the changes of both driving forces and intrinsic diffusion fluxes of Cu and Sn in the phases. It should also be noted that if solid state aging follows soldering, the resulting morphology after aging is dependent on the initial morphology formed during the solid–liquid contact. In addition, it must be noted that the growth of the phases in a given multiphase reactive diffusion couple is not independent, but is affected by other growing layers [41].

Hence, Cu_3Sn consumes some of the Cu_6Sn_5 as it grows, and this effect is enhanced when the flux of Sn through Cu_6Sn_5 is reduced, as shown in previous publication [42]. This effect can be seen by comparing the diagrams in Figs. 9.23, 9.25, as not only the thickness of Cu_3Sn surpass that of Cu_6Sn_5 earlier that in the case of pure Sn–Cu couple, but also the total IMC thickness is somewhat reduced in comparison with Sn–Cu couple.

This is because as the flux of Sn through Cu_6Sn_5 is decreased, also the growth of Cu_3Sn is slowed down due to the reduced availability of Sn at the Cu_3Sn/Cu_6Sn_5 interface. Finally, it is to be noted that as the solubility of Ag in solid Sn is very small it has only minor effect on the activity of Sn [32]. Therefore, the difference between the growth kinetics of IMCs in Sn11Ag–Cu and 100Sn–Cu systems do not display marked differences. One can also ask whether the formation of Ag_3Sn platelets at the interface may play some role in the intermetallic compound formation by physically blocking the diffusion of Sn to Cu_6Sn_5 layer at specific locations. However, this effect is anticipated to be of much less of significance than the effect of Ag on the activity of Sn. This is because there still remain plenty of paths for Sn diffusion to Cu–Sn IMC layers despite the Ag_3Sn precipitates and it can even be argued that the phase interface between Ag_3Sn precipitates and Sn-based solder matrix offers fast diffusion paths for Sn atoms along the interface, thus acting against the above-mentioned “blocking effect”.

9.5.4.2 Effect of Ni

Figure 9.26 shows the interface between Sn1Ni (at%) and Cu after annealing at 150°C for 2560 h. The plot of IMC thickness as a function of square root of time for this couple is shown in Fig. 9.27. When Figs. 9.25, 9.27 are compared, it can be seen that: (a) total IMC layer thickness is smaller than in the Sn–Cu case, (b) its growth does not follow parabolic kinetics and (c) especially the thickness of Cu_3Sn is drastically reduced. As the growth kinetics related to this system have been described in detail (both in liquid and in solid state) [42, 43] together with lengthy list of appropriate references, only a brief discussion is presented here.

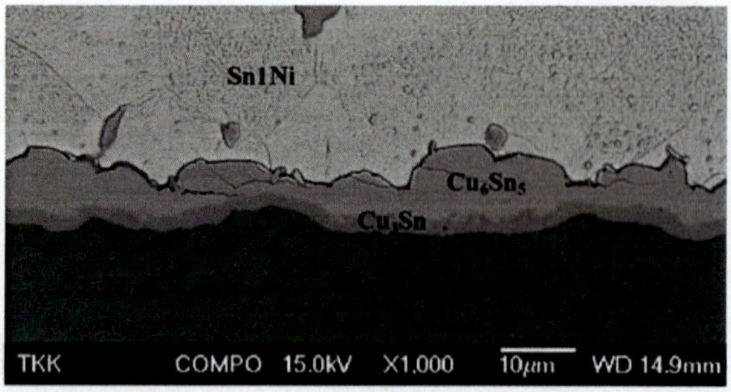

Fig. 9.26 Reaction layers formed in the reaction between Sn1Ni and Cu after annealing at 150°C for 2560 h

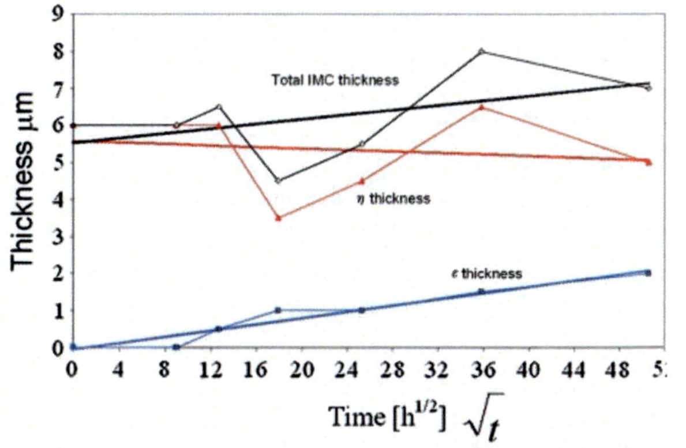

Fig. 9.27 The plot of IMC thickness as a function of square root of time (hours) for the Sn1Ni–Cu couple

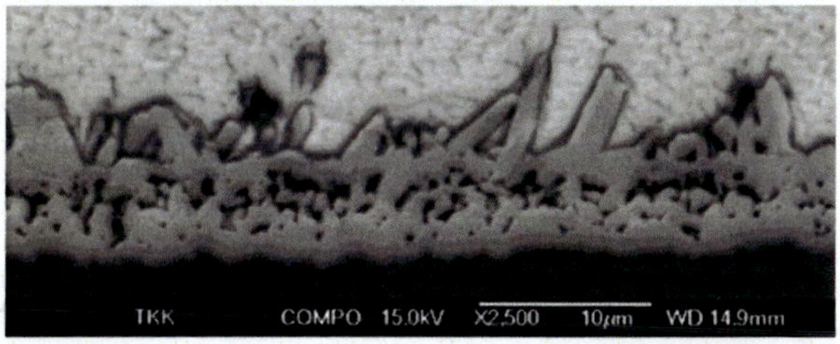

Fig. 9.28 Cu/Sn1Ni interface after reflow soldering

The deviation from the parabolic growth law of the total IMC comes mainly from the $(Cu,Ni)_6Sn_5$ layer, as it is seen that Cu_3Sn follows nicely diffusion-controlled growth kinetics (Fig. 9.27). The reason for the abnormal growth of $(Cu,Ni)_6Sn_5$ most probably has its origin in the liquid stage reaction. Because of the effect of Ni on the interaction between Cu and Sn atoms in the liquid (and thus also on the solubilities), the formation of a relatively thick and highly porous two-phase structure $[(Cu,Ni)_6Sn_5+Sn]$ takes place (see Fig. 9.28), as explained in [43].

Thus, during the solid state annealing, sintering of this porous structure takes place first, and only after it transforms to the dense $(Cu,Ni)_6Sn_5$ structure, can the "normal" parabolic growth continue. Based on the experimental results, it seems that the sintering of the porous $(Cu,Ni)_6Sn_5$ layer has barely reached its end within the time frame of the annealing. On the other hand, the decreased growth of Cu_3Sn can be explained with the help of thermodynamic-kinetic considerations about the Cu–Sn–Ni system [44]. During soldering process, the solder is in molten state for about a minute. During such a short period of time, the ternary compounds that have been observed or proposed to be present in the system [45–47] cannot form. Therefore, the metastable phase diagram without these compounds can more realistically describe the equilibria between solder alloys and conductor metals [24, 41, 43]. Using the optimized dataset [32], the Gibbs free energy curves corresponding to this metastable phase diagram were calculated. When 1 at% of Ni is added to Sn, the Ni content of $(Cu,Ni)_6Sn_5$ is about 20 at% (measured by EDS) and its stability is increased markedly. The Cu_3Sn being in equilibrium with it is practically free of Ni, and so its stability will remain practically the same as that of pure Cu_3Sn. As a result, the driving forces for the diffusion over the intermetallic compounds will change. The driving force for the diffusion of Sn through the $(Cu,Ni)_6Sn_5$ layer is now about ten times the driving force for the 100Sn–Cu couple. On the other hand, the driving force for the diffusion of Cu through Cu_3Sn reduces almost to zero. This means that through the Nernst–Einstein relation the fluxes through the Cu_3Sn and the $(Cu,Ni)_6Sn_5$ layers are markedly altered, thus resulting into the observed growth behavior (i.e. the marked decrease in the layer thickness of Cu_3Sn layer), as also previously shown [42, 43].

9.6 Summary

The ongoing trend toward ever smaller electronic products forces to larger scales of integration and to the use of smaller and finer pitch components, such as (wafer-level) chip scale packages and flip chips. Because of the small-scale interconnections, components become closer to the printed wiring boards and the strains and stresses experienced by solder interconnections are increased. These miniaturized interconnections must be able to withstand variety of different loadings out of which the sudden mechanical shocks are perhaps the most severe threat to their reliability. Hence, understanding why the different tests produce different failure mechanisms and, ultimately, different reliability performance is of importance. This can only be achieved by knowing how the stress states are produced, how the materials respond to different types of loading, and how the microstructures of solder interconnections affect the failure mechanisms. Therefore, the emphasis of this paper was placed on describing the loading condition under mechanical shock loading and how it differs from that of thermal cycling.

The failure modes under mechanical shock loads are different from those typically observed in the thermal cycling, where solder interconnections fail by cracking through the bulk solder assisted by the localized recrystallization. Under the mechanical shock, loading the strain-rate hardening of the solder forces cracks to propagate in the intermetallic layers on either side of the interconnections instead of the bulk solder.

The drop reliability of the high-density electronic assemblies was studied by applying standardized test procedures for test vehicles that represent the technologies and lead-free materials typically used in portable electronic products. The component studied in this work was a 12 mm × 12 mm Chip Scale Packaged Ball Grid Array component, which today is the most commonly used package in portable electronic products. The assembled component boards had optional under bump metallizations (Ni or Cu), printed wiring board coatings (Cu|OSP or Ni(P)|Au) and solder compositions (SnAgCu or low-silver SnAgCuNi). The reliability of interconnections on the Ni(P)|Au was inferior to that of the interconnections on the Cu|OSP due to the formation of the complex and mechanically poor P-rich reaction layers between the Ni(P) coating and the solder in the

Ni(P)|Au interconnections. Interconnections on Cu|OSP coatings, in turn, fail by cracking of the intermetallic compound layers on the component side. The binary Cu_6Sn_5 formed on bare Cu under bump metallization showed good structural integrity but the $(Cu,Ni)_6Sn_5$ formed on the Ni metallization is weaker. Solder interconnections with the low-silver nickel containing composition (SnAgCuNi) showed higher average number of drops-to-failure.

Three different failure modes were identified in the drop-tested component boards: (a) cracking of the interconnections in the component side intermetallic layer, (b) cracking of the copper trace on the PWB and (c) cracking of the bulk of solder interconnections on the component side. The mode (c) was found only in the SnAgCuNi interconnections, and the cracks were not observed to have propagated entirely through the interconnections. The SnAgCu interconnections failed by the mode (a) while the SnAgCuNi interconnections failed by the mode (b).

Because the reliability of electronic assemblies under mechanical shock loading is very often dependent on the ability of intermetallic layers to withstand the stresses produced in their operation conditions, the formation and properties of the three metallurgically different interfacial regions (Cu|SnAgCu, Ni|SnAgCu, Ni(P)|SnAgCu, and Cu|SnAgCuNi) were discussed in detail. Further, it was shown that alloying and impurity elements can have strong effects on the growth and properties of IMC layers. The binary Cu–Sn system was used as the base reaction couple. Based on the solubility of the alloying and impurity elements in the IMC layers, these elements can be divided into two categories: (a) alloying elements that do not dissolve in either Cu_6Sn_5 or Cu_3Sn and (b) elements that exhibit significant solubility in (usually) Cu_6Sn_5 and (possibly) Cu_3Sn. It was shown that the latter group of elements have markedly stronger effect on the growth behavior of IMC's in the Sn–Cu system than those belonging to the first group, because these (that do not dissolve in IMC's) can influence diffusion fluxes in the layers only indirectly through the activity of Sn. On the other hand, if the element dissolves in IMC layer it can alter its stability and probably also its microstructure, thus influencing the growth kinetics directly.

References

1. Mattila TT, Kivilahti JK (2004) Impact of printed wiring board coatings on the reliability of lead-free chip scale package interconnections. J Mater Res 19(11):3214–3223
2. Mattila TT, Kivilahti JK (2010) The role of recrystallization in the failure mechanism of SnAgCu solder interconnections under thermomechanical loading. IEEE Trans Compon Packaging Technol 33(3):629–635
3. Mattila TT (2005) Reliability of high-density lead-free solder interconnections under thermal cycling and mechanical shock loading, Espoo, HUT-EPT-13, http://lib.tkk.fi/Diss/2005/isbn9512279843/, Otamedia, p 202
4. Mattila TT, Laurila T, Kivilahti JK (2007) Metallurgical factors behind the reliability of high density lead-free interconnections. In: Suhir E, Wong CP, Lee YC (eds) Micro-and opto-electronic materials and structures: physics, mechanics, design, reliability, packaging, vol 1. Springer, New York, pp 313–350

5. Mattila TT, Marjamäki P, Kivilahti JK (2006) Reliability of CSP components under mechanical shock loading. IEEE Trans Compon Packaging Technol 29(4):787–795
6. Mattila TT, Kivilahti JK (2005) Failure mechanisms of lead-free CSP interconnections under fast mechanical loading. J Electron Mater 34(7):969–976
7. Mattila TT, James R, Nguyen L, Kivilahti JK (2006) Effect of temperature on the drop reliability of electronic assemblies. In: The proceedings of the 57th electronic component and technology conference, Reno, CA, May 29–June 1, 2007, IEEE/EIA CPMT
8. Mattila TT, Kaloinen E, Syed A, Kivilahti JK (2006) Reliability of SnAgCu interconnections with minor additions of Ni or Bi under mechanical shock loading at elevated temperatures. In: The proceedings of the 57th electronic component and technology conference, Reno, CA, May 29–June 1, 2007, IEEE/EIA CPMT
9. Mattila TT, Šimeček J, Kivilahti JK (2006) Failure modes of solder interconnections under mechanical shock loading at elevated temperatures. In: The proceedings of the 1st electronics systemintegration technology conference, Dresden, Germany, September 5–7, 2007, IEEE/EIA CPMT, pp 195–202
10. Mattila TT, Marjamäki P, Nguyen L, Kivilahti JK (2006) Reliability of wafer-level chip scale packages under mechanical shock loading. In: The proceedings of the 56th electronic component and technology conference, San Diego, CA, May 30–June 2, 2006, IEEE/EIA CPMT, pp 95–101
11. Mattila TT, Kivilahti JK (2006) Reliability lead-free interconnections under consecutive thermal and mechanical loadings. J Electron Mater 35(2):250–255
12. Marjamäki P, Mattila TT, Kivilahti JK (2006) A comparative study of the failure mechanisms encountered in drop and large amplitude vibration tests. In: The proceedings of the 56th electronic component and technology conference, San Diego, CA, May 30–June 2, 2006, IEEE/EIA CPMT
13. JESD22-B111 (2003) Board level drop test method of components for handheld electronic products. JEDEC Solid State Technology Association, p 16
14. Marjamäki P (2007) Vibration test as a new method for studying the mechanical reliability of solder interconnections under shock loading conditions, Doctoral dissertation. Helsinki University of Technology, (to be published)
15. Reinikainen TO, Marjamäki P, Kivilahti JK (2005) Deformation characteristics and microstructural evolution of SnAgCu solder joints. In: The proceedings of the 6th EuroSimE conference, Berlin, Germany, April 18–20, 2005, IEEE, pp 91–98
16. Nikander R (1999) Characterization of the mechanical properties of the dilute tin based solder alloys, Espoo, master's theses, Helsinki University of Technology, p 79
17. Reinikainen T, Kivilahti JK (1999) Deformation behavior of dilute SnBi(0.5 to 6 at. pct) solid solutions. Metallurgical Mater Trans A 30:123–132
18. Amagai M (2006) A study of nano particles in SnAg-based lead free solders for intermetallic compounds and drop test performance. In: The proceedings of the 56th electronic components technology conference, San Diego, CA, May 30–June 2, 2006, IEEE/CPMT, pp 1170–1190
19. Amagai M, Watanabe M, Omiya M, Kishimoto K, Shibuya T (2002) Mechanical characterization of Sn–Ag-based lead-free solders. Microelectron Reliab 42(6):951–966
20. Che FX, Luan JE, Baraton X (2008) Effect of silver content and nickel dopant on mechanical properties of Sn-Ag-based solders. In: The proceedings of the 58th electronic components technology conference, Orlando, FL, May 27–30, 2008, IEEE/CPMT, pp 485–490
21. Garner L, Sane S, Suh D, Byrne T, Dani A, Martin T, Mello M, Patel M, Williams R (2005) Finding solutions to the challenges in package interconnect reliability. Intel Technol J 9(4):297–308
22. Tanaka M, Sasaki T, Kobayashi T, Tatsumi K (2006) Improvement in drop shock reliability of Sn-1.2Ag-0.5Cu BGA interconnects by Ni addition. In: The proceedings of the 56th electronic components technology conference, San Diego, CA, March 31–June 2, 2006, IEEE/CPMT, pp 78–84

23. Laurila T, Vuorinen V, Kivilahti JK (2005) Interfacial reactions between lead-free solders and common base materials. Mater Sci Eng R 49(1–2):1–60
24. Bader WG (1969) Dissolution of Au, Ag, Pd, Pt, Cu and Ni in a Molten Tin-Lead Solder. Welding J Res Suppl 48(12):551–557
25. Bader WG (1980) Dissolution and formation on intermetallics in the soldering process. In: Proceedings of the conference on physical metallurgy and metal joining, St. Louis, MO, Oct 16–17. Warrendale, USA
26. Haimovich J (1989) Intermetallic compound growth in Tin and Tin-Lead platings over nickel and its effects on solderability. Welding J Res Suppl 68(3):102–111
27. Vuorinen V, Laurila T, Yu H, Kivilahti JK (2006) Phase formation between Lead-free SnAgCu Solder and Ni(P)/Au Finished on PWB. J Appl Phys 99(2):3530–3535
28. Bae K, Kim S (2001) Interdiffusion analysis of the soldering reactions in Sn-3.5Ag/Cu couples. J Electron Mater 30(11):1452
29. Choubey A, Yu H, Osterman M, Pecht M, Yun F, Younghong L, Ming X (2008) Intermetallics characterization of lead-free solder joints under isothermal aging. J Electron Mater 37(8):1130
30. Moon K, Boettinger W, Kattner U, Biancaniello F, Handwerker C (2000) Experimental and thermodynamic assessment of Sn-Ag-Cu solder alloys. J Electron Mater 29(10):1122
31. IPMA (2009) The thermodynamic databank for interconnection and packaging materials. Helsinki University of Technology, Helsinki
32. Bhedwar H, Ray K, Kulkarni S, Balasubramanian V (1972) Kirkendall effect studies in copper-tin diffusion couples. Scr Metallurgica 6:919
33. Onishi M, Fujibuchi H (1975) Reaction-Diffusion in the Cu-Sn system. Trans Jpn Inst Metals 16:539
34. Oh M (1994) Growth kinetics of intermetallic phases in the Cu-Sn binary and the Cu-Ni-Sn ternary systems at low temperatures. Doctoral Dissertation, Lehigh University
35. Tu KN (1973) Interdiffusion and reaction in bimetallic Cu-Sn thin films. Acta Metall 21:347
36. Tu KN, Thompson R (1982) Kinetics of interfacial reaction in bimetallic Cu—Sn thin films. Acta Metall 30:947
37. Tu KN (1996) Cu/Sn interdiffusion reactions: thin film case vs bulk case. Mater Chem Phys 46:217
38. Revay L (1977) Interdiffusion and formation of intermetallic compounds in tin-copper alloy surface coatings. Surf Technol 5:57
39. Paul A (2004) The kirkendall effect in solid state diffusion. Doctoral Dissertation, Technical University of Eindhoven
40. Laurila T, Molarius J (2003) Reactive phase formation in thin film metal/metal and metal/silicon diffusion couples. Crit Rev Solid State Mater Sci 28:185–230
41. Vuorinen V, Laurila T, Mattila T, Heikinheimo E, Kivilahti JK (2007) Solid state reactions between Cu(Ni) alloys and Sn. J Electron Mater 36(10):1355–1362
42. Vuorinen V, Yu H, Laurila T, Kivilahti JK (2008) Formation of the intermetallic compound between liquid Sn and different CuNix metallizations. J Electron Mater 37(6):792–805
43. Yu H, Vuorinen V, Kivilahti JK (2007) Effect of Ni on the formation of Cu_6Sn_5 and Cu_3Sn intermetallics. IEEE Trans Electron Packaging Manuf 30(4):293–298
44. Oberndorff P (2001) Lead-Free solder systems: Phase relations and microstructures. Doctoral Dissertation, Technical University of Eindhoven
45. Lin C-H, Chen S-W, Wang C-H (2002) Phase equilibria and solidification properties of Sn-Cu-Ni alloys. J Electron Mater 31(9):907
46. Schmetterer C, Flandorfer H, Luef Ch, Kodentsov A, Ipser H (2009) Cu-Ni-Sn: A key system for lead-free soldering. J Electron Mater 38(1):10
47. Vuorinen V (2006) Interfacial reactions between Sn-based solders and common metallizations used in electronics. Doctoral Dissertation, Helsinki University of Technology

Chapter 10
Impact of Humidity and Contamination on Surface Insulation Resistance and Electrochemical Migration

P.-E. Tegehall

10.1 Introduction

Adequate evaluation of the impact of humidity and contamination on the reliability of electronic circuitry must be based on acceleration of the crucial failure mechanisms that will occur in field conditions. When electronic products are used in humid environments, water will be adsorbed on most surfaces. The amount of water adsorbed will depend on the temperature and the relative humidity but also on the properties of the surfaces. More water will be adsorbed on hydrophilic surfaces compared to hydrophobic surfaces. When polymeric materials are used, these materials may also absorb water. The adsorbed and absorbed water will affect the surface insulation resistance (SIR) on the circuitries. On clean surfaces, the decrease in SIR will normally be too small to affect the functionality of the product, but if ionic contamination is present, the drop in SIR may be large enough to cause failures.

Humidity and contaminants do not only affect SIR. They may also cause electrochemical reactions resulting in oxidation of anodic conductors. Metal ions formed by the oxidation reaction can under certain conditions migrate towards cathodic conductors and there be reduced back to metal. The precipitated metal will form tree-like metal filaments called dendrites. These will grow towards the anodic conductor where the metal ions come from and, with time, may cause a short circuit when they reach the anodic conductor (Fig. 10.1). This failure mechanism is called electrochemical migration (ECM). There is also another failure mechanism closely related to ECM resulting in the formation of conductive anodic filaments (CAFs). The main difference between these two failure mechanisms is that formation of CAFs occurs within a laminate, whereas formation of

P.-E. Tegehall (✉)
Swerea IVF AB, Box 104, 431 22, Mölndal, Sweden
e-mail: per-erik.tegehall@swerea.se

G. Grossmann and C. Zardini (eds.), *The ELFNET Book on Failure Mechanisms, Testing Methods, and Quality Issues of Lead-Free Solder Interconnects*,
DOI: 10.1007/978-0-85729-236-0_10, © Springer-Verlag London Limited 2011

Fig. 10.1 Dendrites formed on an assembly exposed to condensation. The dendrite to the right has caused a short circuit and has been partly blown away by the resulting electric discharge

dendrites normally occurs on the surface of a laminate or a component. However, only failures due to degradation of SIR and formation of ECM will be covered here.

10.2 Definition of Terms

Surface insulation resistance testing has long been used to evaluate the impact of contaminants on current leakage and dendrite formation. However, it may be difficult to interpret the results from such testing since misconceptions and poor understanding of the objectives of SIR testing are common [1]. The difficulties with using SIR testing as a tool for reliability evaluations of assemblies are caused by several factors. The most important are as follows:

- Lack of definitions of expressions used or inconsistent use of defined expressions
- Insufficient knowledge of the failure mechanism
- Poor understanding of the objective with SIR testing
- Difficulties in testing functional assemblies. Specially designed test patterns are required for adequate testing

Surface insulation resistance is defined in IPC-9201 [2] as "a property of the material and electrode system. It represents the electrical resistance between two electrical conductors separated by some dielectric material(s). This property is loosely based on the concept of sheet resistance, but also contains element of bulk conductivity, leakage through electrolytic contaminants, multiple dielectric and metallisation materials and air". That is, SIR is defined as a materials system property.

However, the main purpose of almost all SIR test methods used today is not primarily to measure SIR. Instead, the objective is to evaluate the long-term reliability of materials systems in humid environments [2]. One of several failure mechanisms that may affect the long-term reliability is electrochemical migration. It is defined in IPC-TR-476 [3] as "the growth of conductive metal filaments on a printed board under the influence of a DC voltage bias. This may occur at an external surface, an internal interface, or through the bulk material of a composite. Growth is by electro-deposition from a solution containing metal ions which are dissolved from the anode, transported by the electric field and redeposited at the cathode. We are thus excluding phenomena such as field induced metal transport in semiconductors and diffusion of the products arising from metallic corrosion".

Other failure mechanisms in humid environments that may affect the long-term reliability are degradation of solder masks and conformal coatings, degradation in dielectric strength and corrosion [2]. Most of these failure mechanisms are not or only indirectly related to SIR [4]. Tests to evaluate the impact of humidity are often called temperature–humidity (TH) tests or, if a bias is applied during the tests, temperature–humidity-bias (THB) tests. Many of these tests do not include SIR measurements.

10.3 Description of Failure Mechanisms

10.3.1 Effect of Humidity and Contaminants on Surface Insulation Resistance

Between two conductors of different potential separated by a dielectric material, a current leakage will occur. If the conductors are embedded in the dielectric material, the current leakage will depend on the resistivity of the bulk material. Besides the properties of the dielectric material, the resistivity will also be affected by the amount of absorbed humidity and contaminants present.

When conductors are located on the surface of a dielectric material, the current leakage is determined by the integrated effect of both surface and bulk resistivity. Some current leakage may also occur through the air. Since measurement of the resistance between two conductors on the surface of a dielectric material inevitably will include both bulk and surface resistance, the measured resistance is usually denoted by surface insulation resistance [5, 6]. A special case occurs when the conductors are located between two materials, for example conductors in an inner layer or conductors under a solder mask. The current leakage will then occur through the two bulk materials but also in the interface between the two materials. If contaminants are present in the interface, a large fraction of the current leakage then may take place there.

Dielectric materials that have been properly qualified for production of PCB assemblies have high surface and bulk resistivity. Low surface and/or bulk resistivity is almost always caused by either improper processing (not fully cured or properly mixed) or contaminants. Furthermore, it is normally the result of ionic conduction caused by non-reacted constituents, by-products and/or ionic contaminants. Since migration of ions is involved in the current leakage, the presence of water is normally a prerequisite for ions to take part in the process. Without water, ions will not be formed. Therefore, ionic contamination has a negligible effect on the resistivity in dry conditions [7].

In humid conditions, the bulk resistivity is normally much higher than the surface resistivity [6]. The reason for this behaviour is that a film of adsorbed water is formed on the surface. On α-alumina substrates, roughly one monolayer of water is adsorbed at 35% RH and more than 5 layers of water molecules at relative humidities higher than 70% [8]. Since the surface of the alumina substrate is hydrophobic, thicker water films can be expected to be adsorbed on hydrophilic organic substrates, but it is more difficult to determine the thickness of adsorbed water films on organic substrates such as epoxy- and polyimide–glass laminates since these materials also absorb water.

Due to the intrinsic ionisation of water into hydrogen and hydroxide ions, an adsorbed water film has always some conductive properties. This may be further enhanced by adsorption of some gases from the atmosphere that form ionic compounds with water, for example carbon dioxide. Nevertheless, the impact of a pure water film on the surface resistivity does not normally pose a problem. The conductivity of an adsorbed water film is much less than for bulk water [8]. A water film five layers of molecules thick has a conductivity approximately two orders of magnitude lower than that of bulk water, and a water film of 20 layers has a conductivity which is still one order of magnitude lower than bulk water. The reduced conductivity is due to a strong interaction between the adsorbed water molecules and the substrates, which affects the properties of the water film. Probably, it is also affected by a non-uniform distribution of the water on the surface.

Hence, SIR is to a very large extent affected by the relative humidity but also by the temperature. An increase in the relative humidity by about 20–30% or an increase in the temperature by 20–30°C have been found to decrease SIR by about one decade on fairly clean boards [6, 9–11].

However, it is the function of the water film as a medium for ionisation of ionic compounds that renders adsorbed water films hazardous. A large impact on SIR will be observed if hygroscopic contaminants are present. At or above a critical relative humidity, many salts absorb water forming a saturated solution of the salt. Thereby, the thickness and electrical conductivity of the absorbed water film is increased tremendously. In Table 10.1, the critical relative humidities at which this happens are given for a number of halide-containing inorganic compounds.

As shown in Table 10.1, the critical relative humidities vary considerably for the presented compounds. Furthermore, the data for calcium chloride and potassium bromide show that the critical relative humidity is temperature dependent, the compounds become more hygroscopic at higher temperatures.

Table 10.1 Critical relative humidities for a number of inorganic compounds at which saturated solutions are formed [12, 13]

Compound	Temperature (°C)	Relative humidity (RH) (%)
$LiCl{\cdot}H_2O$	20	15
KF	100	22.9
NaBr	100	22.9
$CaCl_2{\cdot}6H_2O$	24.5	31
$CaCl_2{\cdot}6H_2O$	5	39.8
KBr	100	69.2
NaCl	10–40	75
KCl	5–25	84
KBr	20	84
NaF	100	96.6

Anderson et al. [14] have shown that the resistivity of a surface contaminated with a hygroscopic compound decreases dramatically when the relative humidity is increased to values above the critical relative humidity for the hygroscopic compound. The resistivity is then rather little affected by a further increase in relative humidity. For an assembled PCB, the surface would be contaminated with a mixture of contaminants from various process steps having varying critical relative humidities. Therefore, the surface resistivity for an ordinary production assembly can be expected to decrease more or less gradually with increasing relative humidity.

Not only ionic compounds may have hygroscopic properties but also polyglycols and many other types of organic non-ionic surfactants common in some types of fluxes and fusing oils are very hygroscopic even at relative humidities down to almost 0% [15]. When present alone, they contribute to a decrease in the surface resistivity mainly by increasing the thickness of the adsorbed water film. Since they do not contribute to ionic conduction, the decrease in surface resistivity is rather small. What makes them hazardous is the fact that they promote a dissolving medium for ionic contaminants with low hygroscopicity that would otherwise be rather harmless except at very high relative humidities. Thereby, strong synergistic effects may be observed when hygroscopic non-ionic compounds are mixed with ionic compounds of low hygroscopicity. For example, as shown in Table 10.2, a mixture of polyethylene glycol and adipic acid has the same impact on SIR as sodium chloride [16]. The figures in Table 10.2 also show that ionic compounds have negligible impact on SIR at relative humidities below the critical relative humidities for the compounds. The reason why NaBr and KCl do not have an impact on the surface conductivity despite a critical RH below that at which the measurements were performed may be due to that surface dispersed chemicals have lower critical RH than the bulk salts [16].

Fortunately, not all contaminants are hazardous. Some are even beneficial. Rosin, a common base in many fluxes is hydrophobic, i.e., water repellent [15]. Therefore, rosin residues decrease the amount of water adsorbed on the surface and thereby improve the surface resistivity. Rosin residues may also encapsulate

Table 10.2 Surface conductivity measured at 35°C and 90% RH for copper comb patterns on FR-4 substrate contaminated with various compounds [16]

Compound	Added amount (μg/cm^2)	Surface conductivity (ohm^{-1})	Critical RH[a] (%)
NaCl	2.00	1.3×10^{-8}	76
NaF	1.44	4.0×10^{-11}	97
NaBr	3.52	3.8×10^{-11}	84
KCl	2.55	3.1×10^{-11}	84
$MgCl_2$	1.63	2.9×10^{-8}	44
$CaCl_2$	1.90	9.5×10^{-8}	29
Adipic acid	5.00	2.7×10^{-11}	99.6
PEG 400[b]	13.70	3.2×10^{-10}	0
Adipic acid + PEG 400	5.00 + 13.70	1.4×10^{-8}	–
None (Reference)	0.00	3.5×10^{-11}	–

[a] Calculated values
[b] PEG 400 = polyethylene glycol with a molecule weight of 400

ionic contamination and thereby immobilise ions. In addition, water that condenses on a hydrophobic surface tends to form isolated droplets as water does on a greasy surface. Thus, even if condensation occurs, rosin residues improve the situation.

10.3.2 Effect of Humidity and Contaminants on Electrochemical Migration

Current leakage due to ionic conduction is usually accompanied by a reduction process at the cathode and an oxidation process at the anode. The cathode reaction may be:

$$2H_2O + 2e^- \rightarrow 2OH^- + H_2 \tag{10.1}$$

or

$$\tfrac{1}{2}O_2 + H_2O + 2e^- \rightarrow 2OH^- \tag{10.2}$$

Whereas the anode reaction may be:

$$H_2O \rightarrow \tfrac{1}{2}O_2 + 2H^+ + 2e^- \tag{10.3}$$

The cathodic reactions give rise to an increase in the pH at the cathode, whereas the anodic reaction brings about a decrease in the pH at the anode. The anode reaction may also involve oxidation of metals forming the anode. All common metals used in conductors can be oxidised including gold, platinum and palladium. However, the three latter can only be oxidised in the presence of contaminants forming strong complexes with the metal ions, for example chloride, bromide and iodide [17] of which the first two are common ingredients in fluxes. The oxidation of metals at the anode can be written:

$$\mathrm{Me} \rightarrow \mathrm{Me}^{z+} + z\mathrm{e}^{-}$$

Under favourable (or rather unfavourable) conditions, the dissolved metal ions can migrate to the cathode and there be reduced back to metal:

$$\mathrm{Me}^{z+} + z\mathrm{e}^{-} \rightarrow \mathrm{Me}$$

These two latter reactions are the same reactions that occur during electroplating. However, in contrast to electroplating, the metal precipitated in this case is not plated as a metal film on the cathode. Instead, it is plated in tree-like structures called dendrites (see Fig. 10.1). With time, the dendrites may completely bridge the space between the conductors, causing short circuits. Since the dendrites are very thin, short circuits usually burn off part of or the whole dendrite. Thus, failures caused by the formation of dendrites are normally intermittent in nature, and it may be very difficult to track the reason for such failures, especially if the dendrites have been burned off or if they are formed under components where it is difficult to visually observe them.

It has been proposed that the propensity of a metal to form dendrites can be assessed from the standard reduction potential: the lower the standard potential, the larger propensity to form dendrites [18]. Thus, the propensity to form dendrites should decrease in the following order: Ni, Sn, Pb, Cu, Ag and Au. However, the fact that silver has been found to be the metal that easiest form dendrites [3], indicates that it is not only the standard reduction potential that determines the propensity to form dendrites. If dendrites to form, the metal ions produced at the anode must be able to migrate all the way to the cathode without being precipitated as insoluble compounds. As already indicated, acidic conditions can be expected to prevail at the anode and alkaline at the cathode. However, the actual pH of the surface will also depend on contaminants present on the surface. Some common contaminants are alkaline, for example sodium and potassium carbonate from alkaline solder mask developers or residues from alkaline cleaning solutions. Examples of acidic residues are weak organic acids, such as adipic acid, forming a part of many no-clean fluxes.

Rather few, if any, of the metal ions causing growth of dendrites are soluble through the entire pH range that may prevail between the conductors, especially at neutral and high pH. For example, copper dendrites require a pH of less than 5 if they shall be able to form [19]. Silver(I) ions are among those metal ions that are soluble at a rather high pH [20] which is likely one of the reasons why silver has been found to be the metal that easiest form dendrites. When tin is oxidised, tin(II) ions are first formed which may be further oxidised to tin(IV). Tin(II) and especially tin(IV) require very low pH if they shall not be precipitated as hydroxides in water solutions. Lead(II), which is formed when lead is oxidised, is less soluble than silver(I) but more soluble than tin(II) and tin(IV). Thus, lead can expected to be more prone to form dendrites than tin under right conditions. In addition, for an assembly exposed to field conditions, exposure to high humidity levels will usually be followed by dry-up periods. Soluble metal salts may then be converted to more or less insoluble compounds, which will delay the growth of dendrites.

As already mentioned, gold can only be oxidised in the presence of certain anions, notably chloride, bromide and iodide. These are anions that form very strong soluble complexes with gold ions. In the presence of chloride ions, the anode reaction can be written:

$$Au + 4Cl^- \rightarrow AuCl_4^- + 3e^-$$

The strong complexes formed with chloride also make gold ions soluble in solutions with rather high pH, thus facilitating the migration of gold ions towards the cathode and thereby the formation of gold dendrites. Chloride and bromide ions form strong complexes also with cations of silver, copper, tin and lead. In fact, that is the reason why they are added to fluxes. Their strong tendency to form complexes with the metals commonly used as surface finishes facilitates the breaking of the oxides on these surfaces during the soldering process. Consequently, many of the substances added to fluxes in order to facilitate soldering can be expected to form strong complexes with metals common in surface finishes of printed boards although not necessarily as strong as those formed with chloride and bromide. The presence of complex-forming compounds is highly favourable for the formation of dendrites and is likely a prerequisite for dendrites to form in most cases.

An interesting fact is that most complexes formed with chloride and bromide ions have a negative charge. For example, the complexes that may form between silver and chloride, AgCl, $AgCl_2^-$, $AgCl_3^{2-}$ and $AgCl_4^{3-}$, are either neutral or negatively charged. The concentration of each species depends on the concentration of chloride ions in relation to silver ions. The larger the surplus of chloride ions, the higher the concentration of the species with three and four chloride ligands. Thus, in the presence of chloride and bromide ions, migration of silver ions from the anode to the cathode will not be driven by the potential difference since it will work in the opposite direction. Migration would be driven by difference in concentration. Formation of highly negatively charged complexes may also explain the finding by Bumiller et al. that as chloride contamination increases, the failure mechanism shifts from ECM to uniform corrosion of the anodic conductor [21]. Hence, it is possible that dendrites would form faster if power-off periods were included in the test, under condition that the humidity level is preserved.

10.4 Performance of SIR and ECM Measurements

It is far more complicated to measure SIR than it may seem at a first glance. Moreover, as discussed previously, the purpose of most "SIR test methods" is not to determine the SIR but to determine the risk for electrochemical migration. Therefore, the complexity and purpose of SIR measurements and the connection to ECM testing will first be discussed before test methods for SIR and ECM tests are reviewed.

10.4.1 SIR Measurements

SIR is determined using Ohm's law $R = U/I$. In humid conditions, most of the current on contaminated boards is generated by migration of ions. This will cause a depletion of ions in the area between the biased surfaces, which will result in a decrease in the current, i.e., an increase in SIR. Thus, measuring of SIR will affect the measured value. As shown by Chan [22], the increase in SIR on a contaminated board is fast at the beginning and then getting slower and slower. When a board was contaminated with 10 mg/in^2 (1.56 mg/cm^2) table salt, the initial SIR was 5.3×10^6 ohms at 30°C and 90% RH compared to about 1×10^{10} ohms for a clean board. When a bias of 100 V DC was applied to the contaminated board, SIR increased to 3.7×10^7 ohms after 60 s, 1×10^9 ohms after a few hours and 8×10^9 ohms after 100 h (Fig. 10.2). The clean test board showed almost no drift in SIR values. Thus, the SIR on the contaminated board approached the value of the clean board after 100 h of testing.

Obviously, in order to get reproducible data, one must define a stabilisation (electrification) time after the voltage has been applied before a measurement is taken. Many standards prescribe a stabilisation time of 60 s, but that may be a little too short since the SIR may still drift rather fast after 1 min [10, 22]. On the other hand, waiting too long will cause a too large drift in SIR. Furthermore, a measurement of SIR will be affected by previously performed measurements. This influence will be larger the longer the stabilisation time is. Another implication of the ionic contribution to current leakage is that the measured SIR and the optimal

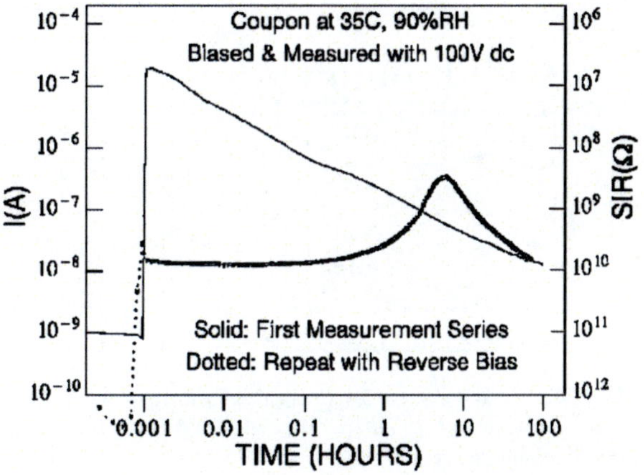

Fig. 10.2 Current and SIR for coupon contaminated with 10 μg/in^2 table salt [22] (© 1996 IEEE)

stabilisation time will depend on the applied voltage but also on the geometry of the test pattern.

The test patterns most frequently used for SIR and ECM testing are Y and interdigitated comb patterns of various designs, of which some are shown in Fig. 10.3.

In order to get test patterns that are more representative for the circuitry on PCBs, interdigitated solder lands have also been used. Three examples of such patterns are shown in Fig. 10.4. The two first types of test patterns have also comb patterns beneath the component bodies. In order to minimise current leakage between the test patterns for the interdigitated solder lands and the comb patterns, the connectors connecting the solder lands and the comb patterns should be coated with a solder mask.

The distance between the conductors and their number and length will all affect the SIR of a test pattern. Therefore, it is difficult to compare SIR values measured for patterns of different design. Ideally, this can be handled using the ohms/square concept [6]. If SIR is assumed to obey Ohm's law, the SIR for a square will be independent of the size of the square. By calculating the total number of squares in a test pattern (dark grey area in Fig. 10.5) and then dividing the measured SIR with the total number of squares, SIR can be expressed as ohms/square. If SIR had obeyed Ohm's law, this value would have been independent of test pattern geometry. However, in reality, it has been found that SIR does not obey Ohm's

Fig. 10.3 Examples of comb patterns and a Y-pattern from Test Board IPC-B-25A used for SIR and electrochemical migration testing

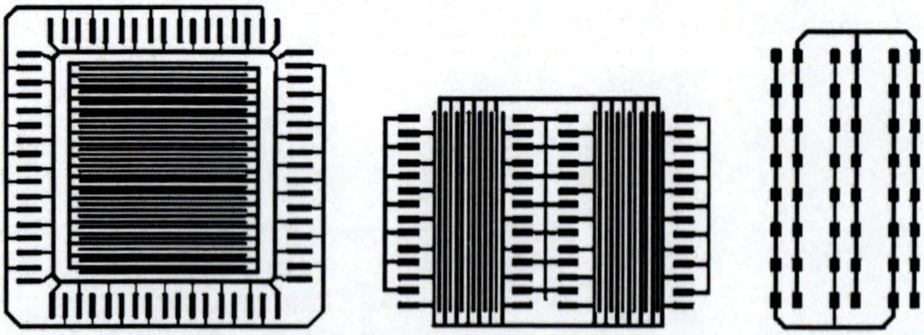

Fig. 10.4 Test patterns for measuring SIR between solder lands to an LCCC68 or PLCC68 component, two SO28 components and 0805 chip capacitors. Comb patterns are located beneath the IC components

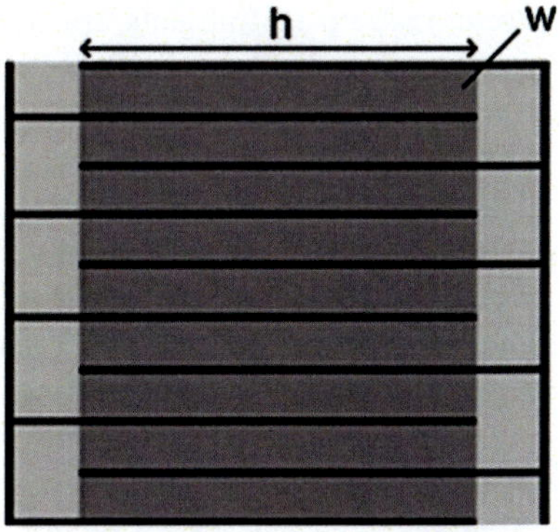

Fig. 10.5 The number of squares for the area of interlap in a comb pattern (*dark grey*) is calculated by dividing the length (h) of the conductors in the interlap area with the conductor spacing (w) and then multiplying by the number of conductors (n) minus one, i.e., $h(n-1)/w$

law due to the ionic conduction, so this concept should be used with caution [2, 23]. Furthermore, the current leakage will not be restricted to the area coloured dark grey in Fig. 10.5. Some current leakage will also occur in the light grey area. This will not have any significant impact on the calculated ohms per square values as long as the length of the conductors is much larger than the insulation distance between the conductors, as in comb patterns, but it will have a large impact on the values for interdigitated solder lands.

10.4.2 Purpose of SIR Measurements and Connection to ECM Testing

In many SIR test methods, a bias is applied to the test patterns during part of or for the whole test period, which may have a duration of up to 56 days. Although SIR is measured during these tests, the purpose of such extended tests is not to determine the SIR. Instead, the objective is to assess to the long-term effect of exposure to humidity and then especially the risk for electrochemical migration. Thus, a more appropriate name of such a test is electrochemical migration test.

The fact that SIR is measured during these tests may give the impression that the risk for electrochemical migration can be measured and determined from the measured SIR values. That is not usually the case. Although electrochemical migration causes migration of metal ions, it is not possible to separate electric current due to migrating metal ions from current due to other migrating ions. Thus, it is not possible to determine from a measured SIR value if, and to what extent, migration of metal ions have occurred. Moreover, the formation of a dendrite does normally not have any impact on SIR except for a few seconds when a short circuit is formed [3]. Due to the very small cross-sectional area, a part of the dendrite is burned off almost instantly when the short circuit occurs. Since SIR in most tests is measured once a day, or even less frequently, and each measurement takes less than a second, it is very unlikely that a very short drop in SIR would be detected.

Some modern instruments are designed to automatically measure SIR at a high frequency [24]. By using these instruments, the chance to detect an intermittent short circuit increases, but there is still a risk for missing failures that will occur. This could be mastered by using an event detector similar to the type used for detecting intermittent open circuits in solder joints, but in this case designed to detect intermittent short circuits. The time elapsed before a short circuit occurs would be a good measure of the propensity for dendrite formation.

As long as event detectors are not used to detect short circuits, dendrites have to be detected using some other method. Today, this is done by visual examination of the test boards using an optical microscope after the test has been run. If test patterns are located beneath components, it may be necessary to remove the components before the inspection. Using visual inspection, only the number, size and location of dendrites can be evaluated. It is not possible to determine when they were formed during the test. Furthermore, current-limiting resistors must be connected in series with each test pattern during the test in order to prevent that dendrites are completely burned off when a short circuit is formed.

If electrochemical migration cannot be detected by SIR measurements, then why measure SIR at all during an electrochemical migration test? Although there is no clear correlation between SIR and the inclination for dendrite formation, a high SIR indicates that electrochemical migration is unlikely to occur and a low SIR indicates an increased risk for dendrite formation. However, it is not possible to exclude that dendrites will form if SIR is above a certain value, nor will dendrite formation be inevitable if SIR is below a certain value. A large number of parameters not directly correlated to SIR have a large impact on the inclination for dendrite formation, for example, as previously discussed, the pH of the surface and the presence of substances forming strong complexes with metal ions.

There are also other factors that will affect the results from SIR and ECM tests. Originally, SIR measuring was done with an instrument using 500 V DC as measurement voltage [4]. Several standards still prescribe that a measurement voltage of 100–500 V DC should be used. Today, with much shorter insulation distances and lower voltages, this is a far too high voltage for many applications. Even if the applied polarisation bias generally is lower, a polarisation bias of 50–100 V DC is common. Such a high polarisation bias may induce failure mechanisms that would not occur in field conditions. Not only will the increased electrical field cause an increased migration rate of ions, but it will also increase the rate of the oxidation and reduction reactions at the anode and the cathode, respectively. This will cause concentrations of ionic species that would not be found on a true product. Furthermore, the pH gradient between the conductors would be larger than normal, i.e., the pH would be lower at the anode and higher at the cathode. Thus, the surroundings for migrating metal ions may be quite different from those experienced in field conditions. Therefore, it is advisable to not use polarisation and measurement voltages that differ too much from the voltage to be used for the final product. To be on the safe side, polarisation and measurement voltages not higher than twice the service voltage gradient have been recommended [4].

Due to the continuous increase in SIR caused by depletion of ions in the area between biased conductors during an ECM test, some standards prescribe that measurements of SIR should be taken using a voltage with reversed polarity and a stabilisation time of 60 s. As expected, by reversing the polarity, the SIR will first drop as ions start to migrate in the opposite direction and then increase again after some time when the area between the conductors once more becomes depleted of ions. The drop in SIR may begin immediately but, as shown in Fig. 10.2, it may also take more than 1 h before it occurs and 10 h to reach a minimum [22]. Therefore, using reversed bias when measuring SIR only makes it even more difficult to interpret the results of ECM tests. Since also a change of voltage with retained polarity will cause an unpredictable drift of SIR, measurements should preferably be done using the same voltage and polarity as the polarisation bias and without any interruption of the bias.

However, unbiased periods during the measurement could be used for another reason, but then they should be of some length and not performed directly before measurements are taken. As discussed previously, metal ions are probably in many cases mainly present as negatively charged complexes. Thus, the migration of these to the cathode must be by diffusion against the electrical field. A test consisting of periods with applied bias followed by periods without any bias to allow for "free" diffusion would then perhaps be a more adequate test producing dendrites in a shorter time.

Zou and Hunt have proposed a fast method to predict the propensity for electrochemical migration of solders and PCB finish materials based on Electrochemical Impedance (EI) measurements [25] (see Chap. 4). Using this method, the ionic resistance between two electrodes can be separated from the overall impedance of the system. The former was found to be dependent on the metal corrosion rate and solubility of the metal hydroxide. The propensity for dendrite formation for various materials was shown to have an inverse correlation with ionic resistance.

10.4.3 Test Vehicles

Production of test boards will inevitably cause contamination of the boards from processing solutions. Figure 10.6 shows the change of SIR during an ECM test for printed boards with different surface finishes produced by one printed board manufacturer using ordinary production processes [26].

Initially, the boards with HASL (hot-air solder levelling) finish have about five decades lower SIR than the boards with OSP (organic solderability preservative) and ENIG (electroless nickel/immersion gold) finishes. The much lower SIR values for the HASL finish is due to residues from the flux used in the HASL process. Due to the fact that it is impossible to clean away all flux residues [27, 28], HASL boards will always have rather low SIR. However, the impact of flux residues from different printed board manufactures may vary greatly.

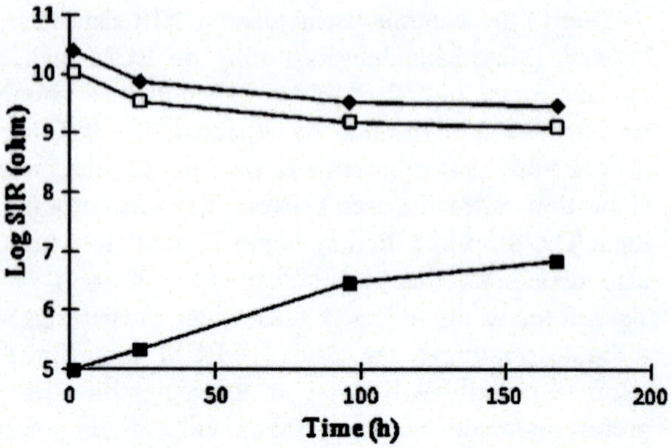

Fig. 10.6 Surface insulation resistance measured during an electrochemical migration test performed at 85°C, 85% RH and 5 V DC polarisation and measurement voltage for comb patterns on printed boards with the following surface finishes: HASL (*filled square*), ENIG (*open square*) and OSP (*filled diamond*)

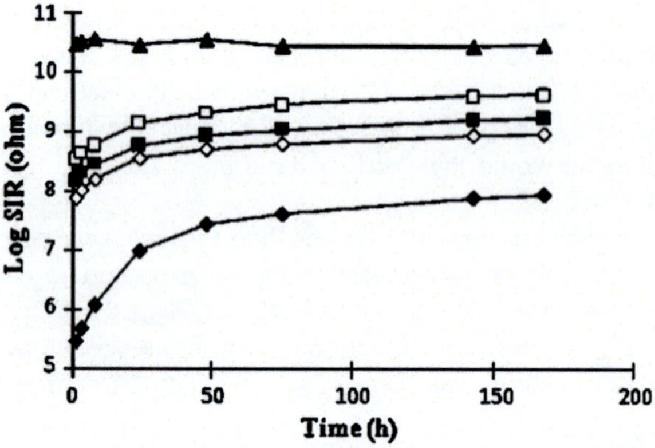

Fig. 10.7 Surface insulation resistance measured during an electrochemical migration test performed at 65°C, 85% RH and 5 V DC polarisation and measurement voltage for comb patterns on printed boards from four printed board manufactures: A (*filled square*), B (*open square*), C (*filled diamond*) and D (*open diamond*). For reference, results are given for a board from manufacturer A cleaned for 2 h before the test (*filled triangle*)

Figure 10.7 shows the results from SIR measurements of HASL printed circuit boards from four printed board manufacturers [27]. Initially, the SIR values differ three decades between the best and the worst boards. With 2 h of extensive cleaning using ultra-sonic agitation of boards from Manufacturer A, the SIR was improved by about two decades, i.e., to about the same value as OSP and ENIG boards if the SIR is adjusted for the difference in test temperature. However, when a thoroughly cleaned board went through a soldering process without any added flux, the SIR decreased almost back to the original value indicating that absorbed flux residues diffused up to the surface during the heat treatment [27].

To minimise the effect of contamination, test boards with bare copper patterns, without any solder mask applied and extensively cleaned are often used.

Thereby, the effect of a single material, chemical or process can be evaluated. Passing such a test is mandatory but not sufficient for approval of a material. When various materials and processes are combined, synergistic effects may occur. For example, as shown by Adams et al. [16], polyethylene glycol which is a common ingredient in fluxes used for HASL gives a strong synergistic effect when combined with adipic acid, a common ingredient in many no-clean fluxes used for soldering of assemblies (see Table 10.2).

The cleanliness of a printed board will affect the SIR of an assembled board whether cleaning of the assembly is performed or not. Figure 10.8 shows the results from measuring SIR on assemblies soldered at some assemblers having different types of soldering processes using printed boards from the four manufacturers presented in Fig. 10.7 [27]. Although the absolute measured SIR values varied considerably for the various assemblies, the variation in SIR on the printed boards was reflected in the SIR of the assemblies independent of the type of flux and cleaning process used. It can be noted that rosin residues have a very benign influence on SIR even when cleaning is performed.

Thus, one has to choose between a comparatively clean test board with a copper pattern and no solder mask, and a test board that is more representative for what will be used in true products. In the first case, it will be possible to evaluate the influence on SIR of a single material, chemical or process and to compare with data from other investigations, but it will not be possible to draw reliable conclusions regarding the SIR on the final product. In the second case, the results will

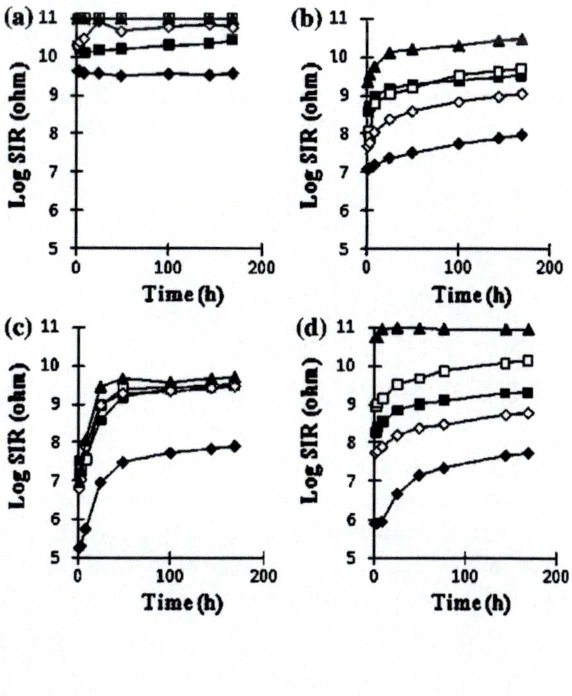

Fig. 10.8 Surface insulation resistance measured between solder joints to wave soldered SO28 components on assemblies manufactured by three different assemblers using their ordinary processes: RA flux and solvent cleaning (**a**), low-solid, rosin-based flux and no cleaning (**b**), water soluble flux and cleaning using pure water (**c**) and non-soldered reference boards (**d**). The printed boards used were produced by Manufacturers A (*filled square*), B (*open square*), C (*filled diamond*), D (*open diamond*) and, as reference, boards from manufacturer A cleaned for 2 h before soldering (*filled triangle*)

show the SIR properties of the final product but they will only be representative for the combination of materials, chemicals and processes used for producing the test boards. Hence, "clean" test boards are mainly used for screening of materials and processes, although it is not desired that material qualification be limited to that. It may lead to approval of materials that later will be found to cause unreliable products. Consequently, test boards used for final verification need to be representative of the materials and processes used in production. That is, final verification can only be done for specific applications. This is reflected in IPC's standard for printed board assemblies (PBAs): J-STD-001D, Appendix C, Material and Process Compatibility Testing, which specifies that the test vehicle should represent the substrate materials, assembly materials and fabrication processes used in production [29].

However, comb patterns are not appropriate for such evaluations. The locations most contaminated and difficult to clean on a PBA are areas between solder joints and beneath components. Furthermore, due to the shadowing effect of components, flux residues underneath components will not be heated to the same temperature as on the rest of the assembly. Therefore, they may not be deactivated or evaporated as they would have been on a free surface. Hence, to truly represent manufacturing processes, it is necessary to mount components on the test boards. By mounting dummy components without any electrical connections inside (open circuit) on test patterns of the type shown in Fig. 10.4, SIR can be measured between solder joints to soldered components. In fact, a test board can be designed and produced exactly in the same way as production PBAs. The results presented in Fig. 10.8 are from measurements of SIR between solder joints to SO28 components soldered to printed circuit boards.

By placing comb patterns under components, SIR can also be checked at such locations, although it must be pointed out that the comb pattern can have some influence on the efficiency of a cleaning process for components with small stand-offs. Ceramic chip capacitors and resistors usually have a very small stand-off, and it is difficult to clean beneath them, especially if they are glued to the boards using glue dots at the sides of the components. The best way to measure SIR beneath this type of components is to use a test pattern as shown in Fig. 10.4, but then only chip capacitors can be used. After the capacitors have been charged, the SIR can be measured.

10.5 Common Contaminants that may Affect SIR and ECM

Contamination on electronic circuits may originate from manufacturing of printed boards and assemblies, handling, storage, transportation, use, rework and repair. Contamination from all these phases of a product's life may affect SIR and ECM.

10.5.1 Contamination from Manufacturing of Printed Boards and Assemblies

The most common laminate used for production of printed boards is FR-4, glassweave-reinforced epoxy. When the epoxy resin in the laminate is cured, small amounts of sodium chloride is formed as a by-product [30]. To render some resistance against fire, brominated flame retardants are added to the laminate. The function of the brominated flame retardants is to liberate hydrogen bromide when exposed to fire. However, some decomposition may already occur during soldering processes [31]; the higher the soldering temperature and the longer the exposure to high temperatures, the more hydrogen bromide is liberated. Furthermore, to render solder masks UV-curable, photo-initiators are added which are not included in the polymerisation reaction [30]. Thus, the board materials themselves contain contaminants of which some are ionic. Also, if the epoxy is under cured, flux ingredients may react with flame retardants at the surface liberating bromide ions [31].

The production of PCBs includes many wet chemical treatments which all may leave residues. Examples of chemical processes are desmearing of drilled holes, etching processes, plating of copper and surface finishes and cleaning processes. It is important that thorough cleaning is performed in order to remove as much as possible of process chemicals.

The most contaminating process used for PCB fabrication is application of solder, whether it is applied using hot-air solder levelling (HASL) or hot-oil fusing. This process involves fluxing of the boards followed by heating to 240–280°C during solder application or hot-oil fusing, i.e., at temperatures far above the glass transition temperature (T_g) for FR-4 laminates and around the T_g for polyimide laminates [32]. The polymer resin in the laminate becomes soft at temperatures around and above T_g, and flux ingredients are absorbed into epoxy resins by this treatment [27, 33] and probably also into polyimide resins. Epoxy and polyimide laminates may absorb flux ingredients to some extent even at temperatures as low as 125°C [34].

One group of ingredients that may be absorbed is polyglycols. It has been known for a long time that polyglycol materials are absorbed into the epoxy resin of FR-4 boards and that it is very difficult to remove these residues by cleaning [35, 36]. Inorganic ionic compounds may also be absorbed. High concentrations of chloride and bromide have been found in FR-4 laminates and solder masks [27, 28]. These residues are not possible to remove completely even if very efficient cleaning processes are used. The halides are believed to originate from fluxes used for HASL or fusing. There are indications that halides may diffuse through several layers in multilayer boards [37].

Fluxes used for soldering of assemblies will also leave residues on the surface, and some of the flux ingredients may be absorbed into the resin.

10.5.2 Contamination from Handling, Storage and Transportation

Common contamination sources during handling, storage and transportation are fingerprints and dust from the environment and packaging materials. Fingerprints contain hygroscopic salts of which a large part is sodium chloride. They also contain organic acids that may form strong complexes with copper, silver, tin and lead ions. Thus, fingerprints have large impact on surface resistivity, and they promote the formation of dendrites. The composition of dust varies considerably between different locations. They may have a high content of ionic materials.

10.5.3 Contamination from Use Environments

Assemblies often become contaminated in the field environment. Especially for assemblies that are cooled by forced-air circulation, this contamination may be significant.

The contamination can be divided into two main classes, gaseous and particulate contaminants. Gaseous contamination consists of gases that will cause corrosion of metals. The primary failure mechanism caused by corrosive gases is contact failures in electrical contacts. Corrosion attacks on conductors, solder joints and component leads on assemblies may affect SIR and the formation of dendrites. However, most corrosion products formed, for example oxides and sulphides, are rather insoluble in water and do not contribute to ionic conduction. An exception is corrosion products formed by hydrogen chloride and chlorine, i.e., metal chlorides. Besides being soluble and thereby contributing to ionic conduction, they are also a source of halide ions, which promote electrochemical migration.

For most applications, particulate contamination is a much larger source of ionic contamination than corrosive gases [38]. Particles are usually divided into two size-groups, coarse and fine particles [38, 39]. Coarse particles (>1 mm diameter) originate mainly from mineralogical or biological sources, whereas fine particles (<1 mm diameter) have mainly anthropological origin. Although fine particles are much more abundant than coarse ones, the total mass in outdoor environments is usually about the same for the two size-groups. Ionic components in fine particles are mainly sulphate and ammonium [40]. The content of ionic species in coarse particles shows large variations between various locations. Sulphate, ammonium, calcium, magnesium, sodium and chloride normally are the most prevailing ionic components with large local variations for the three latter.

Most of the coarse particles are usually trapped by filtration systems but only a small portion of the fine particles is removed unless high efficiency filtration is used [38]. The deposition rates of various ions on a free, horizontal zinc surface in a typical electronic equipment room (telephone office) are presented in Table 10.3 [41]. An investigation of an electronic equipment manufacturing environment

Table 10.3 Average annual accumulations of ions on a horizontal zinc surface [41]

Substance	Concentration (μg/cm^2)
Sulphate	0.20–0.35
Chloride	0.23
Nitrate	0.13–0.26
Sodium	0.11–0.15
Potassium	0.08–0.18
Magnesium	0.03–0.05
Calcium	0.63–0.87

showed a concentration of particles about five times higher in the factory environment compared to electronic equipment rooms [42]. Of course, there can be very large local variations, both regarding the amount of airborne particles and their composition. In equipment without forced-air cooling, the deposition of particles will be much less compared to a free surface. On the other hand, if forced-air cooling is used, the deposition rate can be expected to be much higher [43].

As indicated, indoor airborne particles consist mainly of sulphate and ammonium at most locations. The ammonium/sulphate ratio is normally found to be between one and two, i.e., the formula can be written $(NH_4)_{2-x}H_xSO_4$ where x varies between 0 and 1. The critical relative humidity at room temperature for $(NH_4)_2SO_4$ and NH_4HSO_4 is approximately 80 and 40%, respectively. These values correspond to the observation that deposits tend to be very hygroscopic above 50–65% relative humidity [39].

Contamination may also occur in the form of liquid droplets containing, for examples, oils and fuels. These are usually not ionic or hygroscopic but they may affect the protective properties of conformal coatings.

10.5.4 Contamination from Rework and Repair

When rework and repair are performed on assemblies, this may cause considerable contamination of the boards due to decomposition of board materials and fluxes used for the work. In addition, it is usually not possible to use the same type of efficient cleaning processes as is used in the assembly line, especially if the work is performed in the field. Therefore, reworked and repaired boards normally have higher levels of contamination than boards that have not been reworked or repaired.

10.6 Surface Insulation Resistance and Electrochemical Migration Test Methods

Test methods for SIR and ECM testing may have one of three purposes:

- Qualification of materials

- Qualification of material combinations and manufacturing processes
- Evaluation of performance in field conditions

10.6.1 Test Methods for Qualification of Materials

For a long time, testing was almost exclusively performed in order to qualify specific materials. Some examples of such tests are as follows:

- IPC-TM-650, Test Method 2.5.27, Surface Insulation Resistance of Raw, Printed Wiring Board Material [44]
- IPC-TM-650, Test Method 2.6.3F, Moisture and Insulation Resistance, Printed Boards [45]
- IPC-TM-650, Test Method 2.6.3.1D, Moisture and Insulation Resistance—Solder Masks [46]
- IPC-TM-650, Test Method 2.6.3.3B, Moisture and Insulation Resistance—Fluxes [47]
- IPC-TM-650, Test Method 2.6.14A, Resistance to Electrochemical Migration, Polymer Solder Mask [48]
- Bellcore TR-NWT-000078, Method 13.1.5, Electrochemical Migration [49]

10.6.2 Test Methods for Qualification of Material Combinations and Manufacturing Processes

Since the test boards used for qualification of materials only have Y or comb patterns and usually bare copper finish, interactions between various materials and contaminants that can be found on an assembly will not be evaluated. Therefore, while passing the tests is mandatory for approval of a material, it is no guarantee for that it will not cause reliability problems when mixed with other materials and contaminants that may be present on an assembly.

For this reason, it is required in IPC's standard for soldered electrical and electronic assemblies [29] that "The materials and processes used to assemble/manufacture electronic assemblies shall be selected such that their use, in combination, produce products acceptable to this standard. When major elements of the proven processes are changed (e.g., flux, solder paste, cleaning media or system, solder alloy or soldering system), validation of the acceptability of the change(s) shall be performed and documented. They can also pertain to a change in bare board supplier, solder resist or metallization". It is up to the user and manufacturer to agree how this should be done but an example of a method for accomplishing it is provided in Appendix C to the standard.

The method in Appendix C is based on SIR testing and visual inspection after finished test. It is specified that "The test vehicle should represent the substrate materials, assembly materials and fabrication processes used in the production. The test vehicle circuitry must provide for SIR testing similar to the IPC-B-36 circuitry". The IPC-B-36 circuitry consists of interdigitated solder lands to LCCC68 components as shown in Fig. 10.4. Furthermore, "Components of the type to be soldered in production representative of the "hardest-to-clean" configurations (in term of "shadowing" of the solder connections by component bodies and component-to-substrate spacing) shall be included on the PWA". Test vehicles for products that will be used in non-condensing service environment shall be tested in accordance with IPC-TM-650, Test Method 2.6.3.3 [47], whereas test vehicles for products that will be used in condensing service environments shall be tested in accordance with IPC-TM-650, Test Method 2.6.3, Class 3 [45]. In the latter case, the test vehicles shall be conformally coated using the same coating material/application processes used in "delivered" hardware. In both cases, it is THB tests. When testing according to Test Method 2.6.3.3, the test vehicles are exposed to 85°C and 85% RH for 168 h. SIR measurements are performed after 24, 96 and 168 h. A bias of 45-50 V DC is applied during the test and measurements are taken using a test voltage of −100 V DC (i.e., reversed polarity).

That is, the purpose of the method in Appendix C to IPC J-STD-001D is to qualify the material combinations and manufacturing processes used for production of a specific product. A similar approach is taken in the IEC standard 61189-5, Test 5E02: Surface insulation resistance, assemblies [50]. The test board to be used for the measurements can be equipped with several types of components (QFP, BGA, SO, surface mount and trough-hole mount connectors and surface mount capacitors). These have circuitry similar to those shown in Fig. 10.4. The test conditions for the IPC and IEC test methods are compared in Table 10.4.

Table 10.4 Test conditions specified for IPC test methods 2.6.3 and 2.6.3.3 and IEC test 5E02

	IPC Test 2.6.3.3	IPC Test 2.6.3, Class 3	IEC Test 5E02
Temperature/RH	85°C/85% RH	20 cycles 25–65°C/85–93% RH	40°C/93% RH[a] or 85°C/85% RH
Duration	168 h	160 h	Minimum 168 h
Bias	45–50 V DC	100 V DC	5, 50 or 100 V DC[b]
Test voltage	−100 V DC	Optional[c] but with the same polarity as the bias voltage	5 or 100 V DC[b]
Measurement frequency	After 24, 96 and 168 h	Every third cycle	Depends on bias and test voltage[b]

[a] No-clean processes shall be tested at 40°C and 93% RH

[b] Three test conditions are suggested with the following bias/test voltage/test frequency: 50 V/100 V/24-96-168 h, 100 V/100 V/twice daily and 5 V/5 V/every 20 min

[c] To be specified in procurement documentation

10.6.3 Test Methods for Evaluation of Performance in Field Conditions

It is stated in IPC-9201 that SIR testing has been used as a predictive tool for estimating service life [2] and in Appendix C to IPC J-STD-001D that the intent of the testing is to show that a proposed manufacturing process change can produce hardware with acceptable end-item performance [29]. To my knowledge, the results from SIR testing have never been correlated to field performance. Field conditions are usually quite different from the conditions used in the tests. For most products, exposure to humid conditions will be followed by dry-up periods, which will affect especially electrochemical migration. More important, although the test conditions may seem severe, many field conditions are considerably harsher than the test conditions. If condensation occurs on a biased non-coated assembly, short circuits due to ECM will normally occur within a few minutes even if the assembly is clean. Although IPC's Test Method 2.6.3, Class 3 is specified to be used for condensing service environment, condensation of water does not normally occur in the test when free-standing assemblies are tested. Furthermore, in many environments, and especially when forced-air cooling is used, assemblies will be contaminated with fibres, dust and particles from the environment [26]. Many of these contaminants contain hygroscopic and/or ionic compounds which will degrade SIR and the resistance against ECM. Corrosion of metallic surfaces due to corrosive gas pollutions may also affect SIR and ECM. All these parameters must be considered when assessing how the service life will be affected.

The problem is how to assess how the conditions in the field environment will affect SIR and the risk for ECM. No studies, to my knowledge, have been performed of how dry-up periods affect SIR and dendrite formation. A test method for contamination with ammonium sulphate particles has been developed by AT&T Bell laboratories [51] which could be used to study the impact of ionic contamination from the field on SIR and ECM but that has not been done so far. During the eighties, test methods were developed for evaluating the impact of corrosive gas pollutions on the reliability of connectors [26, 52–54]. These could also be used for evaluating the impact on SIR and ECM.

10.6.4 Acceptance Criteria

Even if the knowledge of how to adequately test SIR and electrochemical migration is far from satisfactory, relatively much has been written about the subject compared to what has been written about acceptance criteria.

There are two aspects on SIR requirements. First, a certain level of SIR is required for the proper function of a PBA. Second, a certain level is believed to be needed in order to avoid electrochemical migration, although this is not well understood. However, there is no standard that makes any distinction between these two aspects, and thus it is not clear to which aspect requirements refer. Obviously, in the first case, the required SIR may differ considerably between different products even if they will be used in the same environment and, therefore, acceptance criteria have to be application specific. The required resistance to avoid dendrite formation will probably be less dependent on the function, i.e., in the second case, the required SIR will be the same for products that will be used in similar environments. However, since humidity and contamination levels in the field environment will have a very large impact on both SIR and ECM, it will be necessary to have different acceptance criteria for various types of applications also when considering the risk for electrochemical migration.

10.6.4.1 Acceptance Criteria for SIR

A low SIR may alone be the cause of failures but the SIR level at which failures will occur depends on the electrical requirements of the circuitry. Thus, SIR measurements are useful for verification of the performance of a printed board assembly, but SIR should then be measured directly with only a short stabilisation period in order to avoid a drift of the measured SIR value. The drift of SIR during measuring and the influence of previous measurements on later measurements may be avoided by using an AC voltage at a low frequency of about 1 Hz [22]. Very little has been done in this field but it may be the most relevant method for SIR measuring, and efforts ought to be made to investigate this alternative more deeply.

As already discussed, SIR is to a very large extent affected by humidity, temperature and contaminants. Especially, hygroscopic contaminant may have a large impact and can cause a sudden drop in SIR when the relative humidity surpasses the critical humidity at which the contaminant becomes hygroscopic. Hence, an informative test would be to screen the relationship between SIR and the relative humidity in order to detect the presence of hygroscopic contaminants and determine critical relative humidities. Since the critical relative humidity decreases for many substances with increasing temperature (see Table 10.1), this screening should preferably be performed at both a low and a high temperature. However, since some residues may be decomposed or evaporated at high temperatures, for example adipic acid and polyethylene glycol, extended exposure to high temperatures should be avoided [55].

A good example of a function-based, application-specific requirement has been described by Mason [11]. The company's most sensitive circuit had a device with 0.5-pA maximum offset current. Two guard tracks are used, each 25 mm long and separated from the guarded track by 1.25 mm. The potential between guarding and guarded tracks was 15 mV. Hence, the minimum required SIR was 3×10^{10} ohms or 1.2×10^{12} ohms per square.

10.6.4.2 Acceptance Criteria for ECM

Although it is not explicit expressed in the tests for SIR and ECM testing, it can be assumed that specified acceptance criteria have been set based on the risk for ECM even when they are called SIR tests.

Most SIR and ECM standards have numerical requirements. The requirements usually differ between various standards. Even if the numerical requirement of the SIR is the same, the test conditions may vary which, in practice, means differing acceptance criteria. Usually there is a requirement that SIR shall not be less than a certain value. Some electrochemical migration tests also require that SIR not decrease more than one decade during the test [49]. Due to the experience that dendrites normally form without any impact on SIR, some later standards also have visual requirements regarding presence of corrosion, dendritic formation and mealing of conformal coating [29].

IPC's standard for soldered assemblies, J-STD-001D [29], is a good example of the many uncertainties that arise when one tries to get a grip on the basis for the acceptance requirement. In Appendix C, the acceptance requirement is that the SIR value shall be not less than 100 Mohm. It is specified that the test pattern must provide for SIR testing similar to that of the IPC-B-36 circuitry which is a leadless ceramic chip carrier with 68 castellations (open circuit) mounted on a pattern with interdigitated solder lands. Components of the type to be soldered in production representative of the "hardest-to-clean" configurations shall be included on the printed board assembly. Then, what does similar circuitry mean? The most likely interpretation is components with open circuit mounted on patterns with interdigitated solder lands. However, any change in the number of connections and/or pitch for a test pattern will, of course, influence the measured SIR and the acceptance criterion ought to be adjusted but nothing is mentioned in the standard about how to deal with that. Ideally, this can be handled using the ohms/square concept [6]. However, since SIR does not follow Ohm's law and contamination is not usually uniformly distributed across the entire test pattern, it is stated in IPC-9201 that normalisation of resistance to an ohms/square value is a questionable practice [2]. Thus, the question remains how to adjust the acceptance criterion in J-STD-001D to differing test patterns.

Furthermore, the reason for setting the acceptance criterion to 100 Mohm in J-STD-001D is not given. When looking in previous IPC standards, it can be noted that acceptance criteria for SIR tests usually are chosen in the range of 100-500 Mohm, although the test conditions and the test pattern may vary considerably. In an IPC technical paper from 1985 [56], it is mentioned that the practice at that time was to require a SIR of 100 Mohm but then only as a requirement for initial resistance. When testing according to J-STD-001D, one can choose between two test methods, one static for non-condensing service environment and one cyclic for condensing service environment. The static test method has requirements for SIR after 96 and 168 h of testing, whereas the cyclic test method has requirements for SIR only after 160 h of testing. There is a tremendous difference between specifying requirement for initial SIR compared to after 100–168 h testing.

As shown by Chan [22], SIR may increase several decades on a contaminated board during a test period of 100 h and approach values of a clean board. Thus, it seems odd to disregard from initial measurements and have requirements on SIR only after some days of measuring, as several standards have. Since no information is given on the background to the SIR requirements in J-STD-001D, it is not possible to draw any conclusion about the relevance of the requirements, nor to justify the disregarding of initial SIR values. The correlation between low initial SIR values and the risk for the dendrite formation needs to be explored more deeply.

It can be noted that the newly developed test for surface insulation resistance of assemblies in IEC standard 61189-5 [50] does not contain any acceptance criteria at all. It only specifies how the test should be performed, how it should be evaluated and what the test report should contain.

References

1. Ellis BN (1994) The determination of flux residue safety: the state of the art. Circuit World 20(4):26–32
2. IPC-9201 (1996) Surface insulation resistance handbook. IPC http://www.ipc.org
3. IPC-TR-476A (1997) Electrochemical migration: electrically induced failures in printed wiring assemblies IPC. http://www.ipc.org
4. Ellis BN (1995) On insulation resistance. Circuit World 21(2):5–18
5. Osborne AG (1983) Modern thoughts on surface resistance. IPC technical paper IPC-TP-498
6. Gorondy EJ (1984) Surface insulation resistance (SIR) part I: the development of an automated SIR measurement technique. IPC technical paper IPC-TP-518
7. Tegehall P-E, Dunn BD (1992) Influence of flux residues and conformal coatings on the surface resistance properties of spacecraft circuit boards. ESA J 16(3):255–273
8. Yan BD, Meilink SL, Warren GW, Wynblatt P (1986) Water adsorption and surface conductivity measurements on α-alumina substrates. In: proceedings of 36th electron compon technol conf, pp 95–99
9. Grossmann G (1996) Contamination of various flux/cleaning combinations on SMT assemblies: a discussion of different analysis methods. Solder Surf Mt Technol 22:16–21
10. Gorondy EJ (1988) Surface/moisture insulation resistance (SIR/MIR): part III analysis of the effect of test parameters and environmental conditions on test results. IPC-TP-825
11. Mason BJ (1994) No-clean flux appraisal—the total process approach. Circuit World 20(2):21–32
12. Alder JF, Isaac CA (1971) Constant humidity. In: Weast RC (ed) Handbook of Chemistry and Physics, 52nd edn. The Chemical Rubber Co., Cleveland, p E-40
13. Wexler A (1995) Constant humidity solutions. In: Lide DR (ed) Handbook of Chemistry and Physics, 76th edn. CRC Press, Boca Raton, Florida, pp 15–23
14. Anderson JE, Markovac V, Troyk PP (1988) Polymer encapsulants for microelectronics: mechanisms for protection and failure. IEEE Trans Compon Hybrids Manuf Technol 11(1):152–158
15. Brous J (1981) Water-soluble flux and its effect on PC board insulation resistance. Electron Packag Prod 21(7): 79-87
16. Adams KM, Anderson JE, Graves YB (1994) Ionograph sensitivity to chemical residues from "no clean" soldering fluxes: comparison of solvent extract conductivity and surface conductivity. Circuit World 20(2):41–44

17. Benson RC, Romenesko BM, Weiner JA, Nall BH, Charles HK Jr (1988) Metal electromigration induced by solder flux residue in hybrid microcircuits. IEEE Trans Compon Hybrids Manuf Technol 11(4):363–370
18. Zou L, Hunt C (1988) Surface insulation resistance (SIR) response to various parameters. NPL report CMMT(A)120. http://www.npl.co.uk
19. Warren GW, Wynblatt P, Zamanzadeh M (1989) The role of electrochemical migration and moisture adsorption on the reliability of metallized ceramic substrates. J Electron Mater 18(2):339–353
20. Steppan JJ, Roth JA, Hall LC, Jeanotte DA, Carbone SP (1987) A review of corrosion failure mechanisms during accelerated tests. J Electrochem Soc 134:175–190
21. Bumiller E, Pecht M, Hillman C (2004) Electrochemical migration on HASL plated FR-4 printed circuit boards. J Surf Mt Technol 17(2):37–41
22. Chan HA (1996) Is the current surface insulation resistance (SIR) methodology appropriate to todaýs manufacturing technology. In: proceedings of 46th electron compon technol conf, pp 234–241
23. Sohn JE, Ray U, Heideman V, Schubert B, Anderson JE, Adams KM, Becka G (1994) How clean is clean: effect of no-clean flux residues and environmental testing conditions on surface insulation resistance. In: proceedings of Surf Mt Int: 1994 391–402
24. Hunt C, Zou L (1999) The impact of temperature and humidity conditions on surface insulation resistance values for various fluxes. Solder Surf Mt Technol 11(1):36–43
25. Zou L, Hunt C (2008) Electrochemical impedance technique to predict circuit reliability with lead-free solders. NPL report MAT 15. http://www.npl.co.uk
26. Tegehall P-E (1998) Reliability verification of printed board assemblies: a critical review of test methods and future test strategy. In: proceedings of Surf Mt 1998 Int 359–382
27. Tegehall P-E (1998) Impact of contamination from production processes on the reliability of printed board assemblies. In: proceedings of int conf electron assembly: materials and process challenges, pp 69–80
28. Sanftleben H (1996) Ionic cleanliness of LPISM circuit boards. IPC Rev 37(2):30–32
29. J-STD-001D (2005) Requirements for soldered electrical and electronic assemblies. IPC http://www.ipc.org
30. Ellis BN (1986) Cleaning and contamination of electronics components and assemblies. Electrochemical Publications, Ayr Chapter 15
31. Lea C (1989) Residues on soldered printed circuit assemblies. Circuit World 16(1):44–46
32. Dunn B (1980) The fusing of tin-lead plating of high quality printed circuit boards. Trans Inst Met Finish 58:26–28
33. Pauls D (2001) Residues in printed wiring boards and assemblies. Circuit World 27(1):32–41
34. Bulletti A, Capineri L, Materassi M, Dunn BD (2007) Surface resistivity of new printed circuit board materials for use in spacecraft electronics. IEEE Trans Electron Packag Manuf 30(2):115–122
35. Zado FM (1983) Effects of non-ionic water soluble flux residues. The Western electr eng, first issue, pp 41–48
36. Brous J (1992) Electrochemical migration and flux residues—causes and detection. In: proceeding of Nepcon West 386–393
37. Ready WJ, Turbini LJ, Stock BA, Smith BA (1996) Conductive anodic filament enhancement in the presence of a polyglycol—containing flux. In: proceedings of 34th IEEE Int Reliab Phys, pp 267–273
38. Sinclair JD (1987) Corrosion of electronics by ionic substances in the environment. In: proceedings of ASM 3rd conf electron packag mater prod corros microelectron, pp 145–154
39. Sinclair JD (1992) The relevance of particle contamination to corrosion of electronics in processing and field environments. In: proceedings of 2nd int symp corros reliab electron mater devices, pp 325–335
40. Sinclair JD, Psota-Kelty LA, Weschler CJ, Shields HC (1990) Measurement and modeling of airborne concentrations and indoor surface accumulation rates of ionic substances at Neenah, Wisconsin. Atmos Environ 24A:627–638

41. Sinclair JD, Psota-Kelty LA, Weschler CJ (1985) Indoor/outdoor concentrations and indoor surface accumulations of ionic substances. Atmos Environ 19(2):315–323
42. Sinclair JD, Psota-Kelty LA, Peins GA, Ibidunni AO (1992) Comparison of electronic equipment room and factory environments. Atmos Environ 26A:871–882
43. Comizzoli RB, Frankenthal RP, Lobnig RE, Peins GA, Psota-Kelty LA, Siconolfi DJ, Sinclair JD (1993) Corrosion of electronic materials and devices by submicron atmospheric particles. Electrochem Soc Interface 2(3):26–33
44. Test Methods Manual IPC-TM-650, Test Method 2.5.2 (1979) Surface insulation resistance of raw, printed wiring board material. IPC http://www.ipc.org
45. Test Methods Manual IPC-TM-650, Test Method 2.6.3F (2004) Moisture and insulation resistance, printed boards. IPC http://www.ipc.org
46. Test Methods Manual IPC-TM-650, Test Method 2.6.3.1D (2000) Moisture and insulation resistance—solder masks. IPC http://www.ipc.org
47. Test Methods Manual IPC-TM-650, Test Method 2.6.3.3B (2004) Moisture and insulation resistance—fluxes. IPC http://www.ipc.org
48. Test Methods Manual IPC-TM-650, Test Method 2.6.14A (1987) Resistance to electrochemical migration, polymer solder mask. IPC http://www.ipc.org
49. TR-NWT-000078, Issue 3, Method 13.1.5, Electrochemical Migration (1991) Generic physical design requirements for telecommunications products and equipment bellcore (now supplied by Telcordia). http://www.telcordia.com
50. IEC 61189-5 (2006) Test methods for electrical materials, interconnection structures and assemblies—part 5: test methods for printed board assemblies, test 5E02: surface insulation resistance, assemblies. IEC http://www.iec.ch
51. Frankenthal RP, Siconolfi DJ, Sinclair JD (1993) Accelerated life testing of electronic devices by atmospheric particles: why and how. J Electrochem Soc 140:3129–3134
52. Abbott WH (1988) The development and performance characteristics of mixed flowing gas test environments. IEEE Trans Compon Hybrids Manuf Technol 11(1):22–35
53. Gore R, Witska R, Chao J (1989) Corrosive gas environmental testing for electrical contacts. Proc Holm Conf Electric Contacts, pp 123–131
54. Zakipour S, Leygraf C (1986) Evaluation of laboratory tests to simulate indoor corrosion of electrical contact materials. J Electrochem Soc 133:21–30
55. Sohn JE, Ray U (1995) Weak organic acids and surface insulation resistance. Circuit World 21(4):22–26
56. Eisenmann ET (1985) On the utility of insulation resistance tests to assess the life performance of printed wiring boards. IPC-TP-542

Chapter 11
Lead-Free and Other Process Effects on Conductive Anodic Filamentation Resistance of Glass-Reinforced Epoxy Laminates

C. Zou, A. Brewin and C. Hunt

11.1 Introduction

Conductive anodic filamentation (CAF) is a failure phenomenon seen within the bulk of glass-reinforced epoxy-printed circuit board laminate, used for the manufacture of circuit assemblies. This distinct electrochemical failure mode was first reported in 1979 [1]. Although the existence of the CAF mechanism has been known for some time, currently there is an increase in concern about these effects.

There are three main drivers for this concern:

Continued requirements for increased circuit density, smaller via geometries and increased layer counts in multilayer boards

- The increased use of electronics in harsh environments and for high reliability and safety critical applications (automotive, avionics, medical)
- The use of lead-free processing in Europe (but affecting global market place) after July 2006, which may affect laminate stability and material choices

The combination of these factors has increased the focus on the CAF phenomenon. The mechanism proposed for CAF growth is based on an electrochemical corrosion of copper ions and the deposition of copper salts that are influenced by electric field, ion concentration and pH gradients. The conductive path is therefore the growth on the anode of a salt, compared to dendrite growth on the surface of the board, which comprises metal ions depositing on the cathode.

In this study, a range of materials, board layout configurations, thermal events and processing steps are investigated, and a robust test method for the assessment

C. Zou (✉), A. Brewin and C. Hunt
National Physical Laboratory, Teddington, Middlesex, UK
e-mail: ling.zou@npl.co.uk

G. Grossmann and C. Zardini (eds.), *The ELFNET Book on Failure Mechanisms, Testing Methods, and Quality Issues of Lead-Free Solder Interconnects*,
DOI: 10.1007/978-0-85729-236-0_11,

of CAF susceptibility for laminates is defined. Furthermore, the data for CAF avoidance form the basis of guidelines for design and material choices.

11.2 The CAF Mechanism

There are sources describing the mechanism for CAF formation [2–5], but brief overviews of the steps that take place in the formation of a CAF are described here.

11.2.1 Initiation

For CAF to occur there must be a source of copper, a electrical bias and a glass reinforcement fibre in the same locality. In dense multilayer PCBs, this can often occur at the via wall where the drilled hole has been seeded and plated, although fibres protruding through the surface of PCBs to the locality of surface tracks have also been reported [6].

11.2.2 Separation of Glass and Epoxy

If glass preparation is of poor quality, separation of the epoxy resin and glass fibres can be more likely. Bonding can also be lost due to hydrolysis of the glass finish or due to residual stress in the weave being relieved [1]. The absorption of moisture into the laminate accelerates these processes, and the de-lamination can then provide a pathway for CAF growth.

11.2.3 Electrochemical Reaction

With moisture present as an electrolyte and a pathway established, current may begin to flow setting up an electrochemical process. The mechanism involves oxidation and dissolution of copper at the anode, and if copper anions reach the cathode, they are reduced back to copper metal. Other ions present from processing or materials such as chloride, sulphur or other metals may also take part in this process, as shown in following equations:

$$\text{Anode:} \quad Cu_{(s)} \rightarrow Cu^{n+}_{(aq)} + n^{e-}$$

$$H_2O \rightarrow {}^1\!/_2 O_{2(g)} + 2H^+_{(aq)} + 2^{e-}$$

$$\text{Cathode:} \quad 2H_2O + 2e^- \rightarrow H_{2(g)} + 2OH^-_{(aq)}$$

$$Cu^{n+}_{(aq)} + ne^- \rightarrow Cu_{(s)}.$$

The process shown here is for low-pH environment; inside the PCB, the process is solubility dependent. The formation of copper at the cathode may not readily happen since the cooper ion at the anode becomes an insoluble salt due to high pH at the cathode. This solubility effect is responsible for CAF deposits growing from the anode, a low-pH environment towards the cathode and a high-pH environment and is illustrated in Fig. 11.1. Other ions present from processing, or materials, such as chloride, sulphide or other metals may also take part in this process.

11.2.4 Salt Deposition

As the electrochemical dissolution of copper continues, a pH gradient is produced due to the H^+ ions produced at the anode and OH^- ions at the cathode. In Fig. 11.2, this is represented by a schematic of two vias, where the anode via wall is the initiation site.

The formation of this pH gradient between the anode and the cathode is key to the mechanism of CAF growth since the solubility of Cu^{2+} is dependant on pH and the electrical potential applied. At the anode, a positive potential and low pH, Cu will corrode to form Cu^{2+} in solution. Under the electric field, Cu^{2+} migrates along the glass fibre/epoxy interface towards the cathode. Since pH increases from the

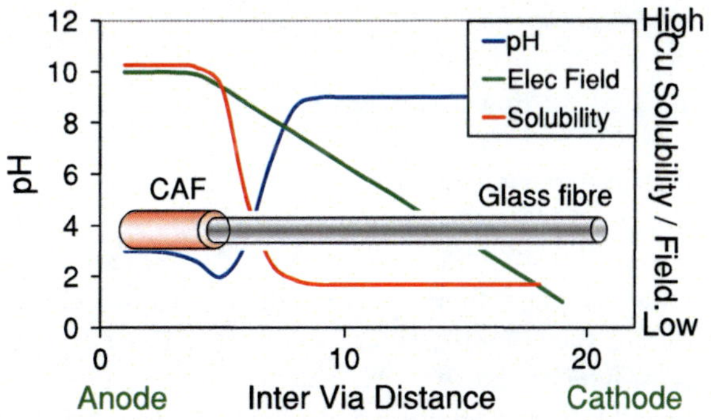

Fig. 11.1 Schematic of pH, electric field and Cu solubility during CAF growth

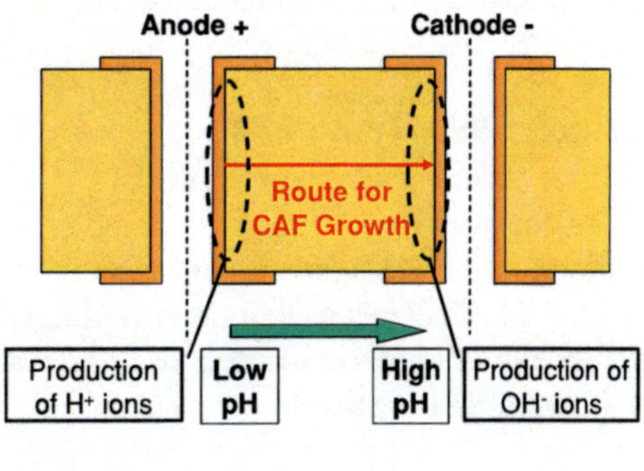

Fig. 11.2 Mechanism for CAF growth (initially proposed by Bell Labs)

anode to the cathode, copper ions Cu^{2+} start to precipitate as an insoluble copper salt. Built-up copper salt forms the conductive path (CAF).

11.2.5 Completion of Conductive Pathway

Catastrophic electrical failure occurs when the filament of copper salts bridge the anode and cathode. Under humid conditions, the salts are conductive and will allow a massive increase in current flow between the previously well-isolated opposite-charged metal conductors in the electronic circuit.

11.3 Experimental

11.3.1 Test Vehicle Design

A The test board used in this study has ten layers and over 6,000 vias with a 32-tab connector layout allowing connection to a test rack system for the application of bias voltage and monitoring of leakage currents, as shown in Fig. 11.3 with the comb area labelled.

11.3.1.1 In-line Test Combs

The in-line test combs A–D comprised alternate rows of either powered tracks or tracks connected to ground as shown in Fig. 11.4. They have via-wall-to-via-wall gap of 300, 400, 550 and 800 μm. The vias are in line with one another orientated with the glass fibre weave. The via-wall-to-via-wall gap is the most common failure site where CAF can occur. The closest point between each via pair is the most likely point for CAF growth (example highlighted in Fig. 11.4). Test pattern G and E were added to combs B and D, but with an orientation at 90° to the other combs. This was done to investigate that any differences in the CAF susceptibility might be sensitive to orientation between the warp and weft glass bundles.

11.3.1.2 Staggered Combs

The construction of staggered comb (comb F) is similar to that the in-line combs; however, the via pairs are arranged at 45° to the glass warp and weft, as shown in Fig. 11.5. This means that the most likely route for potential CAF growth is longer since the orientation of the glass fibres may only permit growth in the horizontal and vertical directions.

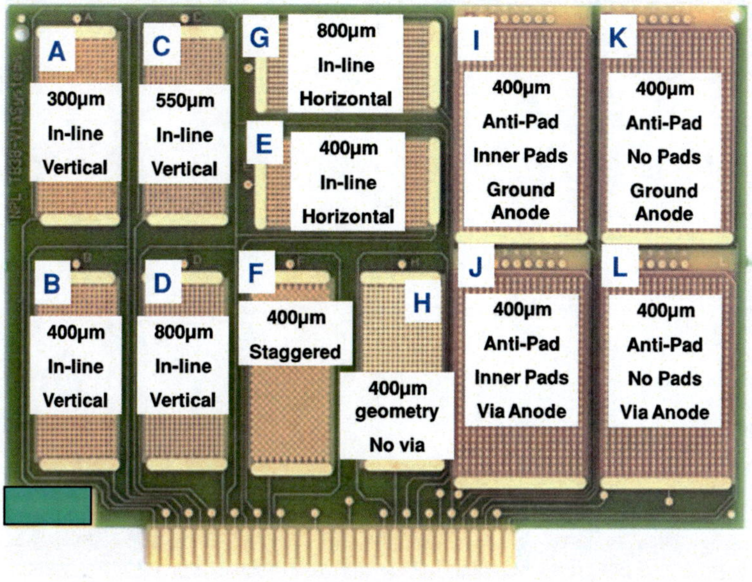

Fig. 11.3 Test vehicle (AuNi finish)

Fig. 11.4 Schematic of in-line test comb, with possible failure site

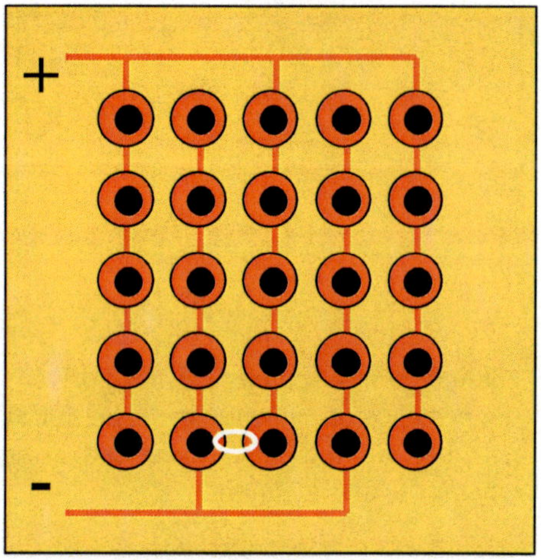

11.3.1.3 Anti-Pad Combs

Anti-pad combs (combs I–L) represent the potential failure site when a CAF may grow between a ground plane in a multilayer board and a nearby via wall. There is a surrounding copper plane at a potential to each via, as shown in Fig. 11.6. There were four anti-pad comb patterns in the test board. Figure 11.7 then presents the

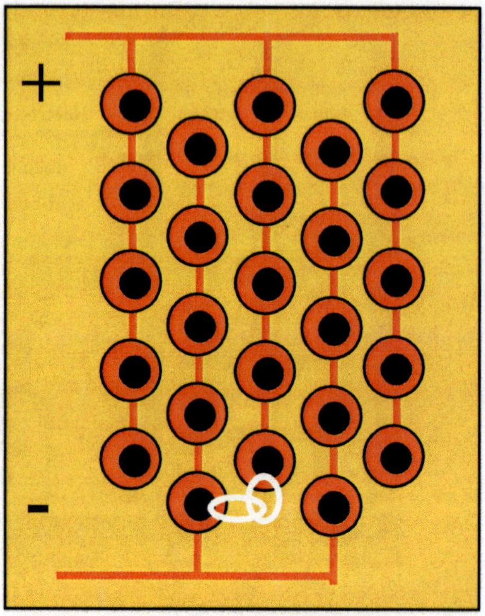

Fig. 11.5 Schematic of staggered test comb, with possible failure site

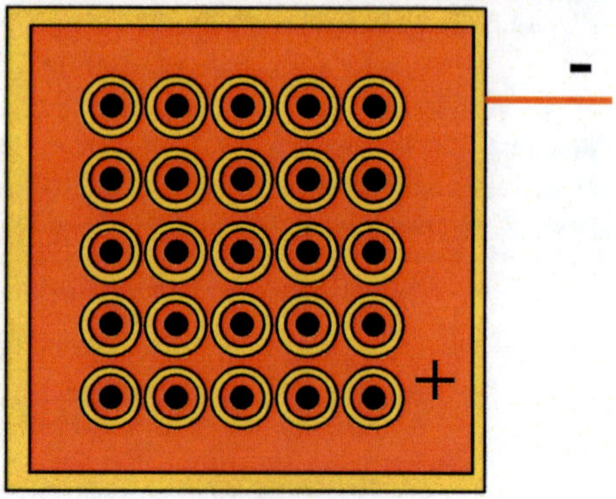

Fig. 11.6 Schematic of anti-pad test comb

configuration of combs I, J, K and L. The dimensions, apart from the changes in the presence of pads, are the same for all four anti-pad style combs. I and J are a pair, but with the polarity reversed, as are K and L. I and J have inner-layer pads, whereas K and L do not.

11.3.2 Conditioning Effects and Variables Investigated

11.3.2.1 Test Voltage

In this study, four test voltages, 5, 50, 150 and 500 V were used to test the samples, to establish the effect of test voltage on CAF formation.

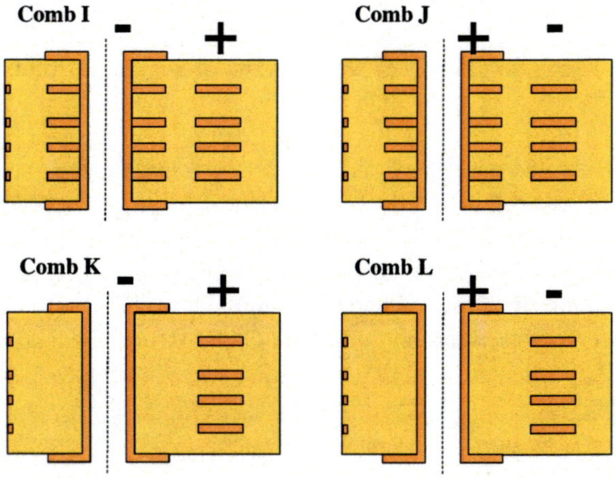

Fig. 11.7 Schematics of configuration for anti-pad combs I, J, K and L

11.3.2.2 Reflow Exposure

Selected test coupons were exposed to three passes through a 5-zone forced convection reflow oven. The boards were allowed to return to room temperature after each pass before being passed through the oven again. Three different reflow conditions are applied:

- No reflow—as a control condition
- SnPb reflow—typical of that used to reflow SnPb solder paste (eutectic 183°C)
- Lead-free reflow—typical of that used to reflow SnAgCu solder paste (eutectic ~217°C).

11.3.2.3 Laminate Supplier

The companies provide the base laminate materials (pre-preg layers, cores, etc.) to the board house. In this study, the materials were from two material supplies L and K.

11.3.2.4 Board House

Board house is the manufacturing site for the PCBs, where the Gerber data are used to etch, drill and plate copper foils. The materials are placed in a press cycle to bond all the layers together forming the multilayer board. A plethora of chemical preparation, etching, drilling, alignment, masking and development stages takes place during the construction of a multilayer PCB; the intention in this work was not to investigate them all but to understand any significance of where boards are produced. There were three board houses, A, B and C, participated this study.

11.3.2.5 Glass Transition Temperature (T_g)

T_g marks the onset temperature of segmental mobility for a polymer (the temperature below which the polymer segments do not have sufficient energy to move past one another). In practice, for FR4 composites, rigidity and co-planarity can be affected once the T_g is exceeded by a large margin, and so some may consider high T_g materials to reduce flex in lead-free production. The epoxy formulation for FR4 is often available as standard T_g or high T_g. High T_g material is often specified for high-temperature applications, although its resistance to thermal degradation is no better than that of standard laminate. In order to see whether the T_g of the laminate epoxy used has any effect on CAF, performance stand and high T_g materials were investigated.

11.3.2.6 Drill Feed Speed

During drilling, the drill bit is driven into the laminate at a controlled feed rate (Note: This is not the speed at which the drill spins). Faster feed speeds mean quicker production; however, it has been postulated that forcing high feed speeds results in more damage to glass fibres during drilling, which in turn could provide a greater number of initiation sites for filaments. In order to investigate any possible trends in CAF resistance caused by the use of different printed circuit drill speeds, four speeds 1–4 were used to make the board.

11.3.2.7 CAF Resistance

Suppliers can offer laminates that use resin cure systems marketed as CAF-resistant. A known contributor to the problem is the curing agent dicyandiamide (DICY), which is often eliminated from such materials. The effect of chemistry in epoxy make-up (CAF- or non-CAF resistant) was investigated in this study.

11.3.2.8 Weave

Weave is the make-up of the glass fibres woven in the composite. There are three generic glass weave types used in this study (2116, 2113 and 1080) and a commercial unknown weave style. Variables include fibre diameter and number of fibres per bundle and can be different for the warp and weft directions.

11.3.2.9 PCB Surface Finish

There are potentially a large number of chemical processes used in the application of surface finish chemistries in board production. Since the CAF mechanism involves the migration of ions and the formation of copper salts, it is possible these residues may impact on CAF formation. In order to investigate any possible trends in CAF resistance caused by the use of different printed circuit-solderable finishes, in this study the following finishes were used:

- NiAu—around 10 μm of Ni as a protective barrier with around 0.1 μm of Au to aid wetting
- Ag—0.2 μm of silver covers the copper
- HASL—(hot air solder levelling)
- OSP—(organic solderable preservative)

11.3.3 Test Method

The samples were placed in an environmental chamber at steady-state conditions of 85°C and 85% RH for 1,000 h. The temperature was increased ahead of the relative humidity to prevent any condensation onto the samples. Throughout the exposure, the insulation resistance (IR) of each comb pattern was measured and logged every 60 min. The resistance measurement system was a 256-channel Auto-SIR using a current limiting resistor of 10^6 Ω.

Two typical IR versus time curve for a failed comb, measured during the 1,000-h artificial ageing, are shown in Fig. 11.8. The decrease in IR as moisture permeates the laminate material can be seen in the first 10–20 h of the test. At about 47 and 136 h (in this example), the sudden catastrophic loss in insulation between the vias in the comb reflected the CAF formation. The IRs do not drop below 10^6 Ω due to the use of a current limiting resistor. Therefore, a important parameter, the time to failure (TTF/hours), can be extracted from the IR–time curve for each comb tested in this work. The IR–time curves and TTF data provided detailed information on CAF performance for all test variables, but it is difficult to see the trend of studied variables.

Fig. 11.8 Typical IR versus time curve for CAF-failed combs

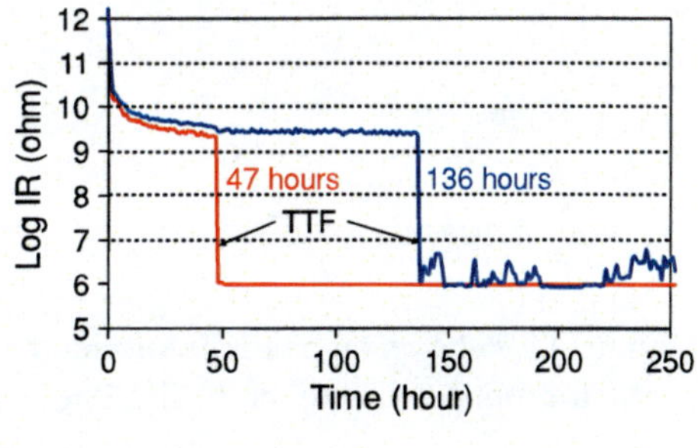

11.3.4 Normalisation of TTF Data for Trend Analysis

In order to study the effect of a particular variable (for example, below 'reflow exposure') on CAF resistance, the TTF data were analysed and normalised in the following procedure:

1. Group data to a set and average the TFT data for samples in which all other materials and processing parameters are identical, except for the key variable, which is then averaged across all test pattern designs. This means that the effects of processes other than the key variable, e.g. reflow exposure, do not influence trends. For example, the effect of reflow exposure (X) in group A sample, the average TTF data for each reflow condition ($TTFA_a$, $TTFA_b$ and $TTFA_c$) can be calculated across all test patterns, and average data from three reflow conditions can be calculated as follows: $TTFA_{mean} = (TTF_a + TTF_b + TTF_c)/3$
2. Normalise data for each group: For each group, the key parameter can be normalised against the average for a particularly reflow condition a, and the TTF data can be normalised using following equation: $Aa = (TTFA_a - TTFA_{mean})/TTFA_{mean}$.
3. Average normalised TTF for all groups: The normalised TTF for each key parameter from different group will be averaged to assess the effect of key parameters on CAF formation. For reflow exposure, average normalised TTF can be calculated as follows: $Xa = Aa + Ba + Ca + \cdots$

11.3.5 EDAX Analysis of CAF

After failures had been established using the electrical monitoring technique, selected combs were sectioned to locate and identify CAF. The area between the failed vias was cut with a slow-speed wet diamond saw, close to the anode via wall. Sequential polishing with 5- to 1-μm particle polish was then used to move through the sample towards the cathode until the CAF was located. A high-resolution optical image of the CAF cross section was acquired, and the polished CAF sample was carbon-coated and mounted for EDAX analysis. This allowed confirmation that the electrical short detected was associated with a CAF failure, and underlined the general approach of using electrical measurements to detect a CAF.

11.4 Results

11.4.1 Analysis of CAF

Figure 11.9 shows an example of elemental digital map produced for correlation with the optical image of CAF. The elemental map shows presence of many

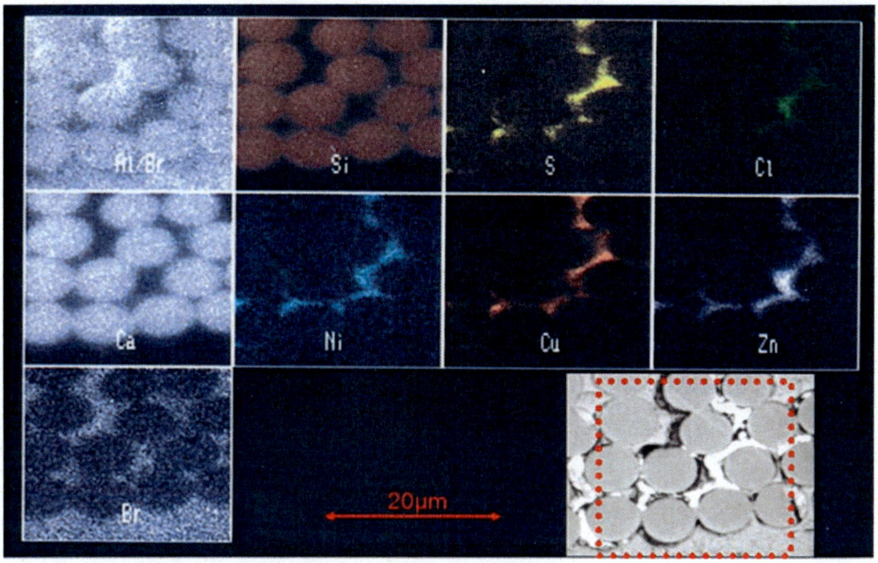

Fig. 11.9 Optical and EDAX images of a CAF sample

elements associated with the CAF. This confirms that electrical short detected was correlated with the physical presence of CAF in the sample.

11.4.2 Normalisation of TTF Data for Trend Analysis

11.4.2.1 Effect of Test Voltage on CAF Failure (In-Line Comb)

Figure 11.10 shows the normalised TTF data for different test voltages. The figure clearly showed that increasing test voltage will significantly shorten the time to CAF failure. This is because CAF formation being an electrochemical process, high test voltage will accelerate copper corrosion at the anode and then CAF formation.

11.4.2.2 Effect of Reflow Conditions on CAF Failure

Previous study indicated that it was high peak reflow temperature that reduced CAF performance, not thermal shock [7]. This suggests that the mechanism for damage in the laminate is not based on TCE mismatch between the materials in the composite FR4, but perhaps a chemical or physical breakdown at a certain temperature. An investigation into reflow conditions was therefore undertaken, and the normalised TTF data are plotted in Fig. 11.11. A significant trend for lower CAF

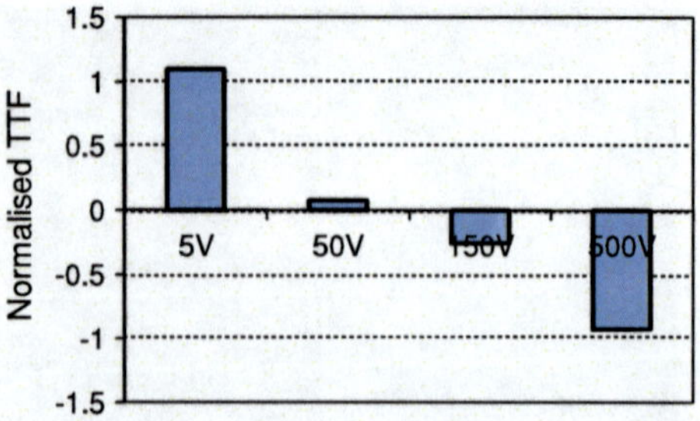

Fig. 11.10 Effect of test voltage on CAF failure

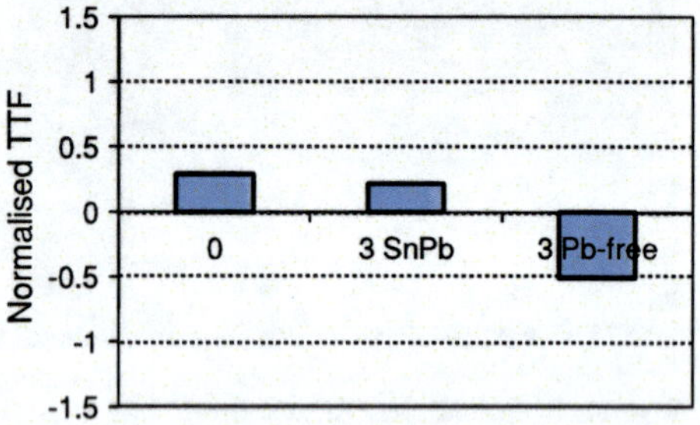

Fig. 11.11 Effect of reflow exposure on CAF failure

performance after exposure to lead-free reflow conditions compared to tin–lead reflow exposure can be seen. Whilst the reflow conditions used can be considered worse case, this could be a concern for those who manufacturing high-reliability multilayer product, especially with high via densities as they switch to lead-free production.

11.4.2.3 Effect of Via to Via Gap on CAF Failure

Figure 11.12 shows the normalised TTF data for different in-line comb geometries. Comparing the vertically aligned combs A to D, the increase in via to via gap from 300 to 800 μm shows an increase in TTF. Thus, larger via to via gaps are, unsurprisingly, more resistance to CAF growth.

11.4.2.4 Effect of Orientation of Vias to Glass Weave on CAF Failure

Figure 11.13 presents the normalised TTF data in different comb orientations. The effect of weave can be seen with staggered vias being more resistant than those in-line and horizontal combs being more CAF resistant than vertical combs. This confirms that CAF is always formed along the fibre. The improvement in CAF resistance for staggered vias is a consequence of there not being a glass fibre running directly between vias, as represented in Fig. 11.5. It is also important to note that the properties of the warp and weft fibres are different.

Fig. 11.12 Effect of the gap between via walls on CAF failure

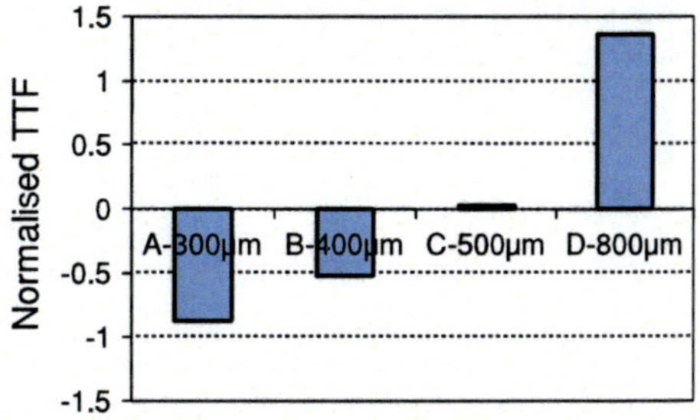

Fig. 11.13 Effect of via alignment on CAF failure

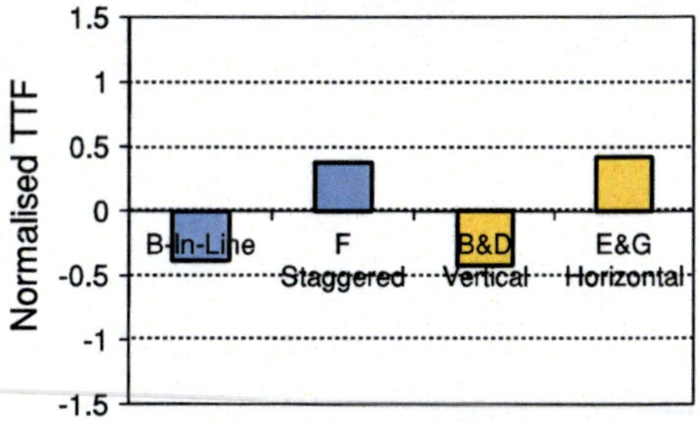

11.4.2.5 Effect of Inner-Layer Pads and Via Polarity on CAF Failure

Figure 11.14 gives normalised TFF data for the anti-pad combs. Firstly, it is clear that if the via is anodic (+ve bias), then the CAF resistance of the system is dramatically reduced compare to a cathodic via system. Put another way, the CAF grows much quicker from a via anode initiation site than from a copper plane anode.

Secondly, one can see that there appears to be no significant change in CAF performance as a result of the presence of inner-layer pads for CAF growing from the via wall (combs J and L). This suggests that the CAF grows from the via wall (associated with the ends of the glass fibres) and not from the edge of the inner-layer pad (held in between the reinforced pre-preg layers away for the fibres). For the case where CAF grows from the copper planes (combs I and K), the failures are earlier for the configuration with inner-layer pads. This is because in this instance the effective distance for the CAF to grow is reduced by the presence of the pads (See Fig. 11.7).

11.4.2.6 Effect of Laminate CAF Chemistry on CAF Failure

The effect of chemistry on the epoxy make-up (CAF- or non-CAF resistant) was investigated. The normalised TTF data are shown in Fig. 11.15. It can be seen that

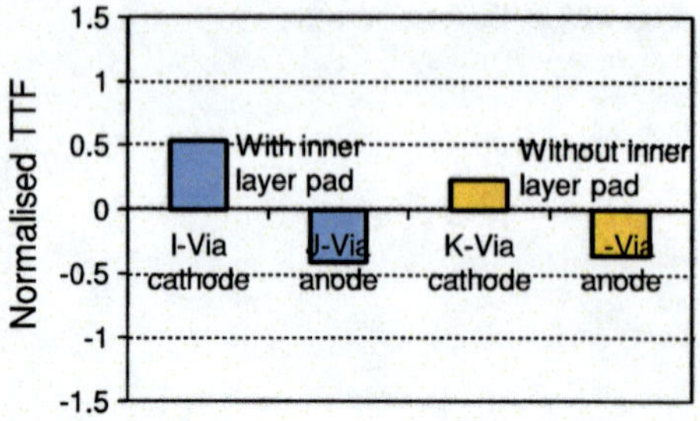

Fig. 11.14 Effect of inner-layer pads and via polarity (anti-pad comb) on CAF failure

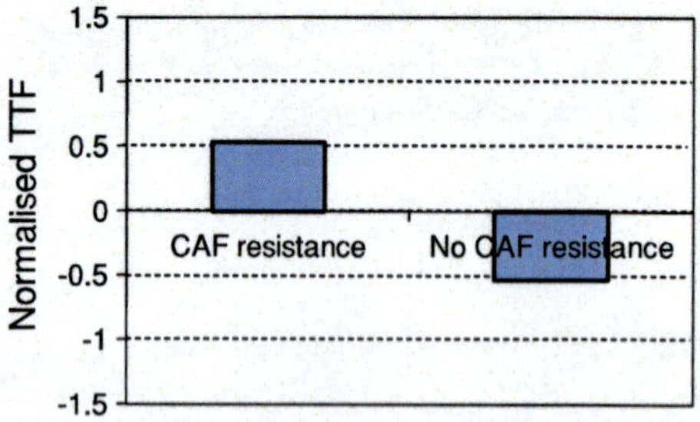

Fig. 11.15 Effect of laminate CAF chemistry on CAF failure

the use of a laminate formulated for 'CAF resistance' does indeed provide an increased resistance to the onset of CAF. It does not eliminate, however, under high-stress conditions.

11.4.2.7 Effect of Board House on CAF Failure

The number of processes and parameters involved in the manufacture of printed boards is immense. Some have been specifically investigated as part of this work, but there are many others that have not. Examples include lamination press cycle (temperature, pressure, time), alignment accuracy, and copper seeding and plating processes. For these processes, the production partners were asked to use their standard production conditions for this type of board.

To establish whether there were any differences in product from different board houses produced under "house" conditions, identical board samples were processed at three locations, A, B and C. The normalised TTF data showing the effect of manufacturing location on the CAF performance are shown in Fig. 11.16. Clearly, the effect of where the laminate is manufactured has significant impact on CAF. It can be seen that for materials manufactured at board house C, boards are more CAF-resistant, and those produced at board house A were the least CAF-resistant. This is a surprising result, and the cause of this is not understood by the project group and could be an effect of one of many processes taking place during the laminate build-up.

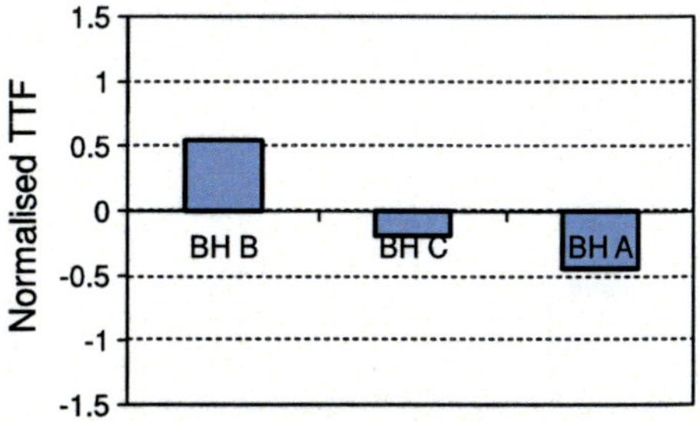

Fig. 11.16 Effect of board house on CAF failure

11.4.2.8 Effect of Glass Reinforcement on CAF Failure

The effect of different styles of glass fibre reinforcement was investigated using the samples with three weave reinforcements. The normalised TTF data are presented in Fig. 11.17. 2,116 weave reinforcement shows higher CAF resistance than 2,113 and 1,080 weave reinforcements, but there is no significant difference in CAF resistance between 2,113 and 1,080 weave reinforcements.

11.4.2.9 Effect of Drill Feed Speed on CAF Failure

Figure 11.18 presents the normalised TTF for the different drill speeds, and it seems that there is no significant effect of drill feed speed for the laminates and

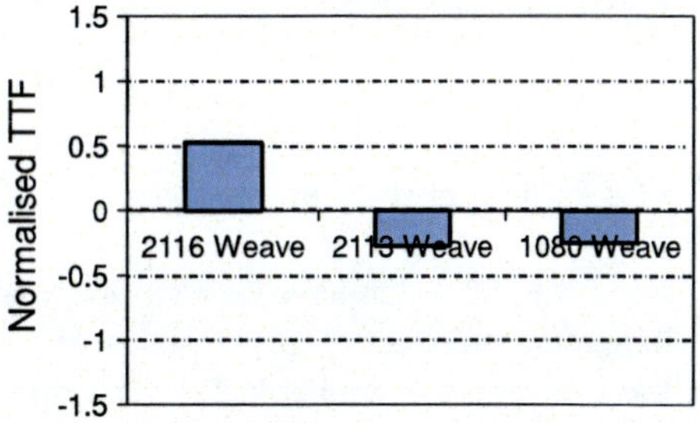

Fig. 11.17 Effect of glass reinforcement on CAF failure

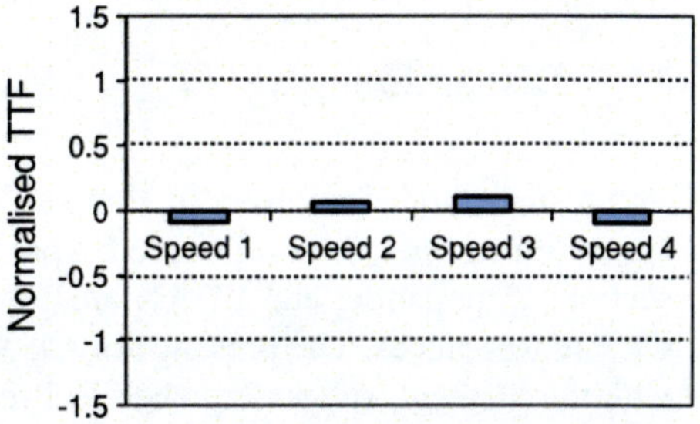

Fig. 11.18 Effect of drill feed speed on CAF performance

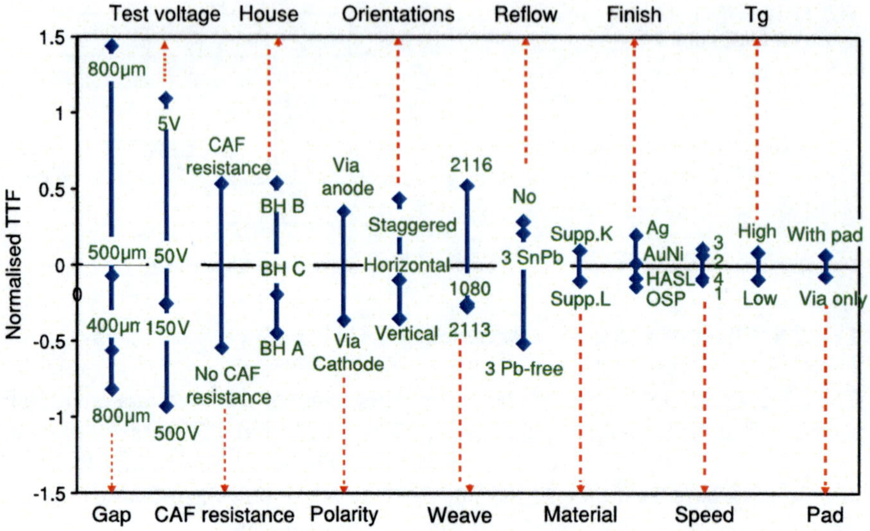

Fig. 11.19 Effect of all variables studied on CAF performance

processing conditions used in the study. Sectioning of the samples did not uncover any significant difference in the quality of the drill hole (i.e. copper plating ingress into glass fibre bundles). Despite the higher feed rates, the drill hole quality appeared to still be acceptable, resulting in little difference between the samples tested. Perhaps the use of blunt or overused drills would give a greater differentiation in drill hole quality.

11.4.2.10 Summary of Findings for All Variables

Some variables that have no significant effect on CAF performance are not detailed here, and the effect of all variables studied are summarised in Fig. 11.19 although the full report is available [7]. The difference in normalised time to failure as a result of change in process and material is clearly shown in the figure.

11.5 Discussion

The test method developed in this study has been successful in measuring the susceptibility of glass-reinforced epoxy laminates to CAF failures. Microsectioning, polishing and EDAX analysis results have confirmed the presence of CAF in test pieces. The presence of Cl, S, Ni, Zn as well as copper are associated with the filament, suggesting chemical residues from many processes or materials can be involved in the electrochemical filament formation (e.g. nickel from ENIG process, sulphur from plating baths, zinc from seeding chemistry, chlorine from acid etch).

It has been demonstrated that CAF can be detected by the measurement of changes in electrical resistance between through-hole via walls, or at other failure sites within the test vehicle. When CAF bridges a gap between two conductors, the loss of electrical insulation is rapid.

The method allows a comparison of material and processing parameters, showing the relative beneficial or harmful effects on product reliability in terms of potential CAF failures. It has been shown clearly that a main concern for those designing and manufacturing electronic laminates must be the significantly earlier failures promoted by decreasing geometries in multilayer board designs. For electronics companies looking at technology roadmaps, this may be an important limiting factor, especially for those with space/weight-sensitive applications, or humid and aggressive end use environments.

After July 2006, electronics manufactured in the EU must be made with lead-free solder, meaning an increase (commonly around 30°C) in peak soldering temperature as a result of replacement of eutectic alloys. Data suggest that the high peak temperatures seen in reflow operations have a detrimental effect on laminate reliability, and so lead-free introduction will have an impact. Whilst laminates can visually darken (when uncovered by solder mask) due to overheating, previous research suggests that the insulation between surface PCB features remains unaffected in lead-free processes [8]. Clearly, it is important that the same assumption for sub-surface failures in multilayer boards is not made, as the risk of CAF initiation and growth will increase.

One of the most interesting findings of this research is that the manufacturing undertaken by the board house (etching, permanganate desmear, seeding, plating, board lay-up, bonding/press cycles, etc.) has a significant impact of the resultant resistance to CAF. Whilst the scope of this study did not cover an investigation into these individual processes, it is clear that specifying materials alone are not adequate, and for better product quality, board house processing needs to be understood and controlled. There is certainly potential for further work in this area.

The trends seen in the accelerated tests support the assertion that CAF always grows from an anodic conductor and preferentially from copper associated with a via wall. CAF can grow from other initiation sites, such as copper planes, but times to failure are much longer. These findings fit well with the mechanism for CAF proposed.

Damage of the glass reinforcement fibres during drilling is thought to be a cause of ‘copper spears’ radiating from the plated via walls, concentrating the electric field and becoming initiation sites for CAF growth. Whilst no significant dependence on drill feed speed was observed in this work, micro-sections demonstrated that the wall quality was good for all the samples. It may be that drill rotation speed and drill quality are more critical in affecting glass damage and subsequent copper ingress than is the feed speed for drilling. From Figure, it can be seen there are no significant points of copper ingress out from the plated via wall, and this was typical of all samples sectioned across a range of materials. As such, this study suggests that drill hole quality can be good even with high feed rates but does not suggest that hole quality has no effect on CAF susceptibility. It is also evident that with high-quality holes, CAF will still occur under high-stress conditions.

There are choices that can be made which can increase product CAF reliability. Design choices include keeping via walls well spaced, and if possible, staggering those that are close at 45° to the glass weave. By choosing 'DICY'-free or CAF-resistant laminate from the supplier, the risks of CAF can be much reduced, but higher T_g materials are not necessarily more resistant to CAF than those with a lower T_g.

11.6 Conclusions

The test method used in this study has proved to be sensitive for monitoring changes in CAF resistance for laminates. The fundamentals of the technique are as follows:

- A multilayer glass-reinforced epoxy laminate test vehicle
- Manufacture and processing of the laminates using actual conditions for product
- Insulation resistance combs incorporating drilled and plated vias in proximity representing a range of geometries
- Application of a DC voltage across vias to provide a driving force for CAF growth
- High temperature and humidity exposure (85°C, 85%RH) for up to 1,000 h
- Frequent monitoring of insulation resistance to determine the time to failure for each test comb

The following parameters have been found to have a significant effect upon CAF resistance:

- Design Geometries
  - Via to Via Gap

    Vias closer together are more susceptible to CAF, even at the same voltage gradient
  - Via and ground planes

    Anodic vias fail faster than geometrically identical cathodic vias
  - Inner-layer pads

    Reduced CAF resistance for geometries where CAF growth is initiated from copper plane, which have an intrinsically higher resistance in any case

- Alignment to reinforcement
  - Warp and weft directions can show markedly different resistances to CAF, both must be examined
  - Vias staggered at 45° to the lay-up have a resistance to CAF greater than those aligned with warp or weft with the same gap
- Voltages
  - Higher voltages decrease time to failure at the same geometries
- Laminate System
  - 'CAF-resistant' (DICY-free) resins reduce the risk of CAF formation
  - Identically specified laminate from different suppliers can have different CAF performance
- Reflow Conditions
  - Peak temperatures of 250°C in reflow are potentially harmful to the CAF performance of laminate
- Manufacturing
  - Processing of laminate at board houses has a significant effect upon CAF performance

The following parameters had less significance with regard to CAF performance:

- Board's solderability surface finish
- Drill feed rate during via drilling
- Presence or size of inner-layer pads
- High/Low T_g designation of laminate

Acknowledgments The work was part of a project in the Materials Processing Metrology Programme of the UK Department of Trade and Industry. NPL gratefully acknowledges the work of the industrial partners: Alcatel Submarine Networks, TRW Automotive Group, Graphic Plc, Prestwick Circuits, Invotec, Isola and Polyclad.

References

1. Lando D, Mitchell JP, Welsher TL (1979) Conductive anodic filaments in reinforced polymeric dielectrics. Formation and prevention. In: Proceedings of the 17th annual reliability physics symposium, pp 51–63
2. Rudra B (1994) Electrochemical migration in multichip modules. In: Surface mount international conference and exposition. Edina, MN, USA. Proceedings of the technical program, NY, USA, Surface Mount Int, IPC: 50–6
3. Bi-Chu WU, Pecht M, Jennings D (1994) Conductive filament formation in printed wiring boards. In: Thirteenth IEE/CHMT international electronics manufacturing technology symposium. Edina, MN, USA, pp 74–79

4. Tsujioka N, Pecht M, Jennings D (1990) Generative mechanisms and preventive methods for micro-defects in PCB. In: Proceedings of the printed circuit world convention, Aylesbury, UK, Hazell Books B2/3–12
5. Mitchell JP, Welsher TL (1981) Conductive anodic filament growth in printed circuit materials. In: Proceedings of the printed circuit world convention II, Wurtt, Germany, Eugen G. Leuze Verlag, vol. 1 of 2 vol., pp 80–93
6. Ready WJ, Stock SR, Freeman GB, Dollar LL, Turbini LJ (1995) Microstructure of conductive anodic filaments formed during accelerated testing of printed wiring boards. Circuit World 21(4):5–9
7. Brewin A, Zou L, Hunt CP (2004) Susceptibility of glass reinforced epoxy circuit laminates to conductive anodic filamentation. NPL Report MATC(A)155
8. Wickham M, Zou L, Hunt C (2002) An assessment of the suitability of current pcb laminates to withstand lead-free reflow profiles. NPL Report MATC(A)91

Chapter 12
PCB Delamination

I. Baylakoglu and E. Hedin

12.1 Definition of Terms

The two light areas on the surface of the PCB are delaminations. They are effective blisters of air inside the board. During reflow, wave soldering or rework, moisture in the board expands causing the blisters (Fig. 12.1).

12.1.1 Blistering

Localized swelling and separation of base material or base and copper.

12.1.2 Delamination

Separation between layers of the base material and/or between the base materials and the copper. A separation between any of the layers of a base material or between the laminate and the conductive foil, or both. Delamination is not measling. Delamination is a rejectable defect, measling is not.

I. Baylakoglu (✉)
TÜBİTAK UZAY, 06531 ODTU Kampüsü, Ankara, Turkey
e-mail: ilknur.baylakoglu@uzay.tubitak.gov.tr

E. Hedin
ERICSSON AB, AB 164 80, Stockholm, Sweden
e-mail: eva.hedin@ericsson.com

G. Grossmann and C. Zardini (eds.), *The ELFNET Book on Failure Mechanisms, Testing Methods, and Quality Issues of Lead-Free Solder Interconnects*,
DOI: 10.1007/978-0-85729-236-0_12,

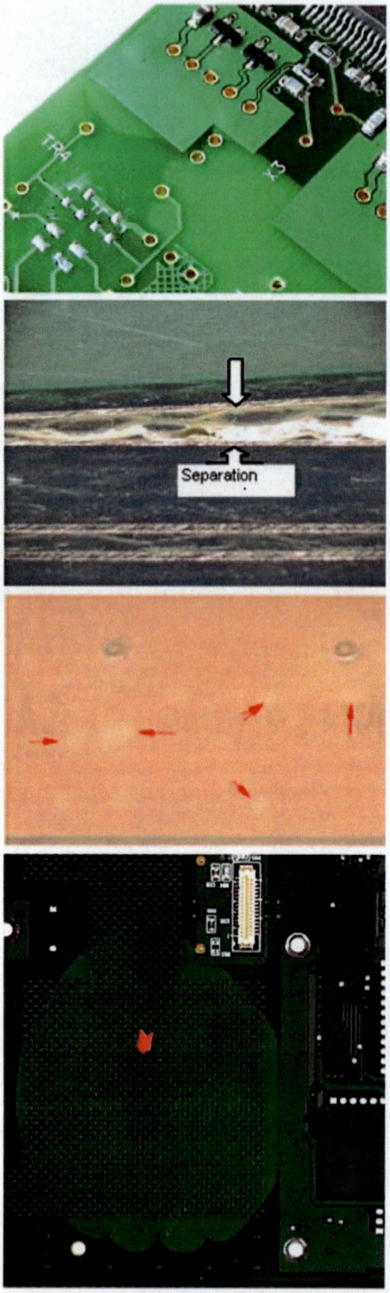

Fig. 12.1 Several optical views of PCB delamination [1]

12.1.3 Measling

Separation of glass fibers from the resin at the weave intersection, measling is seen like white crosses where the fiberglass bundles cross through each other. That becomes the weakest point where fracturing occurs (or slight disbonding). Measling, by definition, occurs only at the intersection of two fiber bundles within the lamina.

12.2 Definition of What is a Failure

A delamination of the composite layers of a circuit board may be a result of overheating encountered during operation environments or soldering by repair personnel (Figs. 12.2, 12.3).

The CTE mismatch in the material during the soldering cycles together with poor adhesive properties are the main reasons. This might occur in large solid Cu planes (bad adhesive) and/or blind vias (large CTE mismatch in the *z*-axis). Absorbed humidity causes high vapor pressure in the laminate, during the soldering process this may also contribute.

Another problem is moisture trapped between the board layers occurring during the PCB production. This will happened if moisture is not fully evaporated or if the material is not correctly cured during production. Large copper plane inside or on the board will raise the risk for delamination.

The delaminations reduce board integrity and can permit moisture to pervade its structure, thereby increasing the probability of metal migration.

Fig. 12.2 Delamination between two layers

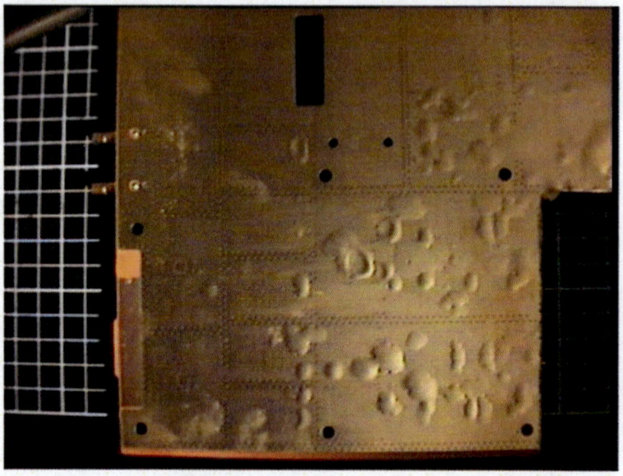

Fig. 12.3 Severe visual seen delamination on backside of a board

12.3 Mechanism

12.3.1 Effect of Td (Decomposition Temperature) and CTE Mismatch

Lead-free peak assembly temperatures can reach point where resin decomposition begins. Higher temperatures result in increased thermal expansion which gives stress on plated holes, buried vias and between copper and laminate as a result.

12.3.2 Effect of Moisture

Vapor pressure of water is much higher at lead-free assembly temperatures. Absorbed moisture can volatilize during thermal cycling and cause voiding or delamination. PCBs that initially pass lead-free assembly testing may exhibit defects after storage in an uncontrolled environment, as a result of moisture absorption.

12.4 Test Methods to Activate the Mechanism

12.4.1 Standard Test

12.4.1.1 Moisture Absorption

Tendency of a material to absorb moisture from the surrounding environment. This can be assessed by more than one method, including water soak or in an increased pressure and humidity environment.

12.4.1.2 Time to Delamination

While not a fundamental property, measures the time for delamination to occur at a specific temperature, e.g., 260°C (T260) or 288°C (T288).

It is related to decomposition temperature and adhesion between material components. Thermal expansion and moisture absorption can also influence results. In multilayer PCBs, the treatment of the internal copper surfaces is also critical.

12.4.1.3 IPC-TM-650, Time to Delamination Method

The specimens for this TMA test are 6.3 mm × 6.3 mm squares cut from the PC board and dried in an oven at 105°C for 2 h to remove moisture. The TMA probe force is 0.05 N. The IPC procedure considers a delamination time greater than 10 min at 260°C as acceptable.

12.4.1.4 Decomposition Temperature, Td

Measures weight loss from resin degradation as a function of temperature. Td is typically defined as the point at which 5% of the original mass is lost to decomposition, but other levels can also be reported, e.g., 1, 2% or "onset." Resin decomposition can result in adhesion loss and delamination. A 5% level of decomposition is severe, and intermediate levels are important for assessing reliability since peak temperatures in lead-free assembly can reach onset points of decomposition. A high Td by the 5% definition does not guarantee performance. Conversely, a low Td by the 5% definition is not necessarily bad if the onset temperature of decomposition is high enough.

12.4.1.5 Peel Strength

The preferred technique for measuring the robustness of an adhesive joint has for many years been the peel strength test. Several methods are available for this measurement in IPC-TM-650. The test is usually performed on a tensile testing machine. The test coupon is composed of a copper foil that is laminated to a PCB substrate. A small width of laminated copper foil is pulled vertically from the sample, and the force required is measured. This force divided by width of the copper foil is referred to as the peel strength. Often this test is performed at an elevated temperature in an attempt to account for the thermal effects of assembly.

12.4.2 *Product-Dependent Test*

Delaminations are design dependent; many layers, double-sided mounting, repairable products, unbalanced copper etc., will raise the risk for delamination. Therefore, it may be necessary with additional tests for advanced PCB to secure the quality of the design with chosen material. Two tests are briefly described below.

12.4.2.1 Test on Bare Board (PCB)

Run the PCB in a standard reflow profile for n cycles. Number of cycles is set from the mission profile point of view. If possible, run the boards until they start to delaminate. This is made to ensure that the PCB has enough margins for production and repair.

12.4.2.2 Test on Fully Assembled Boards

Bare board is not always the same as mounted board in thermal behavior. For high-reliable products, it is sometimes necessary to test the boards with components as well. This is especially important for large boards that have components with large difference in thermal mass.

This test is run if the PCB test above shows out to have no or small margins.

Run the boards in its ordinary soldering profile with components for max 10 cycles or until they delaminate. The test is approved if the boards will pass the number of cycles set in the mission profile.

12.5 Delamination Detection Methods

12.5.1 *Visual Inspection and Microsectioning*

Delamination on the surface layers and severe delamination can be seen with visual inspection.

Since buried vias are one of the critical delamination zones, it is possible to microsection areas with buried vias and look for delamination.

12.5.2 *Terahertz Imaging*

The two modes of imaging available using terahertz radiation are reflection and transmission mode; however, the techniques for acquiring images are similar for both modes. In transmission mode, the acquired image is the result of any radiation that was not completely blocked or absorbed, whereas for reflection mode the image is created from the reflected energy.

Because different materials attenuate terahertz radiation to different extents, the variation in amplitude can be measured to create an image.

Furthermore, since the terahertz pulse is very narrow in time (sub-picoseconds), it covers a broad frequency spectrum. The transit-time of the THz pulse can also be measured. This is also known as a time-of-flight measurement. The time-of-flight measurement effectively measures the change in optical path length that occurs due to either variation in thickness of the sample or changes in the refractive index.

These techniques and imaging modes are collectively characterized as terahertz time domain spectroscopy (THz-TDS) or T-Ray imaging. The application of terahertz imaging for detecting failures in printed circuit boards and electronic components is feasible for many types of failures.

EOS or ESD mechanisms contribute to a large number of integrated circuit failures. A catastrophic failure may result in relatively large-scale delamination or other deformation that still remains hidden during visible light inspection. And non-catastrophic failures often result in modification of substrate or trace properties.

Using reflection mode THz imaging, detection of such damage in integrated circuits is possible down to a spatial scale of approximately 100μ. On the scale of the entire printed circuit board, stress-related cracking and board layer delamination are candidates for detection with terahertz imaging. While IC imaging can be performed primarily in reflection mode, board level imaging can be performed in either reflection or transmission mode. The determining factors are the number of composite layers of which the board is constructed and whether the number and density of metal layers permits adequate information to be acquired. Finally, in situ imaging of PCBs requires interpreting the imaging results of complex boards with multiple components.

12.6 Process Mission Profile

Some designs will experience peak temperatures of around 245°C, while others will experience peak temperatures of up to 260°C, or even higher in a few cases. Some boards may experience 2–3 thermal cycles, others up to 5–6, and more depending on how many reworks are allowed.

It is important to know and set requirements at the PCB that are relevant for the product.

12.7 Design Rules

Below are some rules of thumbs for PBA design.

- Always order PCB as lead-free soldered LFS board with relevant requirements.
- To achieve better adhesion of the copper areas to the laminate.

- Add openings in the Cu area if needed.
- Avoid (if possible) deep buried vias
- Using porous power and ground plane materials in PCBs allows liquids (e.g., water and/or other solvents) to pass through the power and ground planes, thus increasing failures in PCBs caused by cathodic/anodic filament growth and delamination of insulators.
- Components with high thermal mass shall if possible be even distributed over the board.

It is also important to know that the vendors continuously introduce new and improved materials with better thermal performance and filling materials for blind vias with better z-axis property.

12.8 Reference to Existing Standards

Mostly used standard documents for military, commercial and space applications related to delamination are listed below:

MIL-PRF-55110 *Printed Wiring Board, Rigid General Specification For (rev. G) A.3.6.5*

IPC-A-600 *Acceptability of Printed Boards (2.3 and 2.3.3) (rev. H)*

IPC 4101, *Specification for Base Materials for Rigid and Multilayer Printed Boards (rev. C)*

ECSS-Q-ST-70-10C *Qualification of Printed Circuit Boards*

Reference

1. Assurance Technology Corporation (http://www.assurancetechnology.com/FA3.asp)

Chapter 13
Excessive Warpage of Large Packages During Reflow Soldering

Bart Vandevelde

Abstract During a solder reflow process, IC components deform under a rather excessive temperature loading. A too high warpage at temperatures above solder melting can cause either that the joints in the corner are not soldered to the PCB pads or that the solder joints are shorted due to strong compression of the corner joints. In this work, the warpage of a 35 by 35 mm^2 large PBGA package has been measured during a temperature profile, which is similar to a lead-free soldering process. It was found that the warpage becomes very high when the applied temperature is above the glass transition temperature of the overmould material. At that moment, there exists a very large CTE mismatch between the overmould and the BT laminate. The warpage measurements have been successfully verified by Finite Element Modelling, at least when the right material properties are used. This proves that modelling can be used as an estimator of warpage for packages. Also the impact of initial moisture uptake has been experimentally investigated, and it was shown that it has a dominant effect on its warpage behaviour. Finally, a FEM-based parametric study shows the impact of several design parameters.

13.1 Warpage Issues during Reflow Soldering

IC packages consist of different materials, having a different coefficient of thermal expansion (CTE). Under a temperature change, the different mechanical expansions cause internally acting forces and this finally results in internal mechanical stresses and deformations of the package in order to find equilibrium of the internal

---

B. Vandevelde (✉)
Imec, Kapeldreef 75, 3001 Leuven, Belgium
e-mail: bart.vandevelde@imec.be

G. Grossmann and C. Zardini (eds.), *The ELFNET Book on Failure Mechanisms, Testing Methods, and Quality Issues of Lead-Free Solder Interconnects*,
DOI: 10.1007/978-0-85729-236-0_13,

forces. When the package is not symmetric over the cross section, an out-of-plane deformation will also be seen as the forces causing bending moments.

During a solder reflow process, IC components deform under a rather excessive temperature loading. This temperature goes up to 230°C for SnPb soldering, and 260°C for lead-free soldering [1]. A too high warpage at temperatures above the solder melting temperature can cause both open and shorted solder interconnections. Figure 13.1 depicts both cases. The effect of the global bowing is that the corner areas are bending downwards, and as a consequence, it causes compression of the molten solder joints which can shorten when the compression becomes very high. The opposite can also happen. When the package is bowing upward, the corner joints do not make contact at melting temperature resulting in open connections, or even worse, so-called frozen or "cold" connections. These interconnections only make a mechanical contact, giving no failure at room temperature. However, they start to disconnect again at higher temperatures, e.g. in operation mode.

As warpage is typically proportional to the square of the package size, large packages are more vulnerable to these soldering problems. The overmould properties also play a very important role: both its T_g (glass transition temperature) and the mechanical properties above T_g are very important for the warpage at soldering temperature, as will be shown in this chapter.

This phenomenon of interconnection problems due to excessive warpage during solder reflow is strengthened by the change from SnPb to lead-free soldering, which requires a 30°C higher melting temperature.

Also moisture absorption has a large impact on package warpage as during heating up, the swelling (internal vapour pressure) of overmoulds and other polymer-based materials causes extra deformation of the package.

Section 13.2 discusses the warpage problems of a 35 by 35 mm^2 large PBGA package. The warpage of this package has been measured during a temperature profile which is similar to a reflow soldering process. It was found that the warpage becomes very high when the applied temperature is above the glass transition temperature of the overmould material. At that moment, there exists a very large CTE mismatch between the overmould and the BT laminate.

Section 13.3 shows how initial moisture absorption results in a different warpage profile during the soldering process, causing higher warpage at maximum temperature and also package cracking problems after the soldering process [2, 3].

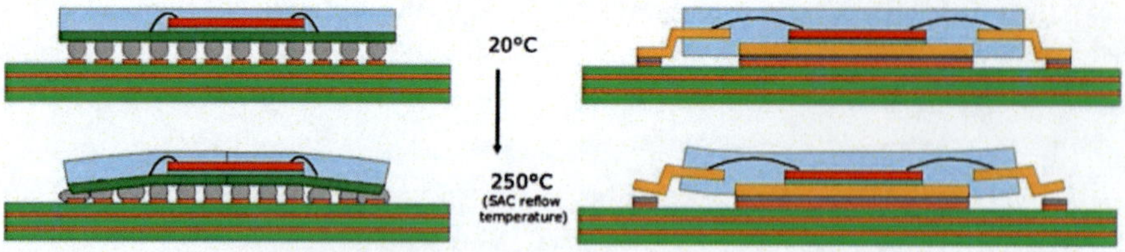

Fig. 13.1 Schematic drawing of excessive warpage at reflow temperature resulting in shorts (*left*) or opens (*right*) after soldering

In Sect. 13.4, a simplified analytical model is derived to calculate the critical warpage as function of ball diameter and ball pitch.

Section 13.5 depicts that finite element modelling (FEM) can predict the measured warpage for this particular PBGA when the right (measured) material properties are used. Using this FEM, a parametric study has been performed showing how dimensional and material parameters influence the warpage of PBGA packages.

13.2 Experimental Study: Warpage of a Regular 35 by 35 mm^2 PBGA Package during Reflow Soldering

In this section, the warpage behaviour is characterized for a particular large PBGA package (Fig. 13.2). The objective is to measure the warpage during a reflow profile, similar to a soldering process profile used in regular Printed Circuit Board (PCB) assemblies. In particular, the warpage at temperatures above the melting temperature of the solder material is of high interest in this study.

Although the results are specific for one package concept with specified dimensions, it very well clarifies the problem and is representative for all large laminate-based packages. Package with metal slugs will need a different study as the metal leadframe is much stiffer than the BT laminate.

13.2.1 Sample Description and Preparation

For this study, a PBGA package is selected with a die glued on a BT laminate and encapsulated by an overmould material. The size of the package is 35 by 35 mm^2, and the die size is 13 × 13 × 0.3 mm^3. The BT laminate has a total thickness of

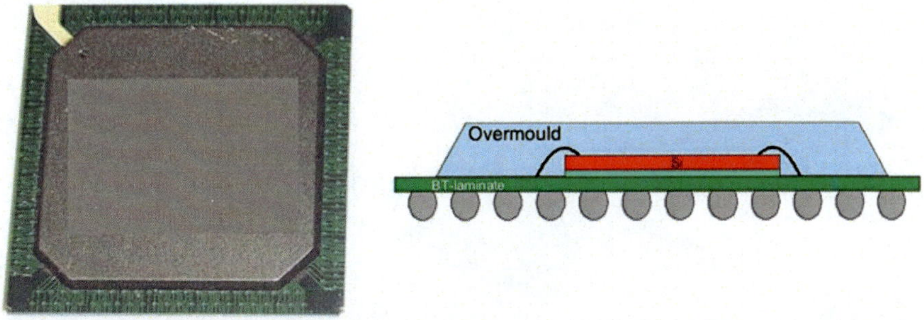

Fig. 13.2 Picture and schematic drawing of the PBGA 35 × 35 mm^2 package

0.47 mm, including 4 copper layers. The overmould cap is 1.15 mm thick and ends at 2.5 mm from the side of the BT laminate.

In order to avoid the impact of any moisture uptake, the samples have been dried for 3 days at 120°C. In total, 5 samples have been measured in order to have a statistical average.

13.2.2 Warpage Measurement Method

In order to measure the warpage as a function of the temperature, the Topography and Deformation Measurement (TDM) system from Insidix has been used [4, 5]. The measurement concept is based on the fringe projection Moiré technique. It measures the topography as the difference between a reference horizontal surface and the measured surface. This reference surface is a calibration sample that is perfectly flat and homogeneous white and is measured beforehand. Known light patterns of stripes are projected on the PBGA sample and viewed by CCD camera. The outcome of one measurement is the out-of-plane profile, as shown in Fig. 13.3 for the PBGA sample. Infrared heaters are warming up the sample, while the temperature is measured using a thermo-couple touching the BGA sample. Figure 13.3 shows the topography of this PBGA package measured at 228°C. By measuring the topography at different temperatures, the out-of-plane expansion versus temperature is characterized.

The warpage is defined as the difference in out-of-plane displacement between the centre and the corner (not the difference in *z*-height!). By definition, the warpage has a positive sign when the corners are deforming downward compared to the central area.

Fig. 13.3 Topography of the 35 × 35 mm^2 PBGA, measured by the TDM equipment at 228°C

13.2.3 Results

A typical reflow temperature profile has been applied to this package, with a maximum of 228°C (Fig. 13.4). In total, 5 samples have been measured and a statistical average is made to determine the warpage values [6].

The first experiment presented in this chapter is the warpage profile measurement for a dry sample. Figure 13.4 shows the out-of-plane profile for this package at 22°C (initial temperature) and at 228°C (=maximum reflow temperature). At 22°C, the warpage is almost zero, while at 228°C, the warpage is visibly high (Fig. 13.5).

The difference between these measured profiles and a perfect non-warped PBGA profile give the out-of-plane deformation, which is shown in Fig. 13.6. At 22°C, the warpage is about −50 μm. At 228°C, the warpage is +450 μm.

The warpage has been measured at different temperatures, both in heating and cooling. The result for this sample is shown in Fig. 13.7.

At room temperature, there is a small warpage of about 50 μm. This warpage is almost constant up to 170°C. Above this temperature, the warpage substantially change and the corner start to bend downward. At that moment, the actual temperature goes over the glass transition of the overmould, resulting in a high CTE mismatch between the overmould and the BT laminate (while they are quite well matched below 170°C). As a consequence, the corners start to warp downward and this becomes very high at the maximum temperature of 228°C. A maximum warpage of 450 μm is achieved.

Five different samples have been measured, and the difference in warpage values between the samples was rather small. Moreover, no significant difference was found between the initial deformation and the deformation after a reflow cycle. Therefore, the deformations were fully reversible and consequently, no nonlinear deformations such as plastic yielding, creep deformation or cracks are noticed. For one sample, a second succeeding reflow temperature step was applied, and the warpage nicely followed the same warpage profile as the first reflow cycle.

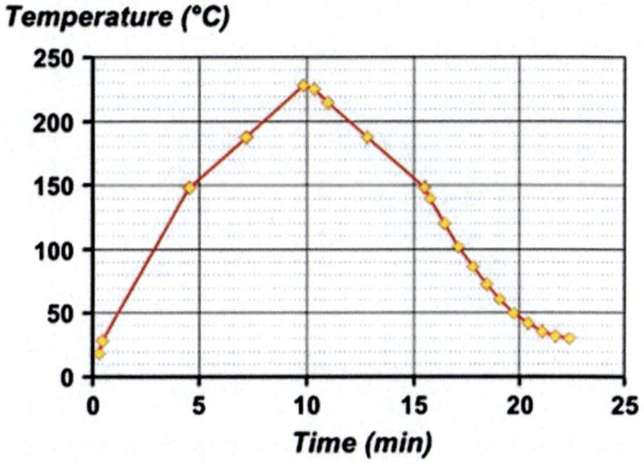

Fig. 13.4 Temperature profile applied to PBGA sample

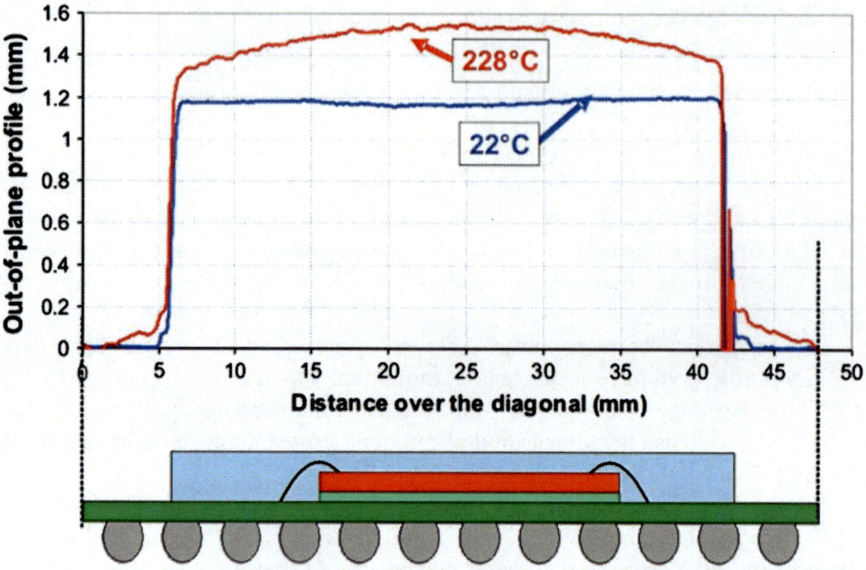

Fig. 13.5 Out-of-plane profile (topography) measured by TDM equipment (along the diagonal)

13.3 Impact of Moisture Absorption on Warpage Behaviour

Although components are supposed to be fully dry, some moisture uptake can not be avoided. In this study, we analysed how two extreme conditions of moisture absorptions have an impact on warpage behaviour during reflow cycling. These two conditions are as follows:

- 168 h at 85% humidity, 85°C
- 18 h at 100% humidity, 120°C

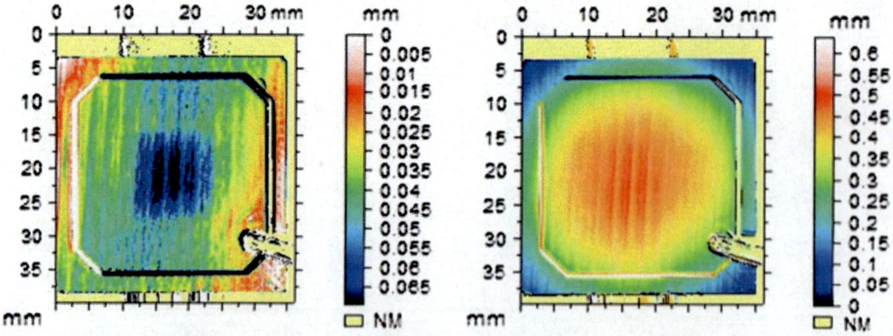

Fig. 13.6 Out-of-plane deformation at 22°C (*left*) and 228°C (*right*)

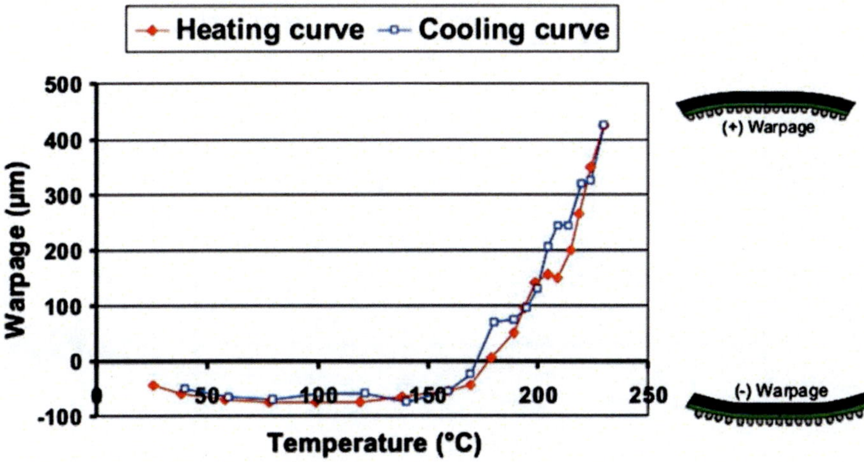

Fig. 13.7 Warpage measured by TDM equipment during a reflow process

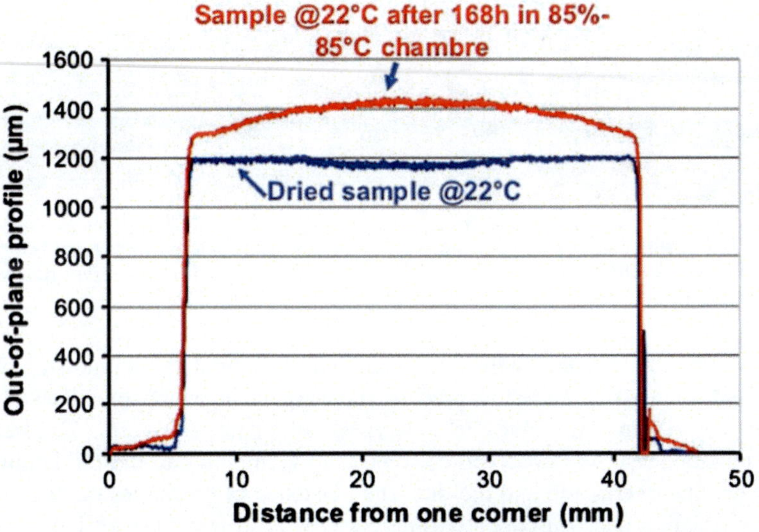

Fig. 13.8 Out-of-plane profile of a dried sample and a sample which has been preconditioned at 85°C/85% humidity for 168 h (both graphs are measured at 22°C)

After being in the moisture oven, the initial warpage is already very different from the dried sample, as shown in Fig. 13.7. The additional bowing is caused by the expansion of the overmould due to the moisture uptake

Figure 13.8 shows the warpage during the reflow profile for the moistured samples versus the dry samples. Due to the moisture uptake, the component has an initial warpage of +200 µm, which is caused by swelling of the overmould due to moisture uptake. There is no difference found between the two preconditions

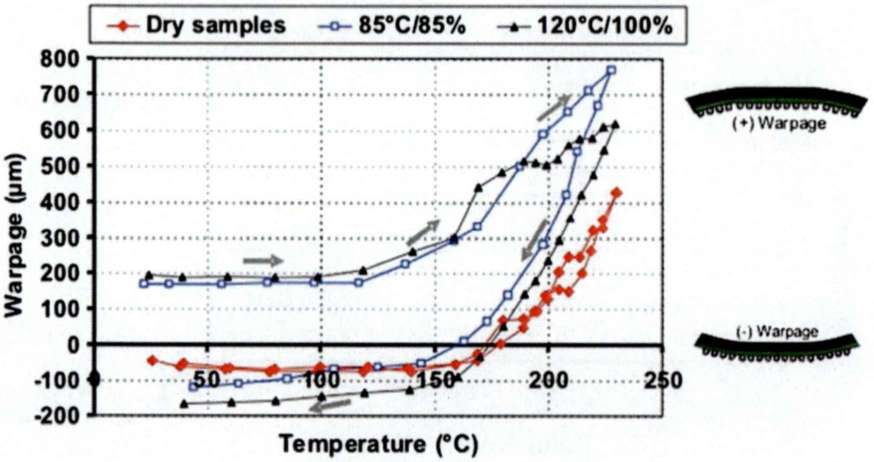

Fig. 13.9 Measured warpage during a reflow temperature profile (from room temperature to 230°C and back)

Fig. 13.10 Cross section of PBGA after a reflow profile. The *top* PBGA was dried before reflow, and the *bottom* PBGA had an 85°C/85% preconditioning and shows after reflow serious delamination

(85% and 100%). Probably the overmould is fully saturated with moisture for both preconditions. After the reflow process, the warpage does not come back along the same curve, but will return to a warpage of about −150 µm. The reason is delamination that was observed between the overmould and the BT laminate and between the overmould and the die. The existence of cracks results in a different stress situation and consequently different bowing (Figs. 13.9, 13.10).

13.4 Analytical Calculation of the Critical Warpage

This section shows a first-order analytical calculation of the critical warpage w_{crit}. This critical warpage is defined as the global bowing of the package needed to shorten the corner joints.

Initially, the solder joint has the shape of a perfect ball or sphere with a radius of r_{ball}. When bowing of the package results in downward bending of the package

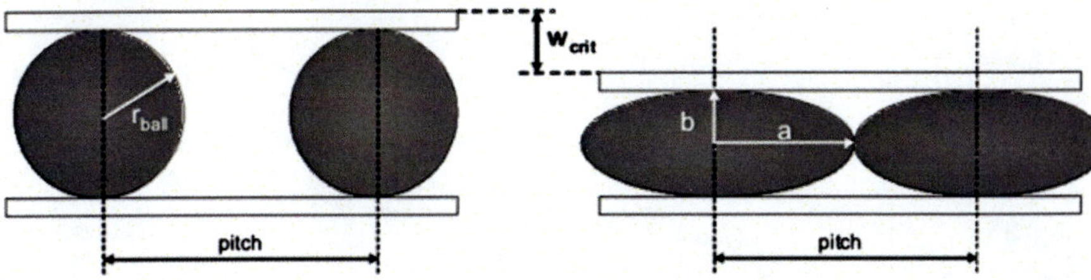

Fig. 13.11 Schematic description of uncompressed "spheres" vs. compressed "ellipsoids". The ellipsoids touches each other when "a = pitch/2". The critical warpage is defined as "$w_{\mathrm{crit}} = 2(r_{\mathrm{ball}} - b)$"

corner, the solder volume becomes an ellipsoid with an equatorial radius 'a' ($>r_{\mathrm{ball}}$) and a polar radius 'b' ($<r_{\mathrm{ball}}$). When 'a' becomes half the pitch, the two neighbouring ellipsoids touch each other (Fig. 13.11).

It is possible to calculate the critical warpage as follows. The volume of the sphere with radius 'r_{ball}' is given by

$$V = \frac{4\pi}{3} r_{\mathrm{ball}}^3 \tag{13.1}$$

The volume of the ellipsoid (compressed solder joint) with equatorial radius 'a' and polar radius 'b' is given by:

$$V = \frac{4\pi}{3} a^2 b \tag{13.2}$$

The critical warpage at which the two ellipsoids touch each other is achieved when following condition is fulfilled:

$$a = \frac{\mathrm{pitch}}{2} \rightarrow b = r_{\mathrm{ball}} - \frac{w_{\mathrm{crit}}}{2} \tag{13.3}$$

As the volume of the ellipsoid is equal to the one of the perfect sphere, the critical warpage can be calculated as follows:

$$V = \frac{4\pi}{3} a^2 b = \frac{4\pi}{3} \left(\frac{\mathrm{pitch}}{2}\right)^2 \left(r_{\mathrm{ball}} - \frac{w_{\mathrm{crit}}}{2}\right) = \frac{4\pi}{3} r_{\mathrm{ball}}^3 \tag{13.4}$$

Solving this equation, the critical warpage where the two ellipsoids touches each other is given by:

$$w_{\mathrm{crit}} = 2r_{\mathrm{ball}} \left[1 - \left(\frac{2r_{\mathrm{ball}}}{\mathrm{pitch}}\right)^2\right] = d_{\mathrm{ball}} \left[1 - \left(\frac{d_{\mathrm{ball}}}{\mathrm{pitch}}\right)^2\right] \tag{13.5}$$

with d_{ball} is the sphere/ball diameter (=two times the ball radius). This relative simple formula clearly depicts that the allowed warpage before ball shorts decreases when the ratio of the ball diameter over the pitch is higher. This is

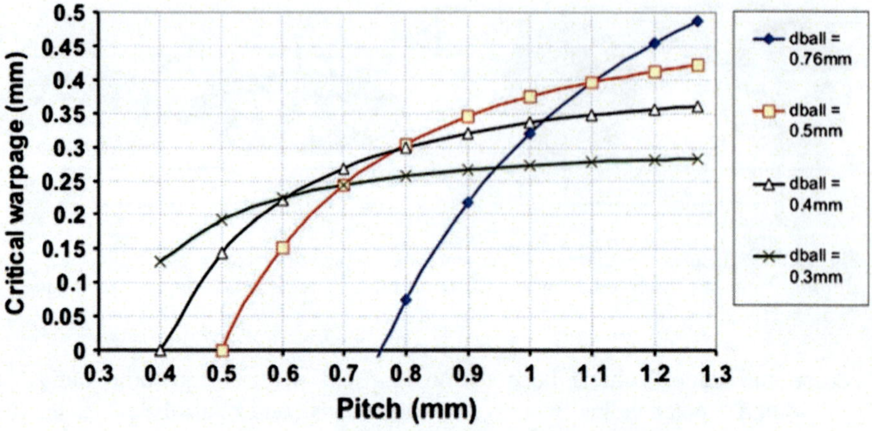

Fig. 13.12 Critical warpage as function of the pitch for different ball diameter sizes

obvious as the free distance between the balls become smaller. Figure 13.12 shows for different ball sizes the critical warpage as a function of the pitch.

In the 35 by 35 mm^2 PBGA package presented above, the pitch is 1.27 mm and the ball diameter is 0.76 mm. Using these numbers, the critical warpage is:

$$w_{\mathrm{crit}} = 0.76 * \left(1 - \left(\frac{0.76}{1.27}\right)^2\right) = 0.488\,\mathrm{mm} \qquad (13.6)$$

This value is only 40 µm higher then the maximum warpage seen for this package (Fig. 13.9), and this shows that warpage could become critical for shorts at the corner balls. Little moisture uptake by the component can result in some higher warpage at reflow temperature with solder shorts as consequence.

13.5 Prediction of Warpage of Large Packages using Finite Element Modelling Simulations

Finite element modelling is a useful tool to simulate thermally induced mechanical deformations of IC packages. In this work, the objective is to simulate the same warpage trend as found in experiments.

13.5.1 Simulation of Warpage during a Reflow Profile

A 3D finite element model is built for this package (Fig. 13.13). Nonlinear material properties are applied both for the overmould and BT laminate, in order to

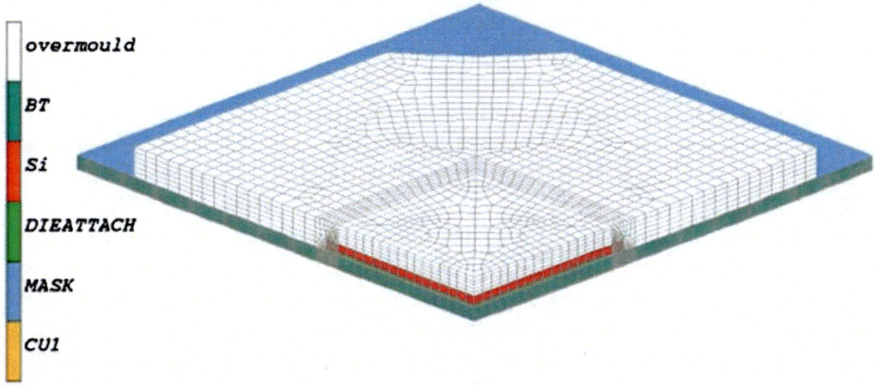

Fig. 13.13 3D Finite Element Model simulating the thermally induced mechanical deformation during a reflow process

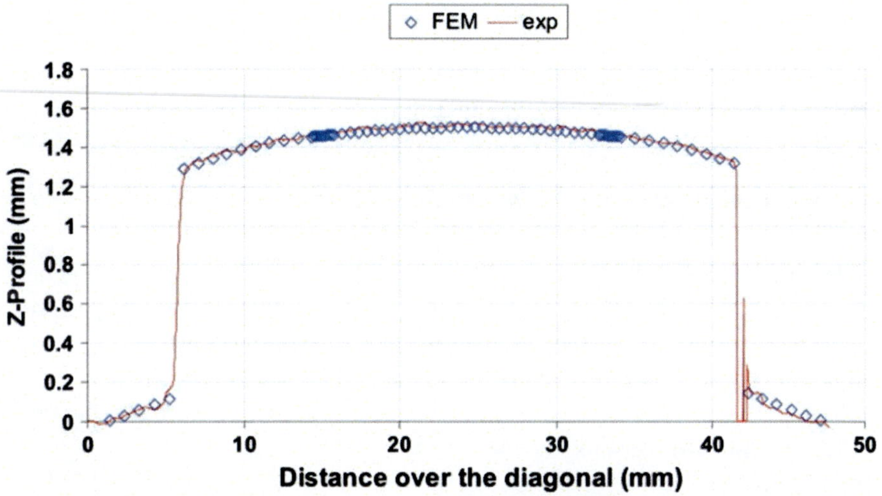

Fig. 13.14 Comparison of FEM and experiment for the PBGA 35 × 35 mm^2 package at 228°C

be able to include the T_g for both materials. For the overmould material, own CTE and E-modulus measured data are used as they differ a lot from the datasheets provided by the material supplier. For example, the measured T_g of the overmould was 170°C, while the datasheet mentioned 210°C. As already shown in Fig. 13.7, the T_g plays a dominant role in the warpage behaviour at the highest temperatures.

With the right material properties, a very good agreement was achieved between the simulated and measured warpage. Figure 13.14 shows a perfect fitting of the bowed top surface of the PBGA at 228°C. Figure 13.15 depicts that the finite element model predicts very well the warpage as function of the temperature.

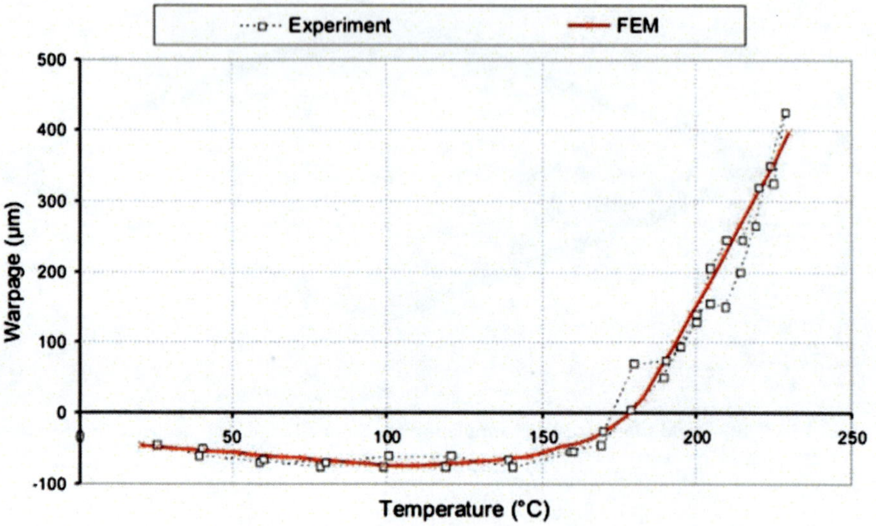

Fig. 13.15 Comparison of FEM and experiment for the warpage versus temperature for the PBGA 35 × 35 mm^2

13.5.2 Parametric Study for PBGA

The Finite Element Model proved to be very reliable tool to predict warpage of the 35 × 35 mm^2 PBGA package. Therefore, a brief parametric study was performed to analyse the impact of some dimensional and material parameters.

13.5.2.1 Chip Dimensions

Figure 13.16 shows that a thicker and larger die decreases the warpage. The explanation is that the die with a high E-modulus of 169 GPa is a stiffener for the package

13.5.2.2 T_g of Overmould and Maximum Reflow Temperature

Figure 13.17 shows that warpage increases with low T_g and higher reflow temperature. This graph also shows the impact of the transition to lead-free assembly. Looking to the middle curve ($T_g = 165°C$), the warpage is below the critical value for SnPb soldering (maximum of 230°C), but appears in the dangerous zone for SnAgCu soldering (up to 260°C)

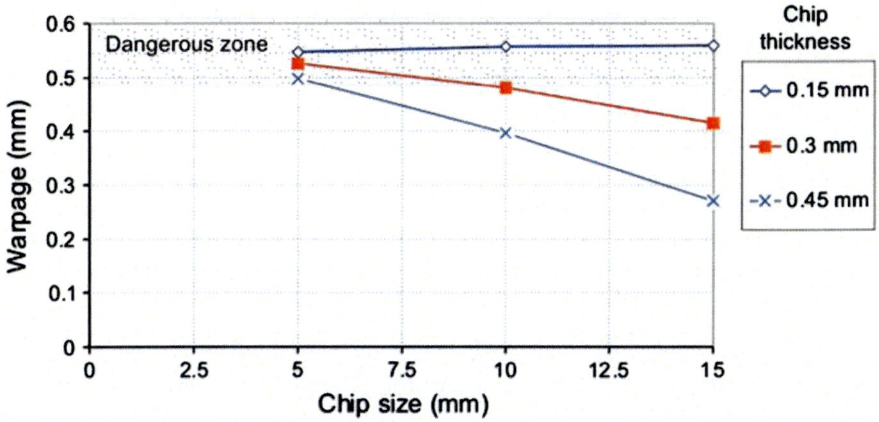

Fig. 13.16 Influence of chip dimensions on warpage at maximum reflow temperature (230°C)

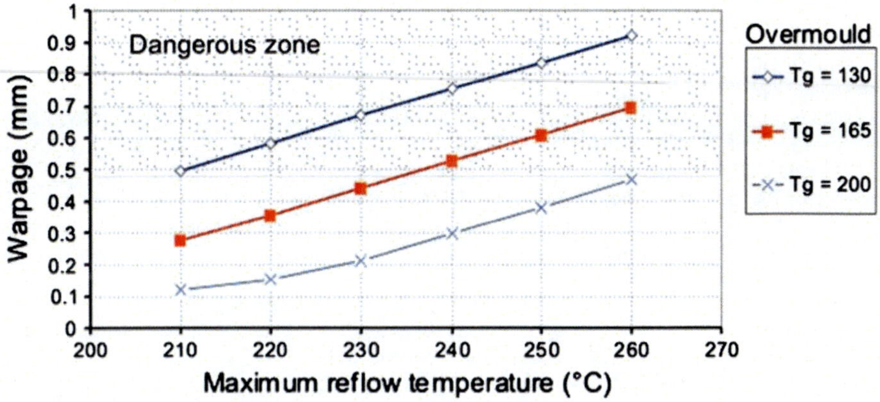

Fig. 13.17 Influence of overmould T_g and maximum reflow temperature

13.6 Guidelines

In order to avoid bad solder interconnections after reflow assembly (induced by strong warpage), following guidelines are summarized below:

It is very useful to measure the warpage of a package during a reflow cycle. It will show you the warpage above solder liquidus, and together with some simple analytical calculations of critical warpage, you can predict whether problems could be expected or not.

Finite Element Modelling can very well predict the warpage as function of temperature, supposing that the right material properties are used. In that sense, it is highly recommended to measure the material properties yourself, in particular the properties of the overmould. Datasheets often overestimate the glass transition temperature of overmould, typically with 30°C.

In general, warpage becomes higher with increased package size and increased temperature. So, if a design change results in such a change, you have to be aware of potential problems.

Moisture absorption has a substantial impact on the warpage of the package. Large moisture absorption can even result in cracks after reflow, but also small moisture absorption already can have an impact, at least if the warpage of a dry sample is already at the limit. In that case, it is highly recommended to dry the components prior to reflow soldering.

Acknowledgments The authors would like to thank Rafael De Weerdt, Veerle Simons, Myriam Vandepeer, Mario Gonzalez, Daniel Vanderstraeten, Guy Brizar and Eddy Blansaer for their support in the measurements and FEM simulations. The work has been supported by the IWT (institute for support of scientific research in Flanders).

References

1. IPC/JEDEC Joint Industry Standard, J-STD-020C (2004) Moisture/reflow sensitivity classification for plastic integrated circuit surface mount devices
2. Vaccaro BT, Shook II, Thomas E, Gilbert JJ, Horvath C, Dairo A, Libricz GJ (2005) PBGA package warpage and impact on traditional MSL classification for Pb-free assembly. SMTA International conference
3. Shook RL, Gilbert JJ, Thomas E, Vaccaro BT, Dairo A, Horvath C, Libricz GJ, Crouthamel DL, Gerlach DL (2003) Impact of ingressed moisture and high temperature warpage behavior on the robust assembly capability for large body PBGAs. In: Proceedings of electronic components and technology conference, pp 1823–1828
4. Richard I, Fayolle R, Smyth A, Lecomte J.-C (2005) New experimental approach for failure prediction in electronics: topography and deformation measurement complemented with acoustic microscopy. EuroSimE conference, pp 305–310
5. Hertl M, Weidmann D, Lecomte J-C (2009) Process optimization: influence of heating and cooling rate on the thermo-mechanical stress generated in components EMPC2009. Rimini, Italy
6. Vandevelde B, Deweerdt R, Duflos F, Gonzalez M, Vanderstraeten D, Eddy B, Guy B, Renaud G (2009) Impact of moisture absorption on warpage of large BGA packages during a lead-free reflow process. Therminic Workshop, Leuven, pp 162–165

Chapter 14
Popcorn Cracking

Rainer Dudek

14.1 Introduction

Interface delamination and popcorn cracking are among the most common reasons for mechanical damage in plastic packaging. Alpern et al. [1] formulate the popcorn mechanism as follows: When a plastic IC package is exposed to normal ambient air (temperature, humidity), moisture diffuses into the molding compound (MC) of the package. The absorbed moisture causes the degradation of adhesion strength of the package's various material interfaces. During printed circuit board (PCB) reflow soldering, an IC package is subjected to solder shock temperatures, typically between 220 and 240°C for conventional SnPb solder, but up to 260°C for lead-free solder. Thermal stresses caused by thermal mismatch between the package's constituent materials may cause delamination of the die-pad/MC interface. Absorbed moisture then diffuses into the delaminated gap resulting in a vapor pressure build-up and subsequently in the doming of the MC underpad layer. When the maximum stress caused by the doming exceeds the MC strength, package cracks occur ("popcorn effect"), see Fig. 14.1.

In general, one distinguishes between three types of popcorn failure modes in relation to the starting interface of the cracks. Mode I refers to cracks emanating from the die-pad backside/MC interface delamination, mode II from the die-attach/die-pad delamination and finally mode III from the chip surface/MC delamination. A polyimide chip coating is normally used to curtail mode III failure. Mode I is the most common failure mode in conventional plastic IC packages.

R. Dudek (✉)
Fraunhofer ENAS, Micro Materials Center Berlin and Chemnitz,
Technologie-Campus 3, 09126 Chemnitz, Germany
e-mail: Rainer.Dudek@enas.fraunhofer.de

G. Grossmann and C. Zardini (eds.), *The ELFNET Book on Failure Mechanisms, Testing Methods, and Quality Issues of Lead-Free Solder Interconnects*,
DOI: 10.1007/978-0-85729-236-0_14, © Springer-Verlag London Limited 2011

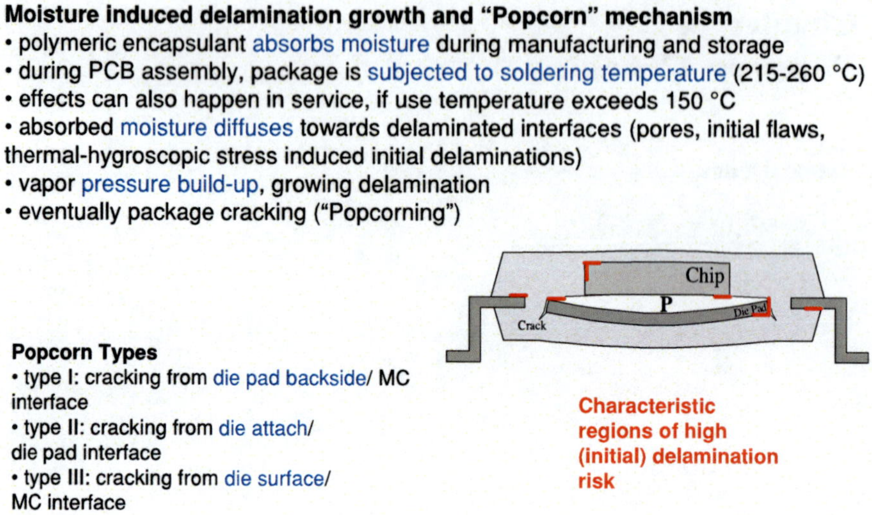

Fig. 14.1 Schematic of vapor pressure-induced delamination and "popcorn effect" in plastic packages

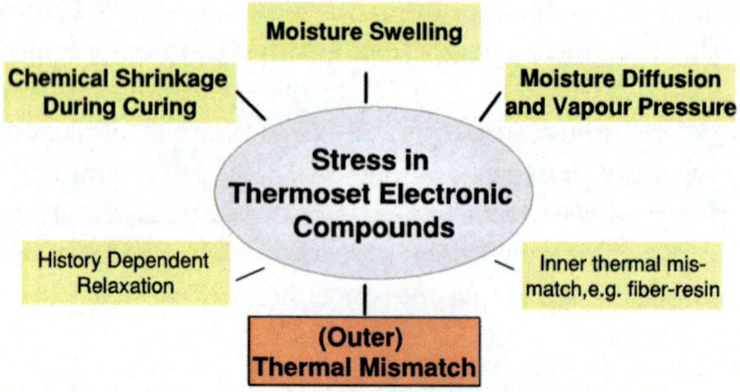

Fig. 14.2 Schematic of stress sources in plastic packages

14.2 Stress in Plastic Packages

Obviously, the phenomenons are related to stress build-up in plastic packages, which might depend on several sources, see Fig. 14.2, the most important one being thermal mismatch. Other sources of package stresses are linked to the action of moisture, as can be observed from Fig. 14.3. According to Koh et al. [2], three property characteristics of the polymers involved are to be considered for reducing the package failures: adhesion strength, moisture absorption, and stress relaxation. The last characteristic is dependent on the material mechanical properties such as the modulus of elasticity, flexural strength, fracture toughness, and the glass transition temperature.

Since package cracking is mainly caused by vapor pressure, moisture absorption and desorption properties of the molding compounds are studied by various authors [3, 4]. They pointed out that both the moisture content and the distribution

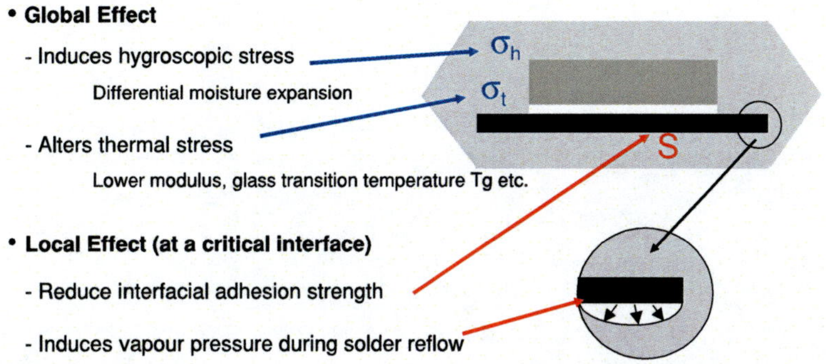

Fig. 14.3 Schematic of effects of moisture in plastic packages

of moisture in regard to the critical interface affect the popcorn resistivity of the packages. Moisture diffusion calculations based upon a one-dimensional simplified diffusion model were applied to calculate the moisture content near the die-pad.

For thin packages investigated, moisture saturation of the encapsulant material is reached fast and concentration gradients become meaningless. Additionally, the typical delamination mechanism at the lower die-pad/encapsulant interface, which is addressed by most of the popcorn models, was seldom observed for thin packages.

14.3 Experimental Observations

The popcorn phenomenon for PQFP packages can be studied by high-speed camera investigations. The already described effect of pressure build-up and popcorn cracking is illustrated in Fig. 14.4.

The effects of moisture preconditioning on package cracking were evaluated in a second experimental step [5]. The specimens were exposed to vapor at a 85°C/85% R.H. environment for 168 h according to JEDEC 112A, level 1. A second batch of samples was exposed to 30°C/60% R.H. for 192 h (level 3 conditions). Both the packages with absorbed moisture and dry packages were subjected to surface mount solder reflow conditions with different peak temperatures up to 215°C.

No package cracking was observed for the dry packages. Again, the unfailed packages were inspected by scanning acoustic microscopy. No inner damage could be detected.

The level 1 preconditioned samples failed in all cases by popcorning. In spite of the commonly expected failure mode with delamination at the lower die-pad/encapsulant interface (type I), all packages exhibited delamination at the upper die-attach/die-pad interface and subsequent package cracking to the outside, see Fig. 14.5. In the case of level 3 preconditioned packages, about 30% of the

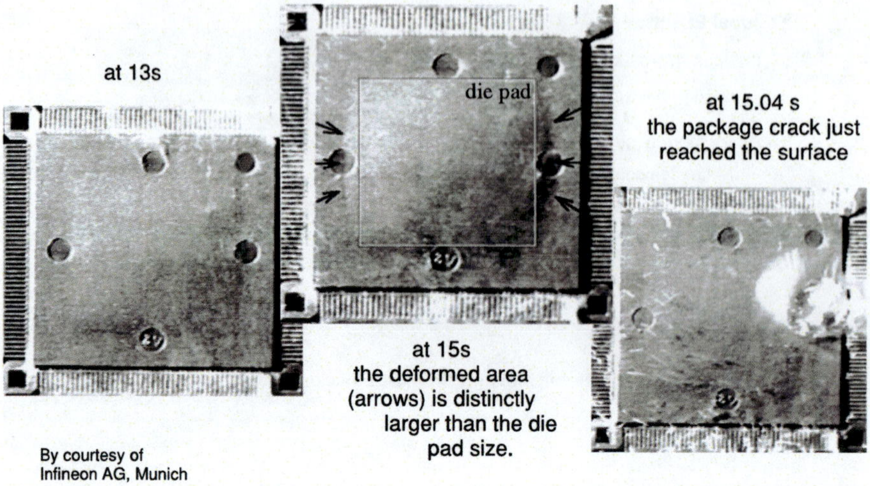

Fig. 14.4 Observation of the development of popcorn delamination growth and MC cracking

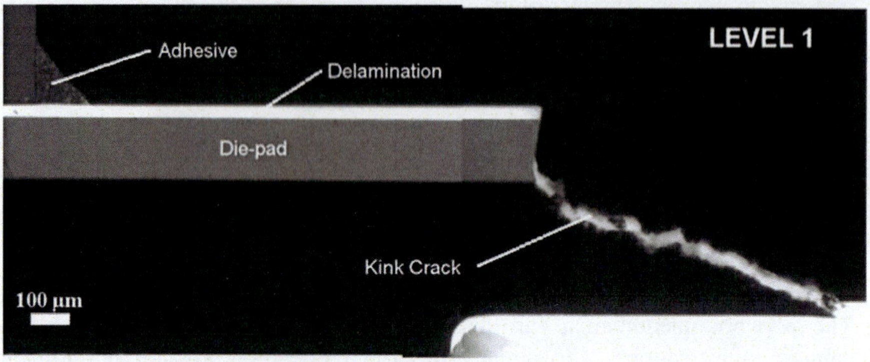

Fig. 14.5 Characteristic popcorn failure mode after level 1 preconditioning (type IV)

cracked samples have shown the same type of cracking, often referred to as type IV cracking, while 70% failed by delamination pad/encapsulant. A typical cross-section of a type I cracked sample is shown in Fig. 14.6. It is obvious from the figures that the kink angles of the package cracks differ for both types of cracking. Cross-sectioning of several samples has validated those cracking directions to be characteristic in connection with the different delaminations. Full delamination of the die-attach and package cracking initiating at the lower edge of the die-pad with crack growth in the characteristic direction shown in Fig. 14.5 is the dominant failure mode with the thin QFP packages at level 1 preconditioning. Thus, this failure mode was studied in further detail by finite element analysis.

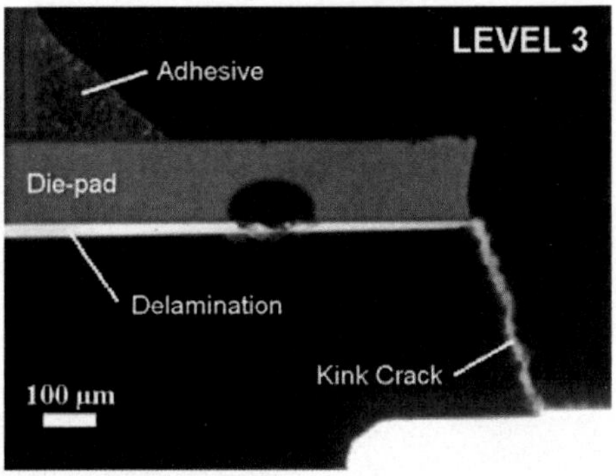

Fig. 14.6 Characteristic popcorn failure mode after level 3 preconditioning (type I)

14.3.1 Characteristics of Encapsulant Materials

It is obvious from the stress phenomena discussed so far that mainly two ways are open to improve package reliability:

- reduce intrinsic stress
- improve interface adhesion

Obviously, the first points become more significant for lead-free soldering, since higher peek temperatures are reached during the soldering process. IC package are subjected to solder shock temperatures, typically up to 260°C for lead-free solder.

To reduce stress, one opportunity in standard application is to use the so-called low-stress encapsulants. According to Koh et al. [2], low-stress encapsulants have commonly been interpreted as having a low CTE: value to be closer to that of the silicon chips. However, since thermal mismatch cannot be avoided, a truly low-stress material must therefore possess additional characteristics that resist crack initiation and moisture penetration. Because the shear stress is proportional to the modulus, lower modulus is desired. Higher flexural strengths, on the other hand, tend to retard crack initiation and propagation.

14.4 Theoretical Analyses

The stress state of the packages, when subjected to moisture, heat, and vapor pressure, can be studied by FE analysis. Different delamination-related types of analysis have to be performed to fully characterize the popcorn phenomenon: moisture diffusion, transient thermal history, and thermo-mechanical stress analyses. However, since a variety of materials data is required and these analyses require non-standard coupling, they are seldom performed in such a detail.

Most important input data is the materials data. It is well known that filled epoxies exhibit dependencies of the mechanical characteristics on moisture,

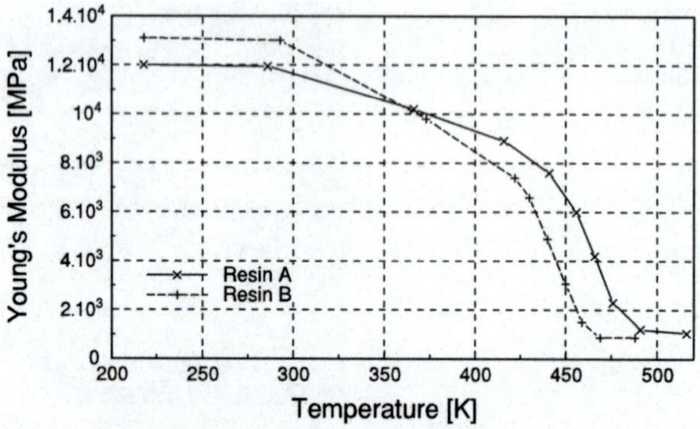

Fig. 14.7 Flexural modulus as a function of temperature

temperature, and time, see e.g. [2, 6, 7]. A drop in mechanical stiffness occurs in the range of the glass transition temperature T_g, while the coefficient of thermal expansion (CTE) usually increases. Additionally, the constitutive response of the filled polymers becomes time dependent, especially at this temperature level.

Characteristic examples for the dependence of the elastic modulus on temperature are shown in Fig. 14.7. The two multifunctional resins shown are the so-called low-stress, highly filled epoxy-based molding compounds with 80 wt% silica content. Both encapsulants reach their T_g at a temperature of about 170°C. Their moisture absorption is low, a value of 0.42 wt% is given by the materials supplier at 85°C/85% R.H. after 168 h. Similarly, the epoxy-based die-bonding adhesive exhibits low moisture content of 0.66 wt% at the same conditions.

Reduction in CTE is achieved through use of optimized filler particle sizes and distributions. The lower limit of approx. 10×10^{-6} mm/mmK has already been reached, and further reduction will be more difficult. Rather than continue to try lowering the CTE, package failures are tried to be reduced more effectively if other properties such as the flexural strength and moisture resistance are improved.

In numerical investigations by means of FEA, either strengths concepts or fracture mechanics concepts are commonly used for design optimizations using sensitivity analyses

In doing so, the most important points in simulation are

- the description of the constitutive behavior of all materials present in the model,
- the measurement and preparation of all relevant geometric and materials properties,
- the definition of all necessary boundary conditions, loading conditions as well as data or additional simulation steps required to define the stress-free state and residual stresses from manufacturing (this could also include special materials models describing the curing process of polymeric materials), and
- (after appropriate simulations) the evaluation of the possibly occurring failure modes.

A variety of investigations on the fracture mechanical treatment of interface mechanical issues has been published by Tay et al., see e.g. [8, 9]. A complex analysis considering all hygro-thermo-mechanical effects is published by Guojun and Tay [9]. The main objective of this paper is to investigate the effects of temperature, moisture diffusion, and vapor pressure on the likelihood of delamination of the interface between the lead frame pad and the encapsulant. In this paper, the entire thermal and moisture history of a plastic IC package is simulated from the start of level 1 moisture preconditioning (85°C/85%RH for 168 h) to subsequent exposure to a solder reflow process lasting about 5 min.

References

1. Alpern P, Lee KC, Dudek R, Tilgner R (2000) A simple model for the mode I popcorn effect for IC packages. In: 11th European symposium, reliability of electron devices, failure physics and analysis (ESREF 2000), Dresden, Germany, 2–6 October Microelectronics Reliability 40, pp 1503–1508
2. Koh W, Tai B, Kolawa A (1995) Low stress encapsulants for reduced failures in plastic packages. In: Proceedings of international mechanical engineering conference, San Francisco
3. Nguyen LT, Chen KL, Schaefer J (1995) A new criterion for package integrity under solder reflow conditions. In: Proceedings of 45th electronic components and technology conference, pp 478–490
4. Wong EH, Koh SW, Lee KH, Rajoo R (2002) Advanced moisture diffusion modeling and characterization for electronic packaging. In: Proceedings of ECTC 52, San Diego, pp 1297–1304
5. Dudek R, Sommer P, Michel B, Alpern P, Birzer CH, Tilgner R (1998) Investigations on popcorn cracking of T-QFP packages. In: Proceedings of ECTC 48, pp 944–951
6. Harper BD, Lu L, Kenner VH (1997) Effects of temperature and moisture upon the mechanical behavior of an Epoxy molding compound. EEP-vol 19-1, Advances in Electronic Packaging, vol 1, ASME
7. Hsu TR, Fawzi KM, Nguyen LT, Kuo A-Y (1997) Fracture strength of Epoxy Molding compounds. EEP-vol 19-1, Advances in Electronic Packaging, vol 1, ASME, pp 261–267
8. Tay AAO, Tan GL, Lim TB (1993) A criterion for predicting delamination in plastic IC packages. In: Proceedings of 31st IRPS, Atlanta, GA, pp 236–243
9. Guojun H, Tay AAO (2004) A numerical study of the effects of temperature, moisture and vapour pressure on delamination in a PQFP during solder reflow. In: EPTC proceedings, Singapore

Chapter 15
Thermal Capability of Components

C. Zardini and J.-Y. Deletage

15.1 Introduction

Since July 1st, 2006, the lead-free technology legally entered in its industrial production phase. For the components manufacturers, the conversion to lead-free required a tremendous work not only to eliminate the banned substances but also to re-qualify the components according to the new soldering processes.

Indeed, for surface mount components, the most used alloy family for reflow soldering, the SnAgCu, shows a liquidus temperature ranging between 217 and 221°C and it is agreed that during reflow the coldest solder joint temperature should be no less than 230°C. Accordingly, the components are submitted to temperatures notably higher than those to which they were submitted with the Pb technology and much of them had to be re-qualified.

The problem of thermal capabilities of components was identified since the Amsterdam meeting by the TEG components members because at that time, there was sometimes confusion between "RoHS compliance" and "lead-free processability".

This topic was thus selected to be included in the prioritized solutions list of the TEG components.

In fact, the high temperatures reached by the components during the soldering processes are an issue for the main following reasons:

- the properties of organic materials are affected (the parts can become deformed in a permanent way, the melting point of some dielectric films can be reached)
- the moisture absorption of molding compounds becomes more critical (increased risk of "popcorning").

C. Zardini (✉) and J.-Y. Deletage
Laboratoire IMS, 351 cours de la libération, 33405 Talence, France
e-mail: christian.zardini@free.fr

G. Grossmann and C. Zardini (eds.), *The ELFNET Book on Failure Mechanisms, Testing Methods, and Quality Issues of Lead-Free Solder Interconnects*,
DOI: 10.1007/978-0-85729-236-0_15,

But the fact that the dissolution of metals in solder alloys is accelerated could also be an issue for the wave soldering.

What can one say today in connection with the thermal compatibility of components with the lead-free soldering processes?

To obtain brief replies, we examined more than 125 web sites of components makers. We also learned from the members of the TEG components.

The obtained results are exposed below by distinguishing active and passive components.

We will not discuss about the through-hole components because, as the lead-free wave solder pot temperatures are in the range of 265–270°C (a slight increase over the 250–260°C of the Pb containing alloys), they are normally robust enough to sustain a wave soldering process.

15.2 Active Components

Plastic package diodes, transistors or ICs are generally made of a semiconductor die, attached to a metallic leadframe or an organic substrate, encapsulated with a molding compound. This compound is a blend of an epoxy resin, a hardener, a filler, a catalyst, a flame retardant, a colorant and a mold-released agent. The most important properties requested from a molding compound are a good moldability, a good adhesion with the leadframe or the substrate, good insulating properties, heat resistance, strength and an excellent resistance to humidity. Moreover, the compound must be RoHS compliant.

To satisfy all these requirements and in particular to improve the heat resistance and the resistance to humidity, the suppliers had to formulate new mold compounds usable with the lead-free technology.

One of the main issues is the moisture absorption by the plastic package. This problem was identified from the beginning of the use of the surface mount technology. During the reflow process, the high value of the vapor pressure of moisture causes delamination of the packaging materials and can lead to well-known "popcorn effect". A moisture sensitivity classification was introduced—for the solid-state devices—in the IPC/JEDEC J-STD-020 standard, which defines eight "Moisture Sensitivity Levels" according to the floor life of the package.

With the lead-free technology, the moisture problem is exacerbated because of the vapor pressure increase with temperature. When the temperature increases from 230 to 260°C, the vapor pressure increases by about 70%. Just so, 260°C is the temperature reached on the top of the package during the reflow process. With such a high temperature level, the package is more prone to internal delaminations, the previous MSL cannot be preserved and the component must be re-qualified.

The new version of the J-STD-020D.1 standard published in March 2008 defines the classification procedure with a reflow profile for Pb-free assembly [1].

The peak reflow temperature T_p measured at the top package surface is in the range of 245 to 260°C according to the package thickness and volume.

Other parameters:

Time maintained above T_L = 217°C	t_L = 60–150 s
Time within 5°C of peak temperature	t_p = 30 s
Average ramp-up time (T_L to T_p)	3°C/s max
Ramp-down rate (T_p to T_L)	6°C/s max.
Pre-heat/soak minimum temperature	T_{smin} = 150°C
Pre-heat/soak maximum temperature	T_{smax} = 200°C
Time from T_{smin} to T_{smax}	t_s = 60–120 s

To be qualified for a given MSL, the packages must be able to sustain three reflow processes. Furthermore, a Pb-free component shall be capable of being reworked at 260°C.

Under these conditions, a J-STD-020D.1 compliant component should not present thermal compatibility problems to the users of the lead-free technology.

When one consults the websites of the device manufacturers, one observes that the home page almost always comprises a “RoHS”, an “Environmental” or a “Support” heading.

Generally present, the FAQ heading is also extremely useful by the answers which it gives.

Under these headings, recommended soldering profiles and data according to J-STD-020D.1 are given. According to the standard, the components are able to sustain three reflow processes. Some of them are also able to sustain four of them.

However, some products have limitations, particularly the optoelectronics components. Some of them cannot sustain more than two reflows.

Some devices such as power components are RoHS compliant but not strictly speaking lead-free because of the Pb containing die attach solder, but this alloy is exempt from current legislation due to the lack of replacement material.

In short, a user of “RoHS compliant” components should not encounter thermal compatibility problems when a manufacturer refers to the J-STD-020D.1 standard.

By prudence, he must, however, consult his supplier for certain product lines.

15.3 Passive Components

As there is not yet a standard similar to J-STD-020D.1 for the passive components, the manufacturers often refer to this one when they present the specifications of their products. So that, RoHS compliant products are supposed to be capable of withstanding lead-free solder processes but it is not the case of all the components.

For some passive components encapsulated in a molding compound, a moisture sensitivity level is defined according to the J-STD-020D.1 qualification process.

15.3.1 Resistors

Chip resistors are constructed by screen printing or depositing a thick or thin resistive film on an alumina substrate. A protective coating is applied on the top of the resistor and solderable terminations on the ends. A nickel barrier is applied under the termination coating to prevent the dissolution (leaching) of the electrodes during the soldering process.

The recommended reflow conditions are generally:

10 s or less at 255/260°C and 60–90 s above 220°C
so thermal capability is not an issue for these components.

Some manufacturers specify two or less reflow processes, probably to limit the resistance drift.

However, these components are very robust and can withstand a wave soldering process.

Trimming potentiometers, which are more complex components, are more sensitive to heat. Nevertheless, the recommended reflow conditions are similar, with a limit to two reflows.

15.3.2 Capacitors

15.3.2.1 Ceramic Chips

Ceramic chip capacitors are normally very robust components but they are prone to cracking if submitted to sudden heating. Care must be taken to limit the heating and cooling rates to about 3°C/s. This is not an issue for the components assembled according to the J-STD-020D.1 reflow profile (provided that the cooling rate is limited to 3°C/s) but could be one during wave soldering if the pre-heating stage is not correctly adjusted in time and temperature.

The heat resistance of the chips, which is tested by dipping in a solder bath, is for example:

260 ± 5°C for 10 s.

In the absence of standardization, the reflow conditions given by the manufacturers are very varied:

- J-STD-020D.1 profile
- 240/260°C peak temperature for 5 s maximum, 30 s max at 240°C
- etc…

The number of reflow cycles is generally limited to two but some products are qualified for three reflows.

The ceramic trimmer capacitors can also withstand two reflow cycles with a peak reflow temperature in the range of 240–260°C following the manufacturers.

15.3.2.2 Film Capacitors

Film chip capacitors made with polyethylene teraphtalate (PET) or polyethylene naphtalate (PEN) metallized films are not well adapted to the lead-free reflow profiles because the highest temperature that they can withstand is in the range of 240–255°C.

In return, the capacitors made with polyphenylene sulfide (PPS) films can sustain a 260°C peak temperature reflow profile including 50 s above 217°C. Moreover, two reflow cycles are allowed.

However, film chip capacitors are not suitable for wave soldering.

15.3.2.3 Electrolytic Capacitors

Aluminum liquid electrolyte capacitors are also not well adapted to the lead-free reflow profile. When their internal temperature increases, at first the liquid electrolyte dilates and if its boiling point (204°C for γ-butyrolactone) is reached, it vaporizes. The inside pressure of the aluminum case increases, leading to the deformation of the cylinder [2]. In the worst case, the case cracks and the capacitor blows.

To avoid such a scenario the recommended peak temperature during reflow is in the range of 230–240°C.

Aluminum solid electrolyte capacitors, the technologies of which were improved these last years, do not have this disadvantage. The reflow profile can have a peak temperature equal to 260°C with 50 s above 220°C. However, they tolerate only one or two reflow cycles.

Tantalum capacitors can also be submitted to reflow profile with a peak temperature in the range of 225–260°C, following the type and size.

As the tantalum and aluminum solid electrolyte are encapsulated in a molding compound prone to absorb moisture, some suppliers specify a moisture sensitivity level as for the active components, for example MSL 3.

15.3.3 Magnetic Components

Since the introduction of the lead-free technology, several problems have been identified with the wirewound surface-mounted magnetic components.

These components are made with enamelled copper wires composed of single or multiple conductors wound on a coil former or directly on a toroid magnetic core. The first identified problem is linked to the ability of the wire insulating enamel to sustain the peak temperature reached during the reflow process. The "enamel" is in fact an organic material applied on copper in multiple passes, each one followed by a curing stage.

Table 15.1 Magnet wires temperature classes

Class	A	B	F	H	200	220	>220
For prolonged life at °C	105	130	155	180	200	220	>220

Theses wires are called "magnet wires" and their specifications are the subject of several standards [3, 4].

According to their insulations properties, the magnet wires can be classified by temperature, from 105°C to more than 220°C (Table 15.1).

To be qualified for a given class, an insulating material should remain effective for a period of 5,000 h when exposed to a temperature 20°C above the class rating, and its extrapolated life must be at least 20,000 h at the claimed rating.

Notice that the temperature to which the enamel degrades is higher than the class temperature but not given by the standards.

Until now, with SnPb reflow profiles, class B magnet wires gave satisfaction, but if one use them with lead-free reflow profiles the enamel can be degraded during the process, leading to defects such as shorts between adjacent turns.

To avoid these problems, experience shows that using class 180 [5] or higher magnet wires is recommended to get components able to sustain for example three reflows.

The same goes for the coil former whose material must be chosen according to the magnet wire class. Liquid crystal polymer reinforced glass coil formers with copper-tin alloy matte tin plated pins give good results.

The winding wires are generally connected to the pins by dip soldering, and this process can be the source of problems which will appear thereafter. It is well known that a piece of metal immersed in liquid solder will dissolve. The main factors influencing the rate of dissolution are the base metal, the solder composition and the temperature. The rate of dissolution of several metals in $Sn_{60}Pb_{40}$ solder was studied and showed that copper dissolves more slowly than silver or gold for example [6]. Until now, with the SnPb alloys, it had not been announced dissolution problems following dip soldering of copper wires to the pins of the coil formers but standardized tests exist to verify it [7].

Problems could appear with the lead-free solder alloys for two reasons. At first, the rate of dissolution of copper in high tin composition alloys such as SAC is much higher than in SnPb alloys. Secondly, the dissolution of metal is a thermally activated process. The temperature of the dip soldering bath and that of the reflow process is higher with SAC alloys than with conventional SnPb ones. As a result, the copper dissolution is more visible with the lead-free processes. If the component is submitted to several reflows, the dissolution is even more sensitive.

This phenomenon is not annoying for wires of large diameter but can be an issue for "litz wires," which consist of a number of individually insulated magnet wires twisted or braided in a uniform pattern to reduce eddy currents losses in high frequency applications. The diameter of elementary wire used to manufacture litz wires starts with 20 μm and can go until the millimeter. If nothing is to be feared with the large diameters, magnetic components designers using litz wire must take into account the copper dissolution issue when choosing the used elementary wires and avoid the too thin ones.

The last issue identified for the surface-mounted magnetics components comes from the insulating material used to separate the layers of wires or to bound the coil. The dielectric films used for the manufacture of the component must also be able to sustain the temperatures encountered during the reflow processes. For example, polyester films whose melting point and service temperature are, respectively, 254 and 130°C cannot be used and must be replaced by better materials such as polyimide, which presents also much better shrinkage properties [8].

At the end of the component fabrication process, a magnetic core is assembled around the bobbin. Most of the magnetic components use a ferrite core and one can wonder whether the properties of this material, in particular the permeability, will not be affected by the temperature reached during the reflow process. In fact, as these materials have been sintered at temperatures in the range of 1,000–1,400°C they are not affected even after three reflows. Finally, if the core is a two-part one assembled by gluing, the adhesive must be carefully chosen to be able to resist three reflows.

The information about the thermal capabilities given on the sites of the SM inductors, filters, transformers or current sense suppliers is ambiguous. The RoHS compliance is always declared and accompanied by the mention "reflow solderable" or "suitable for reflow soldering" but recommended solder reflow profiles are seldom given.

Instead of the profiles, we find specifications or advices such as:

- resistance to soldering heat 260°C for 5 s
- peak temperature 250°C max for 10 s, 50 s max above 230°C
- 260°C compatible.

When a profile is mentioned, it is the J-STD020D.1 one.

Concerning the number of reflows that a component can sustain, the information is generally not available. Sometimes it is mentioned that the component can be submitted to a reflow process "two times or less".

To conceive reliable equipment, the designer using industrial magnetic components will thus be obliged to contact his supplier to obtain this information, which is essential for him.

15.3.4 Connectors

For the connectors' manufacturers who adopted the tin finish, the main RoHS compliance issue was certainly that of the whiskers growth, which had to be solved with the bath suppliers. But moreover, to be reflow compatible, the connector housing material should be capable of withstanding 260°C for 10 s without becoming deformed. This compatibility obliged the connector manufacturers to use thermoplastic materials whose heat deflection temperature is higher than

260°C, and among them liquid crystal polymer (LCP) and polyphthalamide (PPA) which are also most expensive.

The RoHS and reflow compatibility are clearly shown in the datasheets. For some products, a two reflows capability is indicated but generally this point is not mentioned.

The same remarks can be made for the IC sockets.

15.3.5 Crystal Units

Crystal units in ceramic or metal packages are reflow compatible (260°C 10 s and 60 s above 220°C) but the number of reflows is not mentioned.

The same goes for the TCXO, VCXO and ceramic resonators.

15.3.6 Fuses

RoHS compliant resettable and non-resettable surface mount fuses are not all reflow compatible. This capability must be checked on the datasheets.

15.3.7 Relays

RoHS and reflow compatible plastic package surface mount solid state or conventional relays with a specified MSL are available.

15.3.8 Switches

RoHS and reflow compatible rotary and DIP switches are available.

15.4 Conclusion

The RoHS and STD-020D.1 compliant solid-state devices should not pose problems of thermal capability because they are normally able to sustain three reflow cycles and shall be capable of being reworked at 260°C.

For the passive components, the situation is contrasted.

Most of them are reflow compatible but the number of cycles, which they can withstand, is not always given by the manufacturers.

However, some of them like film capacitors or liquid electrolyte capacitors are not recommended for reflow soldering.

As a general rule, even if the manufacturers declare that the RoHS compliant products are supposed to be capable of withstanding lead-free solder processing, this must be checked on the datasheets, specially for the plastic encapsulated components.

References

1. IPC/JEDEC J-STD-020D.1 March 2008 Moisture/Reflow Classification for Nonhermetic Solid State Surface Mount Devices Downloadable from http://www.jedec.org
2. Blattau N, Wright J, Schatz R, Hillman C (2009) Robustness of surface-mount aluminum electrolytic capacitors subjected to Pb-free reflow http://www.dfrsolutions.com/uploads/publications/2005_PCI_Article.pdf. Accessed September 30, 2009
3. IEC 60317 Standards (International Electrotechnical Commission) Specifications for particular of winding wires. http://www.iec.ch
4. NEMA MW 1000-2003 (National Electrical Manufacturers Association) Definitions, type designations, dimensions, construction, performance, and methods of testing magnet wire generally used in the winding of coils for electrical equipment. Downloadable from http://www.nema.org
5. IEC 60317-51, NEMA MW 82 Polyurethane enamelled round copper wire, class 180. http://www.iec.ch
6. Klein Wassink RJ Soldering in electronics, 2nd edn. Electrochemical publications Ltd. http://www.elchempub.com
7. IEC 60068-2-58 Standard Test method for solderability, resistance to dissolution of metallization and to soldering heat of surface mounting devices. http://www.iec.ch
8. http://www.dupont.com/kapton/